Land-based and Marine Hazards

Advances in Natural and Technological Hazards Research

VOLUME 7

The titles published in this series are listed at the end of this volume.

Land-based and Marine Hazards

Scientific and Management Issues

Edited by

M. I. EL-SABH

Université du Québec à Rimouski,
Rimouski, Québec

S. VENKATESH

Environnement Canada,
Toronto, Canada

H. DENIS

Université de Montréal,
Montréal, Canada

and

T. S. MURTY

The Flinders University of South Australia,
Adelaide, Australia

INTERNATIONAL DECADE FOR NATURAL DISASTER REDUCTION

1990 - 2000

KLUWER ACADEMIC PUBLISHERS
DORDRECHT / BOSTON / LONDON

Library of Congress Cataloging-in-Publication Data

Land-based and marine hazards : scientific and management issues / edited by M.I. El-Sabh ... [et al.].
p. cm. -- (Advances in natural and technological hazards research ; v. 7)
"International Decade for Natural Disaster Reduction, 1990-2000."
ISBN 978-94-010-6607-5 (hb : acid free paper)
1. Natural disasters--Congresses. I. El-Sabh, Mohammed I., 1939- II. International Decade for Natural Disaster Reduction (Organization) III. Series.
GB500.L36 1996
363.3'4--dc20 96-17820

ISBN-13: 978-94-010-6607-5 e-ISBN-13: 978-94-009-0273-2
DOI: 10.1007/978-94-009-0273-2

Published by Kluwer Academic Publishers,
P.O. Box 17, 3300 AA Dordrecht, The Netherlands.

Kluwer Academic Publishers incorporates
the publishing programmes of
D. Reidel, Martinus Nijhoff, Dr W. Junk and MTP Press.

Sold and distributed in the U.S.A. and Canada
by Kluwer Academic Publishers,
101 Philip Drive, Norwell, MA 02061, U.S.A.

In all other countries, sold and distributed
by Kluwer Academic Publishers Group,
P.O. Box 322, 3300 AH Dordrecht, The Netherlands.

Printed on acid-free paper

Softcover reprint of the hardcover 1st edition 1996

TABLE OF CONTENTS

PART II: MANAGEMENT ISSUES

PREFACE

Causes for major disasters are many and diverse, and the risks associated with them endanger human lives, property, the environment, the economy, and even the country's political and social well-being. Disasters can also change the fate of a community for decades and the cost of rehabilitation and reconstruction in the wake of disasters is consuming available capital, significantly reducing the resources for new investment. In fact, in recent years the world economy has spent more money to recover from disasters than it has on development. It is clear that, with rapid population growth, environmental degradation, climate change, poorly regulated industries, and continued economic uncertainty, chances are that communities may become even more vulnerable to disasters. The dramatic losses in recent years from volcanic eruptions, earthquakes and landslides, floods and droughts. cyclones and storm surges, wild forest fires attest to the fact that we are still a long way from applying even the knowledge we have today to make communities safe. Tackling this problem requires a sound evaluation of disaster mitigation policies and tools.

As a contribution to the **International Decade for Natural Disaster Reduction, (IDNDR)**, the fifth international interdisciplinary symposium, HAZARDS '93, was held in Qingdao, P.R. China, during the period 29 August - 3 September, 1993. China was an excellent choice to host this event. With its vast territory and large population, China suffers an economic loss, directly from various natural disasters, amounting to about 10 billion US dollars a year, and this loss is increasing with time. China has fought those disasters with massive work projects and with relief efforts for hundreds if not thousands of years. With this in mind, the objective of HAZARDS '93 symposium was to bring together scientists and government officials from 20 countries for the purpose of engaging in a free exchange of knowledge, experience and ideas regarding the scientific and socio-economic aspects of mitigating losses from natural and man-made disasters. It was convened by the International Society for the Prevention and Mitigation of Natural Hazards (NHS), the Commission on Natural Marine Hazards of the International Association for the Physical Sciences of the Oceans (IAPSO) and the Research Committee of Natural Disasters, Chinese Academy of Sciences.

More than 250 participants attended and a total of 180 papers were presented at 28 sessions covering a very broad range of topics which proved to be of significant value for future research. They included, among others, climatic and atmospheric hazards, marine hazards such as tsunami, storm surges, sea level variations and water pollution, coastal erosion, river floods, geological hazards including earthquakes and landslides, environmental hazards and forest fires. Papers were presented not only on the scientific aspects of these hazards, but also on disaster mitigation and management issues and the socio-economic aspects of disasters. The combination provided opportunities for very good interaction between social scientists, economists, planners and managers. Discussions during the Final Concluding Plenary Session were focused on the need to unify communication, and exchange information and data. Specific recommendations were addressed to the IDNDR Secretariat, national governments and scientific experts to increase the effectiveness and efficiency of disaster management. More details of these recommendations are presented at the end of this volume.

Seventy papers were submitted for consideration in a post-symposium publication. Every one of these papers was subjected to rigorous scientific review by at least three referees and twenty-one were accepted for publication in this book volume. In many cases, these papers have undergone substantial revision from the form initially submitted. These papers deal with scientific and management issues of land-based and marine hazards which cause the most severe economic losses, deaths, injuries and environmental degradation in many parts of the world. Given the large number of disasters that China has endured over the years we have made every effort to include papers related to natural disasters in China.

We wish to thank the many reviewers who took time from their already busy schedules to contribute to the technical quality of this book. We also wish to thank the Institute of Oceanology in Qingdao, P.R. China for hosting the meeting and to all members of the Local Organizing Committee, particularly Professor Le Kentang for his untiring efforts in his capacity as the Secretary-General for the symposium. Finally, we wish to acknowledge the different organizations who co-sponsored HAZARDS '93 Symposium: the IDNDR Secretariat, The UNESCO Office of the Intergovernmental Oceanographic Commission (IOC) and Marine Science Related issues, the International Association for the Physical Sciences of the Oceans (IAPSO), the International Association of Geodesy-Special Commission on Marine Position-Working Groups on Practical Applications, the International Tsunami Information Centre of IOC, the Tsunami Society, the Disaster Prevention Research Institute of Kyoto University in Japan, and many Chinese organizations in addition to several institutions and other international agencies supporting the participation of their representatives.

We wish to dedicate this book to the memory of Professor Liu Fengshu who acted as the Co-Chairman of the International Organizing Committee for HAZARDS '93 and died on July 7, 1993, mere days before the meeting.

M.I. El-Sabh
S. Venkatesh
H. Denis
T.S. Murty

PART I: SCIENTIFIC ISSUES

A COMPARATIVE STUDY OF STOCHASTIC MODELS FOR SEISMIC HAZARD ESTIMATION

M. SEMİH YÜCEMEN
Department of Civil Engineering and Earthquake Engineering Research Center, Middle East Technical University, 06531 Ankara, Turkey

and

AYŞEN AKKAYA
Department of Statistics, Middle East Technical University, 06531 Ankara, Turkey

Abstract. Different stochastic models have been developed over the years for the prediction of earthquake occurrences. The mathematical rigor and the extent of input data requirement increase as the stochastic models used in the description of the spatial and temporal dependence characteristics of earthquake occurrences get more realistic. In this study, a review of the most widely used stochastic models, namely: Poisson, extreme value and Markov are presented briefly, together with the random field model proposed by the authors. The shortcomings of each model are discussed. The seismic hazard predictions obtained from these models are then compared among themselves based on the data recorded along the most active portion of the North Anatolian fault zone.

1. INTRODUCTION

The mitigation of earthquake hazard requires first of all the assessment of the potential seismic hazard. Considering the underlying randomness of earthquakes with respect to space, time and magnitude and the uncertainties associated with the utilized models, this hazard can rationally be quantified only within the realm of stochastic processes. In view of this fact a number of stochastic models have been developed over the last two decades for the quantification of seismic hazard.

Most of the earlier models of seismic hazard assessment were based on the assumption that earthquake occurrences are independent events in space and time and utilized the Poisson model (e.g. Cornell, 1968, 1971; Der Kiureghian and Ang, 1977) or the extreme value statistics (e.g. Epstein and Lomnitz, 1966; Lomnitz, 1974; Yegulalp and Kuo, 1974; Burton, 1978; Knopoff and Kagan, 1977; Makropoulos and Burton, 1986). Renewal process models have also been used to describe the occurrence of earthquakes as a sequence of events with independent and identically distributed inter-event times. For example, the Poisson model is a special case of the

M. I. El-Sabh et al. (eds.), Land-based and Marine Hazards, 5–24.

renewal model resulting from the assumption of exponentially distributed inter-event times. Esteva (1970) proposed the gamma distribution for the inter-event times. The Weibull distribution has also been used to model the inter-event times (Hagiwara, 1974; Brillenger, 1982). Later studies consistent with the "elastic rebound theory", considered the temporal dependence of earthquakes based on processes with Markovian characteristics (e.g. Patwardhan et al., 1980; Kiremidjian, 1982; Suzuki and Kiremidjian, 1991). Time-predictable (e.g. Shimazaki and Nakata, 1980; Anagnos and Kiremidjian, 1984) and slip-predictable (e.g. Kiremidjian and Anagnos, 1984) models can be considered as special cases of the Markov model. A combined time-slip predictable model was proposed by Cornell and Winterstein (1988).

In a recent study, the authors have developed a random field model for the occurrence of earthquakes along linearly extending faults and for the prediction of the associated seismic hazard. In this model, the occurrence of earthquakes is treated as a space-time random process and the spatial and temporal correlations are taken into consideration (Yücemen, 1993).

The basic aim of this study is to give a critical review of the conventional Poisson, random field and the most commonly applied non-Poisson stochastic models of seismic hazard, namely extreme value and Markov. The predictions obtained from these four models are compared among themselves based on the past earthquake occurrence data associated with the most active portion of the North Anatolian fault zone (NAFZ). A detailed presentation of these models is not possible due to space limitation. For a comprehensive coverage, the interested reader is referred to the relevant references cited in the text.

2. STOCHASTIC MODELS OF EARTHQUAKE OCCURRENCE

The generation of earthquakes in time and space can be examined as a random process. In this section we will briefly present the most common models of earthquake occurrence (namely: Poisson, Markov and extreme value) together with the random field model.

2.1. Poisson Model

In this model earthquake occurrences are assumed to be independent events. The so called "lack of memory" property of the Poisson process is generally found to be in agreement with the observed seismic activity related to the moderate or large-magnitude earthquakes. According to the Poisson model, the probability distribution of the process is given by

$$P(N=n|\nu,t) = [e^{-\nu t}(\nu t)^n]/n! \qquad n=0,1,2,... \tag{1}$$

where, N = the number of earthquake occurrences in in time interval of [0,t];

ν = mean number of earthquake occurrences per unit time (generally taken as one year). The reciprocal of ν gives the expected time to the next earthquake occurrence.

Statistical data show that the Poisson model is a reasonable assumption especially for large magnitude earthquakes. Several statistical goodness of fit tests are available to check the validity of the Poisson assumption (Ferraes, 1967; Rice, 1975). Since, the Poisson model assumes independence among earthquakes it can neither describe the clustering of earthquakes in the space-time domain in the form of fore- and aftershocks, nor it can reflect the accumulation of strain energy between successive earthquakes. In spite of these shortcomings, the Poisson model is found to yield reasonable results and is widely used due to its simplicity.

2.2. Markov Model

The elastic rebound theory implies a dependence among successive earthquakes which could be described by using the one-step Markov model. According to the elastic rebound theory strain energy builds up inside the earth's crust before a new earthquake occurs. Another earthquake will not occur until strain energy is again accumulated in the region, implying the dependency of an earthquake on the preceding one.

A discrete parameter stochastic process X(t), for t=0,1,2,.... is called a first-order Markov chain if the conditional probability of X(t) depends only on the previous value, X(t-1). Mathematically, this condition is expressed by the following equation:

$$P[X(t)|\, X(1),\, X(2),...,X(t-1)] = P[X(t)|\, X(t-1)] \tag{2}$$

Such a Markov process is said to have a one-step memory. For such processes, the probabilities of the system moving from one state to another are described by a transition matrix, [P]. Considering the events: 0=no earthquake occurs, 1=an earthquake occurs, the one-step transition matrix can be defined as:

$$[P] = \begin{bmatrix} P(0,0) & P(0,1) \\ P(1,0) & P(1,1) \end{bmatrix} \tag{3}$$

where, P(1,1) = probability of having an earthquake in the current period of time given that an earthquake has occurred during the last period;

P(0,1) = probability of having an earthquake in the current period of time given that no earthquake has occurred during the last period.

Similar definitions can be given for P(0,0) and P(1,0). The period of time should be chosen in such a way that no more than one earthquake occurs in a given time period. If the initial state of the system is defined by a probability row vector, $\Pi(0)$, the state of the system at the end of n steps is given by:

$$\Pi(n) = \Pi(0)\,[P]^n \qquad n=0,1,2,... \tag{4}$$

where, $[P]^n$ is the n-step transition matrix.

The closed form of $[P]^n$ can be written as (Parzen, 1962):

$$[P]^n = \frac{1}{2-P(1,1)-P(0,0)} \begin{bmatrix} 1-P(1,1) & 1-P(0,0) \\ 1-P(1,1) & 1-P(0,0) \end{bmatrix} + \frac{[P(1,1)+P(0,0)-1]^n}{2-P(1,1)-P(0,0)} \begin{bmatrix} 1-P(0,0) & -(1-P(0,0)) \\ -(1-P(1,1)) & 1-P(1,1) \end{bmatrix} \tag{5}$$

The above matrix contains the n-step transition probabilities. The first and second portions of this matrix are steady state and the transient components, respectively. For large values of n, the transient part of the matrix approaches to zero and the limiting probabilities can be found directly from the steady-state matrix. After calculating the probability of occurrence of an earthquake in a given year, it is possible to calculate the expected time to the next occurrence, which is found from the following relationship:

$$E(T) = \frac{1}{P_{11}} = \frac{1}{[1-P(0,0)]/[2-P(1,1)-P(0,0)]} \tag{6}$$

where T is in years, and P_{11} is obtained from the steady state part of $[P]^n$ matrix. Note that in this case P_{11} is equal to P_{01}.

In the time-predictable model (Shimazaki and Nakata, 1980) a positive "backward" correlation between the inter-event time and the magnitude of the preceding earthquake is assumed. In the slip-predictable model(Kiremidjian and Anagnos, 1984), a positive "forward" correlation is assigned between the inter-event time and magnitude of the next event. The combined time-slip predictable model proposed by Cornell and Winterstein (1988) allows the existence of both correlations. The time- and slip-predictable models are special cases of the Markov model and are generally classified as semi-Markov models. In this study, the semi-Markov models of earthquake occurrence are not considered.

2.3. Extreme Value Model

In seismic hazard evaluation, the largest earthquake magnitude of a seismic source is an important input parameter. Applying the theory of extreme value distributions, the probability distribution of the largest possible magnitude can be found. The theory of extreme values is formulated under the assumption that the observed largest values are independent of each other. The theoretical bases of the subject are discussed extensively in the work of Gumbel (1958). Here the basic theory will be summarized for the sake of completeness.

2.3.1. Exact Distribution of the Largest Values

Let X be a random variable with a known distribution function, $F_X(x)$. The distribution function of the maximum value of a random sample of $X_1, X_2, \ldots, X_n$, taken from the same distribution $F_X(x)$ will be:

$$\begin{aligned} \Phi(x) &= P[\max(X_1, X_2, \ldots, X_n) \leq x] \\ &= P(X_1 \leq x) P(X_2 \leq x) \ldots P(X_n \leq x) \\ &= [F_X(x)]^n \end{aligned} \tag{7}$$

The function $\Phi(x)$ depends on the initial distribution function $F_X(x)$. Generally, the initial distribution function is not known. In such cases, the asymptotic theory of extremes can be used. The asymptotic theory is valid even if a few neighboring observations are dependent, which is quite favorable in the case of earthquakes.

2.3.2. Asymptotic Theory of Extremes

Asymptotic theory considers the limiting form of the exact distribution function as n increases. Gumbel (1958) has indicated that there are three major mathematically distinct asymptotic distributions of extremes.

The First Asymptotic Distribution (Gumbel Distribution):

Type I distribution is defined as:

$$\Phi^{(1)}(x) = \exp\left[-\exp\left(-\alpha(x-\eta)\right)\right] \qquad -\infty < x < \infty, \quad \alpha > 0 \tag{8}$$

where, α is the extremal intensity function and η is the characteristic largest value. $1/\alpha$ is a measure of dispersion and $\Phi^{(1)}(\eta) = 1/e$. The characteristic largest value η is defined as the value of x at which, in a sample of size n, the expected number of values equal to or larger than η is 1.

Methods for estimating the parameters of the first asymptotic distribution are available in the literature (Gumbel, 1958). For example, in the least squares method, one takes the largest yearly earthquake magnitudes $x_1, x_2, \ldots, x_n$ in a sample of n consecutive years. These magnitudes are arranged in order of increasing size so that $x_{(1)} \leq x_{(2)} \leq \ldots \leq x_{(n)}$. Then the values of $\Phi(x)$ are estimated by using the empirical distribution function:

$$\hat{\Phi}(x_{(i)}) = \frac{i}{n+1} \tag{9}$$

This is also known as the plotting position, p_i, of the ith observation. The values of α and η are then computed based on the following relationships (Gumbel, 1958):

$$\hat{\alpha} = \frac{s_y}{s_x} \tag{10}$$

$$\hat{\eta} = \bar{X} - \frac{1}{\hat{\alpha}} \bar{Y} \tag{11}$$

where, s_x and s_y are the standard deviations of X and Y, and $\bar{X}$ and $\bar{Y}$ denote the mean values of X and Y, respectively. Here $y = -\ln[-\ln \Phi^{(1)}(x)]$ and $x = (1/\alpha)y + \eta$.

The mean and the mode of the n largest earthquake magnitudes according to the Type I extreme value distribution are computed, respectively, from the following relations (Gumbel, 1958):

$$\bar{X} = \eta + \frac{0.577}{\alpha} \tag{12}$$

$$\tilde{X} = \eta + \frac{0.36651}{\alpha} \tag{13}$$

The return period is the mean recurrence time, which gives information concerning the frequency of observing an extreme value and is expressed as follows:

$$T(x_n) = \frac{1}{1 - \Phi^{(1)}(x_n)} \tag{14}$$

<u>The Second Asymptotic Distribution (Frechét Distribution):</u>

The second asymptotic distribution of the largest values is defined by:

$$\Phi^{(2)}(x) = \exp\left[-\left(\frac{\eta-\varepsilon}{x-\varepsilon}\right)^{k}\right] \qquad x \geq \varepsilon, \quad k>0, \quad 0 \leq \varepsilon \leq \eta \tag{15}$$

where, ε is the lower limit of largest values, k is the shape parameter, η is the characteristic largest value and $\Phi^{(2)}(\eta) = 1/e$, $\Phi^{(2)}(\varepsilon)=0$. In estimating the value of maximum earthquake magnitude, the second asymptotic distribution is generally ruled out, because it requires only a lower limit for the variable in consideration.

<u>The Third Asymptotic Distribution:</u>

If X is a random variable whose largest value is limited, the asymptotic distribution of its largest value is suitably represented by the Type III asymptotic distribution and is given by:

$$\Phi^{(3)}(x) = \exp\left[-\left(\frac{w-x}{w-\eta}\right)^{k}\right] \qquad x \leq w, \quad k>0, \quad \eta<w \tag{16}$$

where, w is the upper limit of largest values, k is the shape parameter, η is the characteristic largest value, and $\Phi^{(3)}(\eta) = 1/e$, $\Phi^{(3)}(w)=1$.

The parameters w, η and k of the third asymptotic distribution can be calculated by using the classical method of moments or by the method of least squares (Yegulalp, 1974). In the method of moments either three or two moments are taken into consideration. In the three-moment method, the third sample moment has a large variance and may not be very reliable (Yegulalp and Kuo, 1974). Besides, this method does not guarantee that the upper limit w is greater than the largest earthquake magnitude observed in n years. Thus, the two-moment method is used alternatively.

The two methods of moments mentioned above require that all values of X_i be available, but it is not always possible to have a complete data set, especially for the case where the unit time period considered is one year. In such cases the least squares method seems to be more convenient (Yegulalp and Kuo, 1974). In the least squares method, the main aim is to minimize the error, E:

$$E = \sum_{i=1}^{n} (x_i - \hat{x}_i)^2 \quad for \quad 0 < \eta < w \quad and \quad k > 0 \tag{17}$$

Here x_i's are the annually observed maximum magnitudes in a given region, and $\hat{x}_i$'s are the predicted values of x_i's. Based on the plotting positions (9), $\hat{x}_i$ can be computed from the following expression:

$$\hat{x}_i = w - (w - \eta) z_i^{1/k} \tag{18}$$

where, $z_i = -\ln p_i$.

A computational algorithm and the corresponding software (Yegulalp, 1988) is used to minimize E with respect to the given constraints and estimate the parameters w, η and k. The mean and the mode of the largest magnitude expected in the next n years, according to the assumption of third type asymptotic distribution, are shown to be (Yegulalp, 1974):

$$\bar{X}_n \cong w - (w - \eta)\Gamma(1 + 1/k)(1/n)^{1/k} \tag{19}$$

$$\tilde{X}_n \cong w - (w - \eta)[(1 - 1/k)/n]^{1/k} \tag{20}$$

where, $\Gamma(.)$ is the gamma function.

2.4. Random Field Model

The occurrence of earthquakes in the most comprehensive context can be modeled as a space-time random process and accordingly the seismic hazard is predicted based on the concept of level crossing. The proposed model considers either the strain energy or the stress random field as the basis. The implementation of the stress-based formulation is hindered by the lack of spatio-temporal data to assess the statistical parameters (such as mean, standard deviation, covariance function, etc.) of the shear stress and static friction stress random fields. Energy-based formulation is preferred due to the availability of data on the energy random field through based formulation. However, a detailed study on the stress-based earthquake occurrence mechanism can be found in Yücemen (1989).

The strain energy accumulated at any point of the earth's crust is described by the random function, $e(\underline{u},t)$, where $\underline{u}$ is the 2-dimensional space vector $\underline{u}'$: (x,z)), and t denotes the time. Here it is assumed that a potential rupture segment extends symmetrically between $x_1 = x_0 - w/2$ and $x_2 = x_0 + w/2$, where x_0 is the projection of the shock origin onto the x-axis, and the length of rupture, in the z-direction, is restricted to h (see Figure 1).

The total energy accumulated at a fixed time t over the ruptured portion of the fault plane, which is centered at x_0 along the fault axis and having an area A=hxw, is obtained by the following integration:

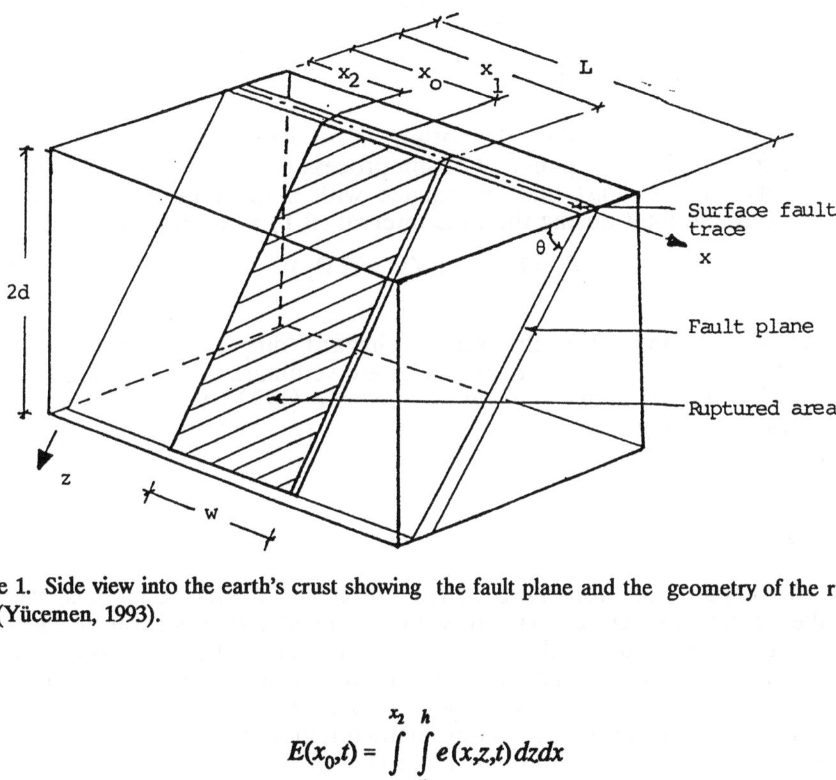

Figure 1. Side view into the earth's crust showing the fault plane and the geometry of the ruptured area (Yücemen, 1993).

$$E(x_0,t) = \int_{x_1}^{x_2} \int_{0}^{h} e(x,z,t)\,dz\,dx \tag{21}$$

Here $E(x_0,t)$ is a "spatial integral process" and is assumed to be Gaussian based on the generalized central limit theorem.

An earthquake having at least a magnitude of m occurs whenever the energy random field E(x,t) exceeds the corresponding energy level e_m at any point of the

2-dimensional region, D: $\{x \in (w/2, L-w/2)$ and $t \in (0,T)\}$, where L is the total length of the fault and T denotes the time period considered. For simplicity a zero-mean Gaussian random field can be obtained by using a transformation, $V(x,t) = E(x,t)-\bar{E}$, where $\bar{E}$ is the mean of the integral process E(x,t). Then the mean number of maxima (excursions) of V(x,t) above "$e_m-\bar{E}$" per unit distance along the fault line and per unit time, based on the concept of scale of fluctuation (Vanmarcke, 1983), is computed from the following asymptotic equation:

$$\Omega_m = \frac{(e_m - \bar{E})}{\pi^{3/2}(2k\alpha A^*)^{1/2}\tilde{V}} \exp\left(-\frac{(e_m - \bar{E})^2}{2\tilde{V}^2}\right) \tag{22}$$

in which, k is a dimensionless constant characterizing the interaction between the principal coordinates; α is the characteristic correlation area; A^* denotes the area of a rectangular window with sides parallel to the x- and time-axes; $\tilde{V}$ is the standard deviation of the random field E(x,t).

For large magnitude earthquakes, the pattern of occurrence of excursions above a high level tends to follow a planar Poisson process (Cox and Isham, 1980). Then the probability of at least one earthquake with a magnitude greater than m occurring along the fault during the time interval (0,T) is approximated by:

$$P_m = 1 - \exp\left(-(L-w)T\Omega_m\right) \tag{23}$$

By the use of equation (23), the seismic hazard due to a linearly extending fault is computed. Here, only the main points are summarized; a detailed coverage of the model described above is given in Yücemen (1993). For more information on the derivation of equation (22) and the theory of random fields the reader is referred to Vanmarcke (1983).

3. CASE STUDY

In order to compare the predictions of the different stochastic models, the past earthquake occurrence data associated with the most active portion of the North Anatolian fault zone are taken into consideration. The data base and the computations required by each model are briefly discussed in the following subsections. A comprehensive study of the North Anatolian fault is beyond the scope of this study; however information on its characteristics can be found in the works of several researchers (e.g. Ambraseys, 1969; Dewey, 1976).

3.1. Data Base

The North Anatolian fault zone is generally considered to extend roughly between longitudes 28°E and 42°E. In this study the most active portion of the fault lying between longitudes 30.31°E and 41.08°E is considered. This portion of the fault extends almost linearly and is modeled as a line source. The zone considered is shown in Figure 2. Earthquakes whose epicenters fall into this zone are compiled from a recently prepared comprehensive catalog (Earthquake Catalog, 1992). The time interval is 1904-1992.

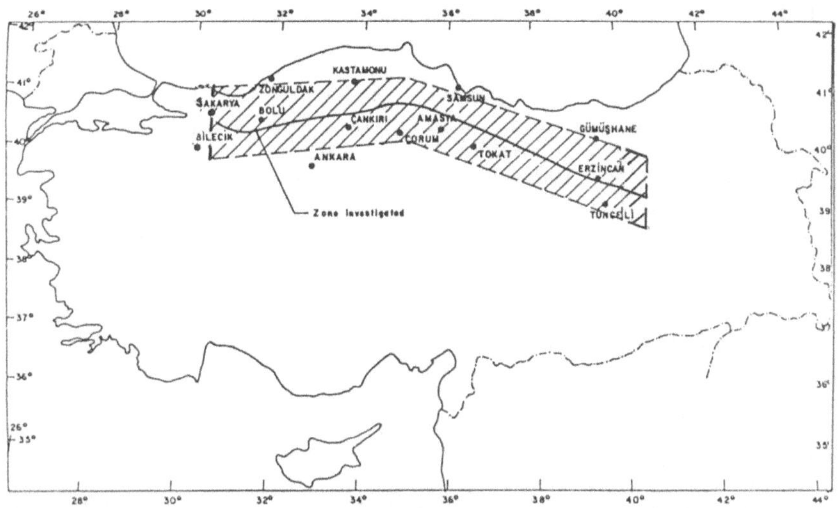

Figure 2. The most active portion of the North Anatolian fault zone considered.

The above-mentioned data are taken as the input for the stochastic models that are presented in the previous section. The raw data require some kind of processing specific to the model before being manipulated. The earthquakes are classified into four magnitude categories as shown below:

Richter magnitude	Classification
6.5 and greater	Large
5.5 - 6.4	Moderate
4.5 - 5.4	Small
Less than 4.5	Very small

This classification is used in all of the earthquake occurrence models.

3.2. Poisson Model

The available data are used first to check the independence of earthquake occurrences in time by using the Chi-square test for independence and then the annual probabilities of earthquake occurrence are calculated. For overall data, earthquake occurrences are found to be dependent at a level of significance of

$\alpha=0.05$ and independent at $\alpha=0.025$. On the other hand, for main shocks ($m\geq6.0$), earthquake occurrences are independent at $\alpha=0.05$. The annual probabilities of earthquake occurrences for the Poisson model are computed from equation (1) and given in Table I.

Table I. The annual probabilities of earthquake occurrence for the Poisson model

Earthquake classification	Time period	ν	$P(N>0\|\nu,t=1)$
Large	1904-1992	0.11	0.10
Moderate	1904-1992	0.30	0.26
Small	1904-1992	1.34	0.74
Very small	1970-1992	1.61	0.80

Here ν represents the mean number of earthquake occurrence for a period of one year. It is to be noted that for very small earthquakes the reliable data starts from 1970.

3.3. Markov Model

The earthquake data obtained along the NAFZ from 1904 to 1992 have been grouped into four magnitude categories as discussed in Section 3.1. For each of the four magnitude categories the one and n-step transition matrices are calculated. From these matrices the probability of future earthquake occurrence can be estimated. The time interval used in the calculation of transition probabilities is taken to be one year and the results are given in Table II.

Table II. One step transition probabilities

Earthquake classification	P(1,1)	P(1,0)	P(0,1)	P(0,0)
Large	0.12	0.88	0.10	0.90
Moderate	0.41	0.59	0.15	0.85
Small	0.61	0.39	0.49	0.51
Very small	0.82	0.18	0.60	0.40

Using the resulting n-step transition matrix, the earthquake occurrence probabilities are computed for any required future time span. The steady state probabilities are attained within a maximum of seven years in all magnitude levels. The steady state probabilities are given in Table III.

Table III. The steady state probabilities

Earthquake classification	P_{11}	P_{10}	P_{01}	P_{00}
Large	0.10	0.90	0.10	0.90
Moderate	0.21	0.79	0.21	0.79
Small	0.56	0.44	0.56	0.44
Very small	0.77	0.23	0.77	0.23

As expected, for large and moderate earthquakes P_{11} is remarkably smaller and P_{00} is remarkably larger than the P_{11} and P_{00} of small and very small earthquakes, respectively, consistent with the observed frequencies. The expected time to the next earthquake occurrence, in years, is also calculated for all categories and are given in Table IV.

Table IV. The expected time to the next earthquake occurrence

Earthquake classification	P_{11}	E(T) (years)
Large	0.10	10
Moderate	0.21	4.8
Small	0.56	1.8
Very small	0.77	1.3

3.4. Extreme Value Model

In this section, the use of the extreme value distributions in forecasting the largest earthquake magnitude along the NAFZ is considered. The largest magnitude in every year for the period 1904-1992 is selected. Since for 23 years either no earthquakes have occurred or the earthquakes with a very small magnitude did occur but were not listed, only 66 yearly maxima are obtained. To estimate the parameters of the asymptotic distributions the software prepared by Yegulalp (1988) is used. The results are summarized in the following sections.

3.4.1. First Asymptotic Distribution

The values of the parameters estimated by the method of least squares are: $\hat{\alpha} = 1.23$, $\hat{\eta} = 4.79$. Thus, the analytical form of the first asymptotic distribution function for the largest earthquake magnitude, m, becomes:

$$\Phi^{(1)}(m) = \exp\ [-\exp\ (-1.23(m-4.79))] \qquad (24)$$

The result of the Chi-square goodness of fit test implies that the observations of largest earthquake magnitude over the NAFZ reveal a good fit to the first asymptotic distribution at a 0.05 level of significance and this fit is observed in Figure 3. The mean and modal values are computed from equations (12) and (13) and found as 5.25 and 5.09, respectively.

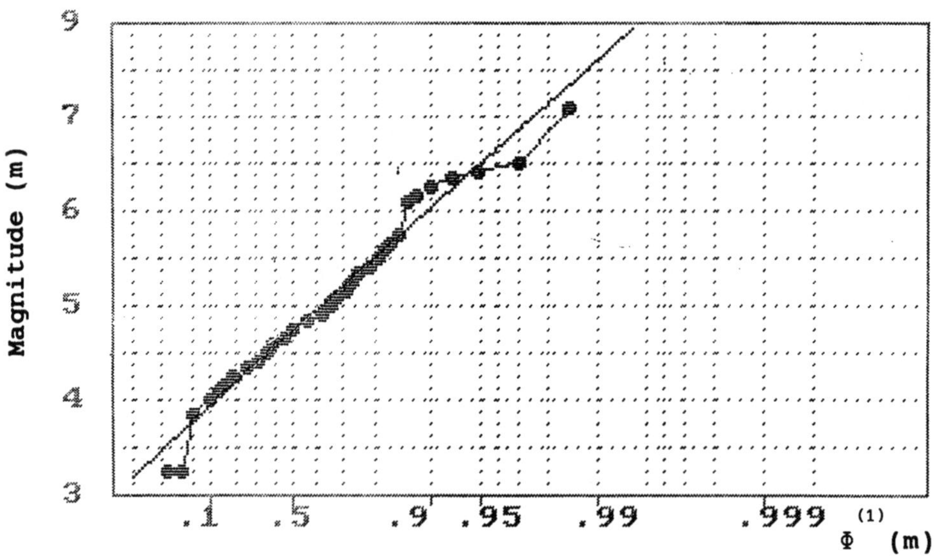

Figure 3. Type I extreme value distribution for maximum earthquake magnitude (North Anatolian fault zone)

3.4.2. Third Asymptotic Distribution

The parameters, w, η and k of the third asymptotic distribution that are calculated by using the three different methods presented in Section 2.3.2 are given in Table V.

Table V. Parameters of the third asymptotic distribution estimated by three different methods

2-moment method	3-moment method	Least squares method
w = 12.05	w = 7.06	w = 10.73
η = 4.85	η = 5.01	η = 4.25
1/k = 0.118	1/k = 0.498	1/k = 0.18

The Chi-square goodness of fit test is applied to the output of the three methods. At a significance level of $\alpha = 0.05$ only the two-moment method yielded to a satisfactory result and this fit is shown in Figure 4. The parameter w, which describes the upper limit is seen to be larger than the observed maximum magnitude of 7.9, in the least squares and the two-moment methods. As a mater of fact, w=12.05 and w=10.73 are physically unrealizable magnitudes. However, w is only a parameter of the distribution and should necessarily not be interpreted as the upper bound magnitude. On the other hand, the value of w=7.06 which is obtained in the case of three-moment method is smaller than the observed largest magnitude of 7.9. The poor fit and the discrepancies observed in the parameter w may be attributed to the large number of missing observations; in fact 26% of yearly maxima are missing.

3.5. Random Field Model

In the random field model, the geometrical parameters for the portion of the North Anatolian fault considered, are set equal to the following:

Fault length:	L = 916 km.
Rupture width:	w = 100 km.
Focal depth:	d = 20 km.
Rupture length in vertical direction:	h = 40 km.

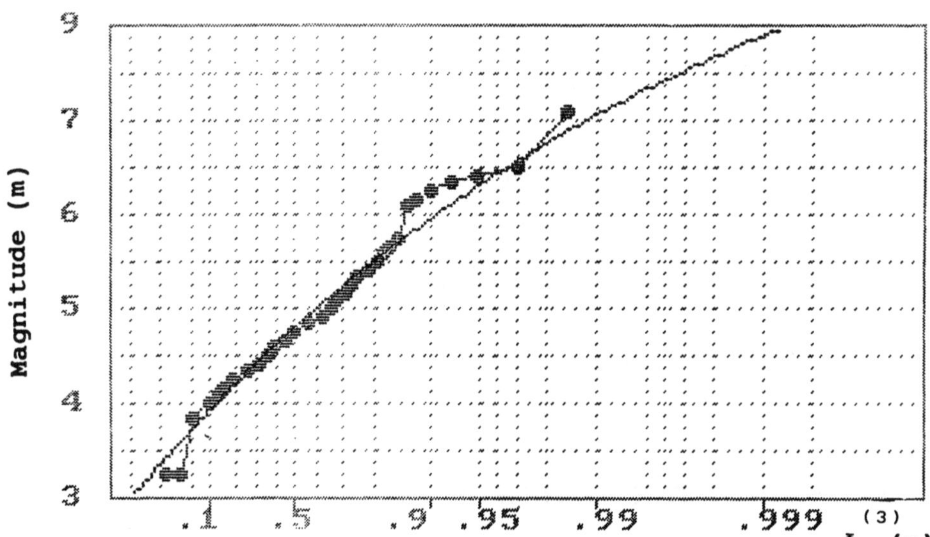

Figure 4. Type III extreme value distribution for maximum earthquake magnitude based on two-moment method (North Anatolian fault zone).

Using the data base described in Section 3.1, the main parameters required to calculate the mean excursion rate from equation (22) are estimated to be:

$\bar{E}$ = 3.064×10^{20} erg/year
$\tilde{V}$ = 1.26×10^{21} erg/year
A^* = 1x100 year.km
λ_x = λ_z = 0.40 w (spatial scale of fluctuations)
λ_t = 0.15 year (temporal scale of fluctuation)

and with the assumption of a separable space-time correlation structure (Yücemen, 1993):

$$k = 1.0$$
$$\alpha = \lambda_t \, \lambda_x = 0.15 \times 40 = 6 \text{ year.km.}$$

The procedures for the computation of these statistical parameters are described in Yücemen (1993).

Assume that we are interested in the probability of an earthquake occurring with a magnitude of at least m=6.8, which is the magnitude of the destructive Erzincan earthquake of March 13, 1992 nucleated from the North Anatolian fault. The energy level corresponding to m=6.8 is 39.8×10^{20} erg (Yücemen, 1993). The mean excursion rate for this large magnitude earthquake is calculated from equation (22) as:

$$\Omega_{6.8} = \frac{(39.8-3.064)\,.\,10^{20}}{\pi^{3/2}(2\,.\,6\,.\,100\,)^{1/2}1.26\,.\,10^{21}} \exp\left(-\frac{(39.8-3.064)^2\,.\,10^{40}}{2(1.26\,.\,10^{21})^2}\right)$$

$$= 0.0002 \;\; maxima \;\; / \; year \; .km.$$

Then, the probability of at least one earthquake with a magnitude of m≥6.8 occurring along the fault in one year is computed from equation (23):

$$P_{6.8} = 1-\exp(-0.0002(916-100)\,) = 0.15$$

4. COMPARISON OF STOCHASTIC MODELS

The results obtained from the four different stochastic models are presented in Table VI for different magnitude levels in terms of annual probability of earthquake occurrence (annual seismic hazard). On the same table the observed annual seismic hazard values are also given. The computation of these empirical values is based on time spans over which the seismic activity data were complete.

As it is seen in this table, for large earthquakes all stochastic models give the same level of annual seismic hazard which is consistent with the observed seismic activity. Also, for moderate earthquakes, all models, except the third type asymptotic distribution derived based on the three-moment method, yield approximately to the same results. On the other hand, for small and very small earthquakes the results differ significantly. The Markov model, which takes into consideration the temporal dependency gives the closest prediction to the observed seismic activity; whereas the Poisson model gives conservative values and the extreme value models underestimate the annual seismic hazard to a great extent. The independence assumption associated with the Poisson model causes the observed overestimation. Among the extreme value models, only the third type distribution, with parameters obtained from the least squares method, reflects the increase in seismic activity with decreasing magnitude levels. The poor predictions obtained from the extreme value models are expected due to the fact that only largest yearly earthquakes are considered and thus, some of the information about small and very small earthquakes is lost. Besides, almost a quarter of yearly maximum values are missing in the data base. Knopoff and Kagan (1977) also concluded that extreme value procedures, which use only part of the data set, give poor predictions of seismic hazard.

5. CONCLUSIONS

Probabilistic methods and theory of random processes provide the appropriate tools for the assessment of seismic hazard due to inherent variabilities and vast amount of uncertainties. In this study different stochastic models are presented for the prediction of seismic hazard, and a comparison of them are carried out based on the seismic activity data compiled for a certain portion of the North Anatolian fault zone. On the basis of this study the following remarks are made:

(i) The stochastic models commonly used to describe the occurrence of earthquakes in the time domain are found to yield different results in predicting the effect of small and moderate earthquakes; but for large magnitude earthquakes they appear to give almost the same level of seismic hazard.

(ii) The random field model appears to be most appropriate one to mimic the existing correlation structure and the occurrence of large earthquakes in the space-time domain; however lack of sufficient data hinders its implementation.

(iii) For the assessment of a realistic estimate of seismic hazard at a single site the use of the stochastic model that the available data base justifies is recommended. On the other hand, for seismic hazard maps requiring the assessment of potential seismic activity at a large number of points (multi-site problem) the simple Poisson model can conveniently be used, since the relative hazard values will be sufficient for such a purpose. This view is also shared by other researchers (e.g. Cornell and Winterstein, 1988).

Table VI. Comparison of different stochastic models based on data from North Anatolian fault zone

Magnitude	Time period	Annual probability of earthquake occurrence	
Large	1904-1992	Poisson	:0.10
		Markov	:0.10
		Type I	:0.11
		Type III	
		i) Two-moment	:0.10
		ii) Three-moment	:0.08
		iii) Least squares	:0.09
		Random field	:0.15
		Observed	:0.11
Moderate	1904-1992	Poisson	:0.26
		Markov	:0.21
		Type I	:0.21
		Type III	
		i) Two-moment	:0.24
		ii) Three-moment	:0.34
		iii) Least squares	:0.16
		Observed	:0.21
Small	1904-1992	Poisson	:0.74
		Markov	:0.56
		Type I	:0.38
		Type III	
		i) Two-moment	:0.38
		ii) Three-moment	:0.31
		iii) Least squares	:0.29
		Observed	:0.50
Very small	1970-1992	Poisson	:0.80
		Markov	:0.77
		Type I	:0.24
		Type III	
		i) Two-moment	:0.22
		ii) Three-moment	:0.21
		iii) Least squares	:0.33
		Observed	:0.78

ACKNOWLEDGEMENTS

This research work is supported, in part, by the Research Fund of Middle East Technical University under grant: AFP 93-03-03-02 and by the Scientific and Research Council of Turkey (TUBITAK) under grant: INTAG-522.

REFERENCES

Ambraseys, N.N., (1969). Some characteristic features of the Anatolian fault zone, *Tectonophysics*, **9**, 143-165.

Anagnos, T. and Kiremidjian, A.S., (1984). Stochastic time-predictable model for earthquake occurrences, *Bull. Seism. Soc. Am.*, **74**(6), 2593-2611.

Brillenger, D.R., (1982). Some bounds for seismic risk, *Bull. Seism. Soc. Am.*, **72**(4), 1403-1410.

Burton, P.W., (1978). The application of extreme value statistics to seismic hazard assessment in the European area, *Proc. Symp. Anal. Seismicity and on Seismic Risk*, 323-334.

Cornell, C.A., (1968). Engineering seismic risk analysis, *Bull. Seism. Soc. Am.*, **58**, 1583-1606.

Cornell, C.A., (1971). Probabilistic analysis of damage to structures under seismic loads, *Dynamic Waves in Civil Engineering*, ed. Howells, D.A., Haigh, I.P., Taylor, C., J. Wiley, London.

Cornell, C.A. and Winterstein, S.R., (1988). Temporal and magnitude dependence in earthquake recurrence models, *Bull. Seism. Soc. Am.*, **78**(4), 1522-1537.

Cox, D.R. and Isham, V., 1980, *Point Processes*, Chapman and Hall, London.

Der Kiureghian, A. and Ang, A, H-S., (1977). A fault-rupture model for seismic risk analysis, *Bull. Seism. Soc. Am.*, **67**(4), 1173-1194.

Dewey, J.W., (1976). Seismicity of Northern Anatolia, *Bull. Seism. Soc. Am.*, **66**, 843-868.

Earthquake Catalog of the General Directorate of Disaster Affairs, Ministry of Public Works and Settlement, 1992, Ankara (unpublished).

Epstein, B., Lomnitz, C., (1966). A model for the occurrence of large earthquakes, *Nature*, **211**, 954-956.

Esteva, L., (1970). Seismic risk and seismic design decisions, *Seismic Design for Nuclear Power Plants*, ed. Hansen, R.J., M.I.T. Press, Cambridge, Mass.

Ferraes, S. G., (1967) Test of Poisson process for earthquakes in Mexico City, *Journal of Geophysical Research*, **72**(14), 3741-3742.

Gumbel, E.J., (1958) *Statistics of Extremes*, Columbia University Press, New York.

Hagiwara, Y., (1974). Probability of earthquake occurrence as obtained from a Weibull distribution analysis of crustal strain, *Tectonophysics*, **23**(3) 313-318.

Kiremidjian, A.S., (1982). A minimum stress level model for large earthquakes, *Proc., 7th ECEE*, Athens, Greece, **2**, 32-41.

Kiremidjian, A.S. and Anagnos, T., (1984). Stochastic slip-predictable model for earthquake occurrences, *Bull. Seism. Soc. Am.*, **74**(2), 739-755.

Knopoff, L. and Kagan, Y., (1977) Analysis of the theory of extremes as applied to earthquake problems, *Journal of Geophysical Research*, **82**(32) 5647-5657.

Lomnitz, C., (1974). *Global Tectonics and Earthquake Risk*, Elsevier, Amsterdam.

Makropoulos, K.C. and Burton, P.W., (1986). HAZAN: A fortran program to evaluate seismic-hazard parameters using Gumbel's theory of extreme value statistics, *Computers and Geosciences*, **12**(1) 29-46.

Parzen,E., (1962). *Stochastic Processes*, Holden-Day, San Francisco.

Patwardhan, A.S., Kulkarni, R.B and Tocher, D„ (1980). A semi-Markov model for characterizing recurrence of great earthquakes, *Bull. Seism. Soc. Am.*, **70**, 323-347.

Rice, J., 1975, Statistical methods of use in analyzing sequences of earthquakes, *Geophys. J. R. Astr. Soc.*, **42**, 671-683.

Shimazaki, K. and Nakata, T., (1980). Time-predictable recurrence for large earthquakes, *Geophys. Res. Letters*, **7**, 279-282.

Suzuki, S. and Kiremidjian, A.S., (1991). A random slip rate model for earthquake occurrences with Bayesian parameters, *Bull. Seism. Soc. Am.*, **81**, 781-795.

Vanmarcke, E.H., (1983). *Random Fields: Analysis and Synthesis*, M.I.T, Press, Cambridge, MA.

Yegulalp, T.M., (1974). Forecasting for large earthquakes, *Management Science*, **21**(4), 418-421.

Yegulalp, T.M. and Kuo, J.T., (1974). Statistical prediction of the occurrence of maximum magnitude earthquakes, *Bull. Seism. Soc. Am.*, **64**(2), 393-414.

Yegulalp, T.M., (1988). *Extreme value statistics package*, Columbia University.

Yücemen, M.S., (1989). A random field model for the estimation of earthquake hazard due to active faults, *Reliability and Optimization of Structural Systems*, Springer Verlag, Berlin, 401- 409.

Yücemen, M.S., (1993) A random field model for the estimation of seismic hazard, *Structural Safety*, **12**, 187-203.

DATA BASE FOR SEISMIC RISK ASSESSMENT IN CUSCO, PERU

MARIO A. CANDIA-GALLEGOS
Vector Engineering, Inc., 12438 Loma Rica Dr., Suite C, Grass Valley, CA 95945, USA.;

INGECO, S.R. LTDA.
Ingenieros Consultores, P.O. Box 865, Cusco, Perú.

KENNETH F. SPRENKE
Department of Geological Engineering, University of Idaho, Moscow, ID 84843, USA.

J. CARLOS PEREZ
GEOMEC Consultores S.R.L., Cusco, Perú.

Abstract: Results from many efforts to better understand the nature of ground motions and the seismic behavior of structures in Cusco, Peru, give now the possibility to combine several sources of information to produce estimates of seismic hazard and risk. A model is presented, which estimates expected damage based on geotechnical characteristics and intensity-damage relations derived for several types of buildings representative of Cusco's constructions. The study area was divided into smaller sections covering the whole city. This information, along with the geographic distribution of soil types and construction density, allows computation of expected losses during a given event for sites distributed throughout the city.

1. INTRODUCTION

Cusco, the oldest city in South America, has suffered several earthquake events during it's almost 1,000-year life time. Records for the last 100 years show severe earthquake-related injuries, death and property loss.

The spatial distribution of damage has demonstrated some correlation between the local geotechnical conditions and the dominant parameters of earthquakes. In order to better assess the seismic risk at different scales, the relationships between ground motion, local geology, and soil behavior should be clarified. Valuable database on geology, hydrology, seismicity, soil behavior, and the output of a regional seismic hazard analysis are placed together along with their geographic location in order to model the geotechnical characteristics of the study area.

M. I. El-Sabh et al. (eds.), Land-based and Marine Hazards, 25–36.

The characterization of the building stock based on age, use, structural type, location, plant configuration, structural design, state of conservation, and material, has been defined. Vulnerability curves were assigned to each category taking into consideration the type, age, natural frequency of vibration, and the known statistics. The correlation of the geotechnical modeling with the vulnerability of buildings in the study area should provide a valuable tool for seismic damage assessment on buildings and, in turn, a solid database is expected to be established in order to make possible practical implementation of better structural design and city planning.

2. GEOGRAPHICAL ENVIRONMENT

The Cusco Valley, is located between 71.9 and 72.3 west longitude, and between 13.4 and 13.6 south latitude. Elevations range from 3,070 meters at the Huatanay River mouth to 4,852 meters that correspond to the Pachatusan Peak.

The floor of the Cusco Valley, 30 km. in length and an average width of 12 km., with an average elevation of about 3,115 meters, is divided by the Angostura and Oropesa narrows into three oval areas of flat-lying land, which are called the Cusco, Oropesa and Lucre Basins. The Cusco Basin is the largest basin where the city of Cusco, including the districts of San Sebastian and San Jeronimo, is located. The Oropesa Basin, has less width and a flatter floor than the Cusco Basin. The Lucre Basin shows the least relief of all three basins. The area of

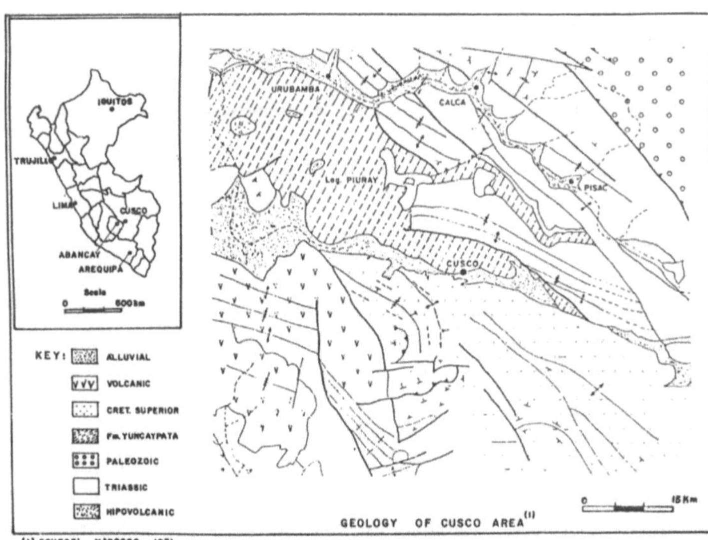

Figure 1. Geology of the Cusco Valley

study is approximately 80 km^2.

3. GEOLOGY AND TECTONICS

The Cusco Basin used to be occupied by an ancient lake. The geology of this area conforms to the lacustrine strata of the San Sebastian formation. Igneous rock masses are located within the Valley. The Rumicolca mass is the largest exposure of igneous rock within the region. Stone from Rumicolca and the Rodadero mass was extensively used in the past, and are the principal source of rock used in modern construction.

Seismic activity in the Cusco area may be related to the active Quaternary "Cusco Faulting System," that divides the Eastern branch of the Andes mountain chain from the Altiplano structural unit (Benavente, R., 1992, Marocco, 1977). This more than 100-km. faulting system crosses the Cusco area, starting from Abancay city, in the northwest, through Urcos city, in the south. About 40 km. south east of Cusco, there is another system, the Vilcanota Faulting System. These faulting systems and Geology of the Cusco Valley is illustrated in Figure 1.

4. SOILS AND HYDROLOGY IN CUSCO VALLEY

The evaluation of about 500 soil site studies is summarized in the soils map shown in Figure 2. These studies were compiled from several sources such as government

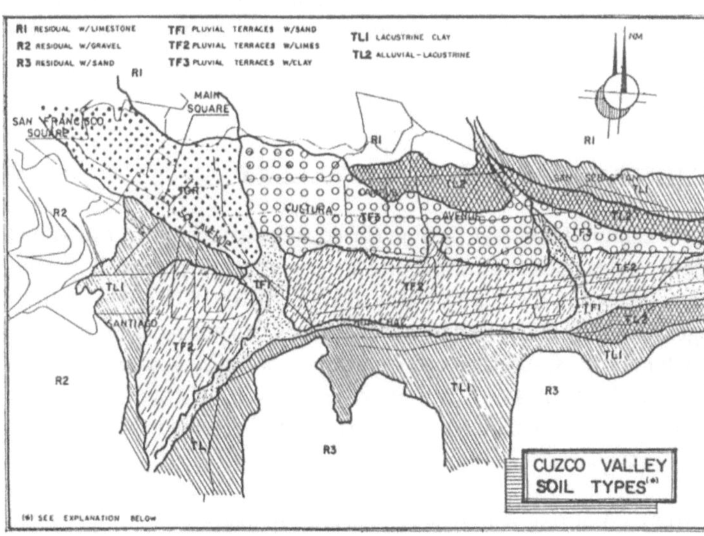

Figure 2. Soil types in Cusco area.

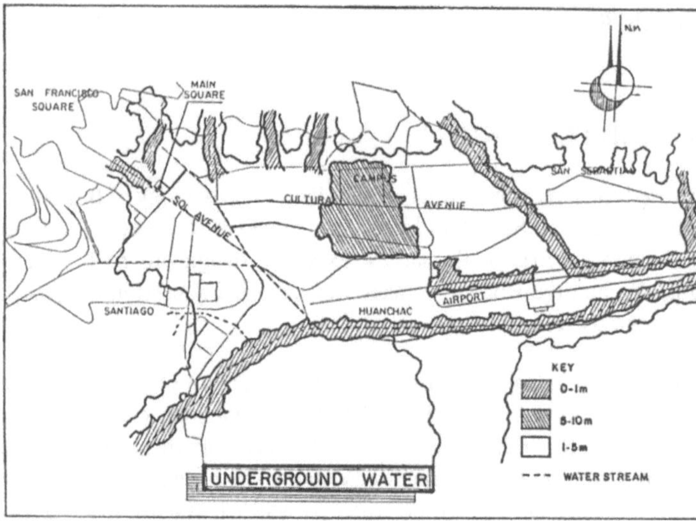

Figure 3. Underground Water in Cusco Valley.

institutions, private companies, and work performed by the authors. The Cusco Valley is drained by the Huatanay river, with a total fall of 50 m. Tributaries to this river vary greatly in amount and permanency of flow. Underground water level, as illustrated in Figure 3, is shallow due to the lacustrine nature of the valley varying from between 0 and 1.0 meter in the lower basin to between 1.0 and 5.0 meters in the upper basin. In upper areas, the water table is more than 5 meters deep. The number of rainy days in a year averages about 175.

5. SEISMIC HISTORY OF THE CUSCO VALLEY

During the present century important data on earthquakes have been recorded by the Instituto Geofísico del Perú, universities and other Peruvian government sponsored institutions. Major earthquakes have occurred in the Cusco Valley in 1823, 1870, 1875, 1943. Other earthquakes occurred damaging seriously some vicinity towns such as Quiquijana in 1717; Paruro, Oropesa, Lucre, in 1746; Chumbivilcas, Abancay and Cotabambas in 1875; Yanaoca and Pampamarca in 1943, Livitaca 1991 (Polo, 1904, Ericksen, et al., 1954). However, the most significant seismic events for the Cusco area are the March 31, 1650, September 18, 1941, May 21, 1950; and April 5, 1986 earthquakes.

On March 31, 1650. A strong earthquake, followed by many strong aftershocks for several days, destroyed nearly all of Cusco. People in Lima, 350 km. northwest of Cusco, felt the strongest shock (Polo, 1904).

On May 21, 1950, more than half the buildings of Cusco were seriously damaged by another shock (Ericksen, et al., 1954). The damage was extensive within an epicentral area of about 12 km^2, with minor damage within a surrounding area of about 500 km^2. The earthquake was considered local in effect since it was felt over a total area only of 16,000 km^2. The depth of focus was considered to be located within 8 to 9 km. The distribution of damage caused by both the March 31, 1650, and the May 21, 1950 (Ericksen, et al., 1954) earthquakes is shown in Figure 4.

The major factors for the extensive seismic damage were poor construction and age of buildings. The maximum intensity in this area was estimated to be VII on the MMI scale, with an acceleration less than 300 gals. (Ericksen, et al., 1954). Colonial churches, some more than 350 years old, and old adobe and combined adobe and rock or brick structures were the most

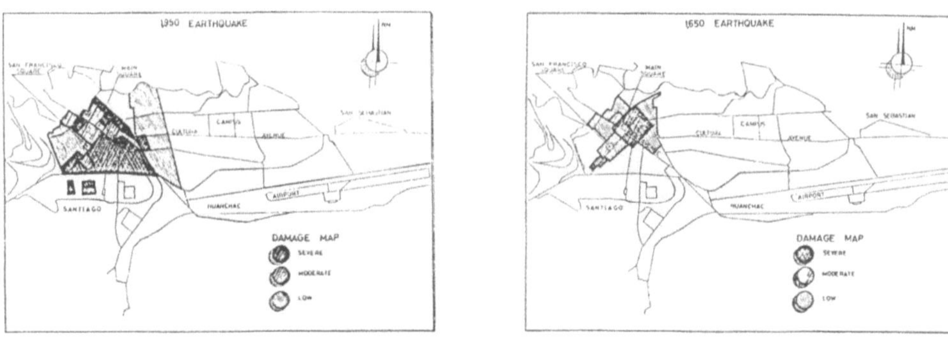

Figure 4. Damage from the March 31, 1650 and the May 21, 1950 earthquakes.

seriously damaged. The most extensive damage was to buildings resting on a thick water-saturated alluvial gravel, the TGR soil deposits, (see Figure 2).

The Cusco earthquake of May 21, 1950 was probably of tectonic origin. An extensive fissure was considered as evidence of a bedrock fault beneath unconsolidated sediments at the south side of the Cusco Basin. Liquefaction and slope stability failure were reported in the same area (Ericksen, et al., 1954).

On April 5, 1986, more than 50,000 people were affected by a earthquake. The epicenter was 20 km. northwest of Cusco, and the magnitude was 5.8 on the Richter scale, the duration was 3 to 5 secs. The shock was felt over a total area of about 12,000 km^2. Isoseismal maps for this event (Huaco, et al., 1986) and for the May 21, 1950 (Ericksen, et al., 1954) phenomenon show MMI values of III about 100 km around Cusco.

The most extensively damaged buildings were the old adobe houses. More than 21,000 adobe houses were affected within the urban area. Inventory of building stock and damage estimation was performed by the University of Cusco and the Instituto Nacional de Cultura. For a more accurate damage assessment, the city area was divided into 22 smaller sections as shown in Figure 5. Geologic risk assessment maps developed by the Instituto Geofísico del Perú, and by the University of Cusco, Catholic University of Lima, and the CERESIS are shown in Figure 6.

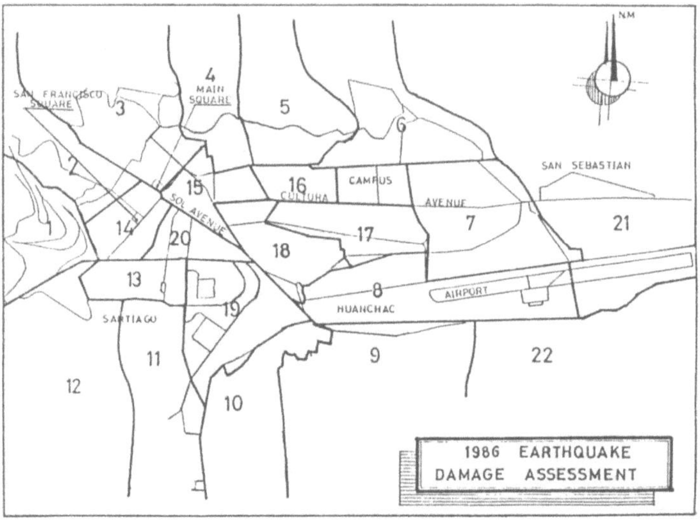

Figure 5. Damage assessment map, April 5, 1986 earthquake (University Cusco, 1986).

6. LOCAL EFFECTS

The duration and frequency content of the ground motion depend on local geological and soil conditions (Mendes, 1991). Landslides in Cusco are very frequent. The rock and gravel displaced by slides vary in amount from a few cubic meters to the enormous mass of material represented by the prominent scar on the Picol Peak, which is visible for a distance of 20 km. Liquefaction, lateral spreading, densification and strength reduction are other local effects reported on past earthquakes for the Cusco area. Figure 7 shows the correlation of seismic intensity and local effects.

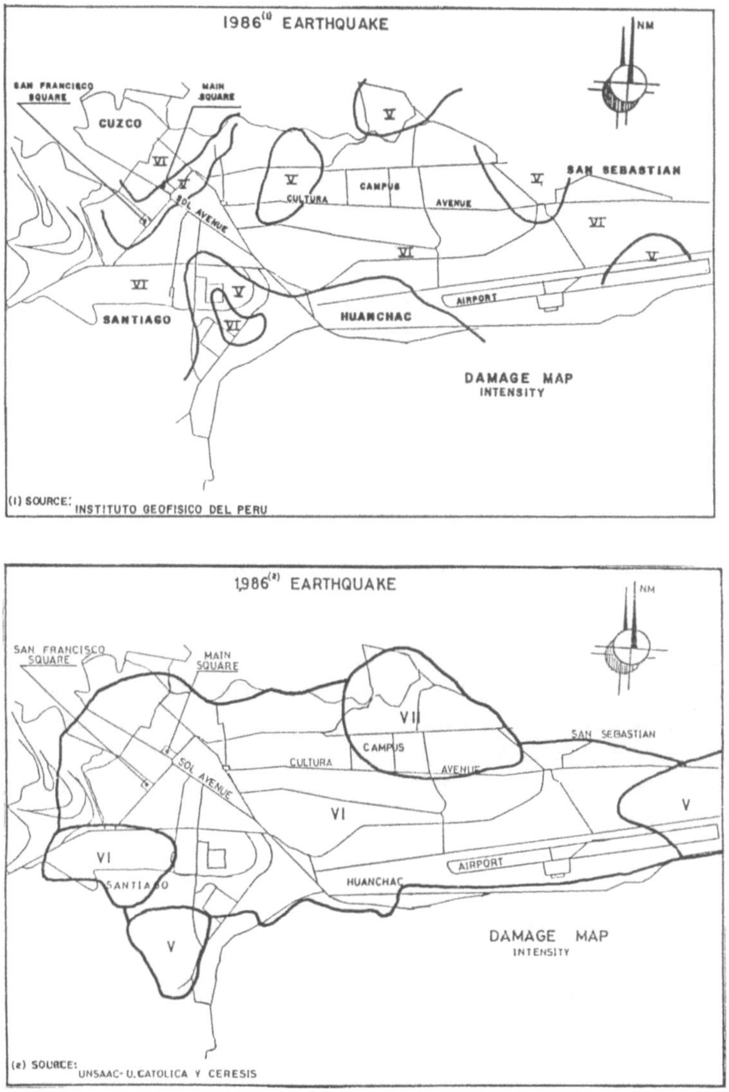

Figure 6. Geologic Risk Maps
(Instituto Geofísico del Perú, 1986, and UNSAAC-CERESIS-U Catolica, 1986)

7. HAZARD ANALYSIS

A methodology, suggested by Thiel and Zsutty (1987), applies peak ground accelerations and soil related factors as independent variables in their seismic damage functions. For this study, PGA having a 10 percent chance of being exceeded over a 50-year period is

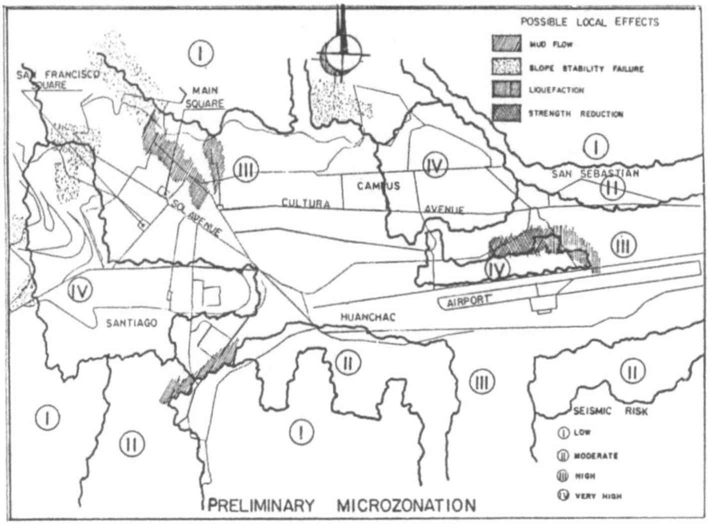

Figure 7. Local Effects and seismic risk.

estimated to be 3.5 g for the Cusco area, as shown in the PGA map of Figure 8, (Candia, 1990, and Candia & Sprenke, 1992). This probabilistic value was calculated using statistical data for the last 100 years with selected attenuation relationships. In a related deterministic study, currently undergoing, using the Tambomachay Fault as a local seismic source for the city of Cusco, Candia and Perez estimated the maximum ground acceleration to be 0.38 g.

8. BUILDING INVENTORIES

Man-made structures in Cusco consist primarily of buildings of low rise importance and middle rise office buildings. In general, buildings can be classified according to their age and structure, configuration in plan and height, as follows:

A. Masonry stone buildings prior to 1900, in bad and regular shape. Freq > 3 Hz.
B. Adobe buildings prior to 1950. Freq. > 3 Hz.
C. Adobe buildings constructed or retroffited after 1950. Freq > 2.5 Hz.
D. Modern RC buildings designed for some lateral load. Freq < 2 Hz.

The characterization of the building stock has been defined based on: use, structural type, location, plant configuration, structural design, state of conservation, and material. Vulnerability curves were assigned to each category taking into consideration the type, age, natural frequency of vibration, and the statistics. Vulnerability is shown in Table 1.

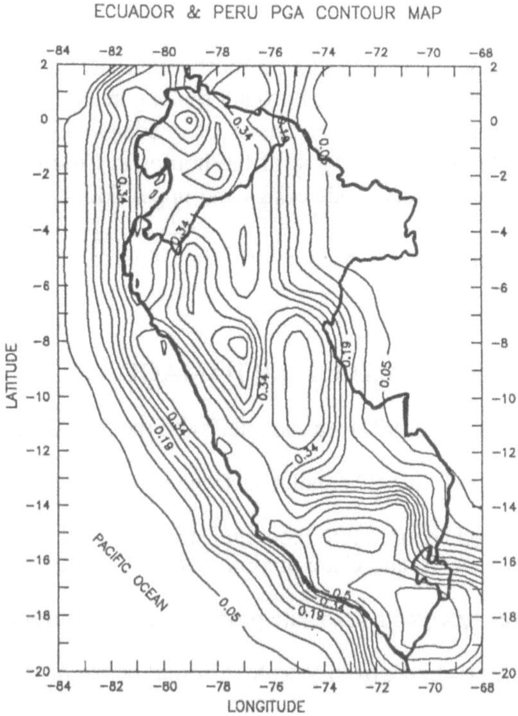

Figure 8. PGA Map: 10% probability exceedance, time period of 50 years. (Candia & Sprenke, 1992).

TABLE 1. Vulnerability of buildings in Cusco

BUILDI NG TYPE	INTENSITY (M.M.)			
	VII	VIII	IX	X
A	1%	5%	10%	50%
B	2%	10%	30%	70%
C	0.5%	1%	5%	10%
D	0.1%	0.5%	1%	2%

9. GENERAL CHARACTERISTICS OF STRUCTURES

The majority of buildings in Cusco were constructed principally of adobe and combined adobe and rock or brick structures. Three major factors make adobe buildings the most important single source of structural damage and associated casualty risk due to earthquake ground motion in Cusco: a) adobe buildings represent the most seismically vulnerable type of buildings in Cusco Valley, b) zones of higher expected ground shaking intensity are occupied mostly by adobe buildings, and c) compared to other classes of buildings, adobe buildings tend to be occupied at relatively higher densities. Most of the houses in Cusco are one or two stories high. Adobe walls are not reinforced, and ceilings and upper floors are supported by wooden logs. Roofs are made with wood and covered with reeds to support the roof tiles. Wooden structures practically are limited to the roof supports; thus, fire hazard is minimum.

10. MODEL TO ESTIMATE EXPECTED DAMAGE

A model developed by Mendes (1991) was adapted with the following assumptions:

1. The study area was divided into 22 smaller sections i (i=1,u); covering the whole city.

2. Buildings were classified according to their age, structure, and configuration, in 4 types j, (j = 1,c). Buildings type A, B, and C are predominantly 2 stories.

3. Seismic sources, k (k=1,cs), with associate probability of occurrence Fk(.)

4. MMI intensity levels, l (l=l,int), from V to X. Ground behavior is incorporated.

5. Four periods were considered for time occurrence of earthquake, m (m=1,P): 6-9, 9-18, 18-21, 21-6 hours.

The modeled functions are given by:

ILF = Individual loss function:

$$ILF_{i,j,k} = C_{i,j}a_i n_i N_{i,j} \Sigma\, V_{i,j,k,l}(a_{l,j})/A_j$$

GLF = Global loss function:

$$GLF_{j,k} = \Sigma\, ILF_{i,j,k}$$

AP = Affected population:

$$AP_{j,k,m} = \Sigma\,\Sigma\, V^{P}_{i,j,k,l}(a_{l,j})/A_j\, N_{i,j}\, P_{j,m}$$

DL = Density of losses index:

$$DL_{j,k} = GLF_{j,k} / \Sigma\, C_{i,j}a_i n_i N_{i,j}$$

DPA = Density of population affected index:

$$DPA_{j,k} = AP_{j,k,m} / P_{j,m}$$

Global loss function for all the sources:

$$GLF_j = \int GLF_{j,k} dF_k$$

where:

$S_{j,k,l}(w)$= power spectrum response
$V_{i,j,k,l}$ = mean vulnerability for $S_{j,k,l}(w)$
$V^{Pi,j,k,l}$ = mean value of affected population
A_j = area of the unit j
$N_{i,j}$ = amount of buildings i in unit j
$a_{l,j}$ = area of MMI intensity l in unit j
$P_{j,m}$ = population in unit j, in time m
$C_{i,j}$ = mean household value/m^2 of building class i in unit j

11. CONCLUSIONS AND RECOMMENDATIONS

An important geotechnical database has been collected and extrapolated for the city of Cusco, Peru.

Control technologies as required for civil structural applications such as building design and construction in the Cusco area clearly need further development and enhanced efforts utilizing better quality materials, updated construction regulations and urban planning involving geotechnical and structural aspects.

It is important to document the repair and retrofitting of damaged structures as well as the design and construction of new ones so that their effectiveness can be valued in the future.

Details of the tectonic environment and geologic processes, must be evaluated in order to estimate reliable boundaries of seismic source zones that can be used for a local seismic hazard analysis of the Cusco area. An improved building inventory is needed as well for further enhance risk assessment in the study area.

REFERENCES

Aparicio, M., O., And Marmanillo E., 1989, "Cusco, Sismo 1986, Evaluación de Inmuebles del Centro Histórico," Instituto Nacional de Cultura, Cusco.

Benavente, R., 1992, "Ocurrencias Sísmicas en la Región Inka," Sp. Report, College of Civil and Geological Engineering, University San Antonio Abad, Cusco, Peru.

Candia, M. A., 1989, "Seismic Hazard Analysis in Peru," Civil Eng. Master's Degree Thesis, University of Idaho, Moscow, Idaho.

Candia, M. A. and Sprenke, K., 1992, "Probabilistic Peak Ground Acc. Map of Peru & Ecuador," Proc. Intl. Symposium on Ethqk. Disaster Prevention, Mexico, D.F., Mexico.

Ericksen, G. E., Fernandez, J., and Silgado, E. (1954),"The Cusco, Peru, Earthquake of May 21, 1950, "Bul. SSA, Vol. 44, #2A, pp. 97-112.

Gregory, H. E. (1916), "A Geologic Reconnss. of the Cusco Valley, Peru," Am. Jour. of Science, 4th Srs., Vol.XXXVII, #220, XXII, pp. 289-298.

Gutenberg, B. and Richter, C. F. (1954), "Seismicity of the Earth and Associated Phenomena," Princeton University Press, Princeton, New Jersey, 165p.

Huaco, P., Minaya, M. and Deza, E., 1986, "Intensidades Causadas por el Sismo de Abril 5, 1986 en la Ciudad del Cusco," Inst. Geofísico del Perú, Sector Educación.

Marocco, R., (1977), "Geólogie Des Andes Peruviennes," Ph.D. Thesis, Université des Sciences et Techniques du Languedoc, France.

Mendes Victor, L.A., 1991, "Elements for Seismic Hazard Evaluation," Ethqk Hazard Assessment, Com. the European Communities, Rep. EUR 13408 EN, pp. 99-118.

Municipalidad Cusco-U. Ingenieria, 1988, "Registro de Microtrepidaciones," Plan Director del Cusco, Geotecnia del Valle del Cusco y Zonas de Expansión.

Polo, J. T., 1904, "Temblores del Peru: Sinopsis de 1647 a 1904," Bol. Soc. Geog. de Lima, anexos del Tomo XVI, pp. 91-118.

Thiel, C. C. JR. and T. C. Zsutty, 1987, "Earthquake Parameters and Damage Statistics," San Francisco: Forell/Elsesser Engineers.

Universidad "San Antonio Abad, Cusco," 1986, "Damage Assessment on Adobe Structures for the April 5, 1986 Earthquake," Sp. Report, C.E. Dept., Cusco, Peru.

Universidad "San Antonio Abad, Cusco," (1986), "Damage Assessment on Adobe Structures for the April 5, 1986 Earthqke." Special Report C.E. Dept., Cusco.

REDUCING THE SEISMIC VULNERABILITY OF EARTH BUILDINGS

JENNY CHRISTIE
School of Architecture, Victoria University of Wellington, P.O. Box 600, Wellington, New Zealand

Abstract: In view of the world housing shortage, and the economic and technological barriers which hamper efforts to alleviate it, the development of traditional techniques of earth construction may be seen as desirable. Certainly earth buildings may be constructed extremely cheaply using simple skills and, provide dwellings which are comfortable and easily modified to suit changing family needs. There are however, many instances where the seismic collapse of earth buildings has caused high death tolls, and building codes generally discourage this type of construction in seismic areas.

Investigation of earthquake damage to earth buildings has shown that their vulnerability may be significantly reduced by incorporating timber or concrete reinforcing elements, and by attention to building layout and detailing.

This paper discusses the results of this research and looks at the approach being taken by engineers in New Zealand to the preparation of a code of practice for earth construction.

1. INTRODUCTION

For more than 10,000 years earth has been used as a construction material, and in many areas including parts of the Middle East, Africa, India and South America it is still widely used, particularly by poor or rural people. In the early 1980's it was estimated (Dethier 1982) that more than one third of the world's population were living in earth dwellings, and in some areas of rural Iran, Pakistan, India and Sudan, almost 90 percent of the houses are constructed of earth or of stone rubble set in clay mortar (El Fadil and Osman 1984, Sharma 1989). Typically the occupant family has a substantial involvement in the construction of earth dwellings (Denker 1989).

Since the 1970's earth construction has also been steadily gaining popularity in developed countries, particularly in Australia and parts of the United States where the climate is hot and dry, and the thermal properties of massive earth walls may be utilised in designing for energy efficient thermal comfort. Increasing awareness of the environmental costs

M. I. El-Sabh et al. (eds.), Land-based and Marine Hazards, 37–48.

associated with many building materials has also assisted this trend by highlighting earth as an "environmentally-friendly" form of construction.

Unfortunately many of the areas where earth construction is common are seismically active and the effect of seismic shaking on earth buildings has resulted in a number of disasters, some of which have involved substantial loss of life. Twenty-five thousand people were killed in the Guatemala earthquake in 1976, of whom 98 percent died as a result of the collapse of adobe houses (Cano 1977, Meli et al 1980). The 1972 Qir earthquake in Iran caused the collapse of 3,200 buildings, most of which were constructed of adobe or unreinforced masonry, and killed 5,300 people, representing 23 percent of the population in the effected area (Ambraseys et al 1972, Razani and Lee 1973). In 1993 the earthquake centred near Latur, India, once again demonstrated the seismic vulnerability of earth construction.

Concern over the hazard posed by earth buildings in seismic areas, coupled with the increasing popularity of this type of construction in developed countries, has emphasised the need for increased understanding of the structural behaviour of earth buildings, and has generated a considerable amount of research in this area over the last ten to fifteen years. This has coincided with greater awareness of the global costs of disasters, and the importance of development strategies in disaster mitigation. In this environment, the need to understand and promulgate seismic-resistant earth construction is urgent in both developed and developing areas.

This paper firstly looks at the seismic vulnerability of traditional earth buildings, and then discusses the approach being taken by earth builders and engineers in New Zealand in preparing a code of practice for earth construction.

2. EARTHQUAKE DAMAGE

Earth, being a brittle material, tends to fail suddenly and completely, giving occupants little time to escape. Like brick masonry or concrete, it is much stronger in compression than in tension, and for this reason, performs best in situations where it is subject only to compressive stresses such as occur in walls supporting the gravity load of a roof or upper storey. During an earthquake the ground accelerates, usually in a predominantly horizontal direction, and building elements are subjected to horizontal forces which are proportional to the mass supported. While traditional construction methods have developed to provide perfectly adequate resistance to vertical gravity forces, there is often no provision for the rare but devastating horizontal loads induced by seismic accelerations.

The degree to which a building is damaged by an earthquake depends on the characteristics of the seismic vibrations it experiences as well as on the resistance provided by the structure. The amplitude and frequency of the earthquake waves depend upon the magnitude and location of the original shock, and are modified by the ground through which the waves travel. Damage caused by shaking is most severe when the natural period of vibration of the building corresponds to the period of the applied earthquake waves and resonance occurs, resulting in greatly increased displacements.

Traditional earth buildings tend to be small heavy rigid structures, which are usually weak in a number of areas, and lack much surplus strength. When additional stresses are imposed by an earthquake, the weakest effected part fails first and ceases contributing to the overall load resistance. The remaining structure is then subjected to increased stresses, and failure occurs more rapidly.

3. ASEISMIC EARTH BUILDINGS

Earth is a comparatively weak construction material, particularly in tension, so it is important that the available strength is well utilised in resisting seismic forces. In this section the main considerations in constructing earth buildings in seismic areas are outlined under the headings of material strength and design issues.

Only the three most common types of wall construction using earth alone are considered here - mud brick (adobe), pressed brick and rammed earth (pisé). There are also a wide range of other techniques, such as wattle and daub, which use earth in combination with other materials. The issues of earthquake damage to buildings constructed using these techniques are similar, but the relatively low mass of earth involved, and the presence of timber reinforcing elements means that a reasonable level of seismic resistance may be more easily achieved.

3.1 Material strength

It is first necessary to consider the factors which effect the strength properties of the material itself:

i) Soil strength is effected by the distribution of particle sizes. Too much fine clay in the soil can result in severe shrinkage upon drying, while the presence of too much coarse sand means it will lack cohesiveness and will tend to crumble.

ii) Moisture content and compaction at the time the element was formed effects the density and thus strength of the material. Blocks formed by mechanical compaction of a fairly dry soil mix are stronger than those formed by pouring a wet mix into a mould. The effect of particle size, compaction and moisture content on soil strength are inter-related, and a change to one will modify the effect of the others (Etzion and Saller 1987, Hughes 1983, Lunt 1980).

iii) Additives such as lime or portland cement help glue the soil particles together, while bitumen emulsion improves the soil's water resistance (Lunt 1980).

iv) The presence of cracks reduces soil strength, and shrinkage cracks need to be minimised, particularly during the drying process. Slow uniform drying, out of direct sunshine is best. Straw is often added to control cracking, although it does reduce the density and therefore compressive strength.

v) Protection from damage by weather or abrasion. Commonly this is achieved with a coating of mud plaster which is regularly maintained. In wetter climates, wide eaves assist in protecting walls from heavy rain.

3.2 Design issues

i) Adequate foundations are required to avoid cracking due to ground settlement as earth, being a brittle material, is particularly sensitive to differential movement. Cracks, whether caused by settlement or by shrinkage, significantly reduce wall strength and increase vulnerability to seismic damage. Foundations are also important in protecting the base of walls from erosion by surface ground water.

ii) The proportions of an earth wall effect its stability and strength, and in order to resist forces perpendicular to its face, a wall should be thick. A long or high wall needs adequate lateral support. This is most easily provided by means of transverse walls and a

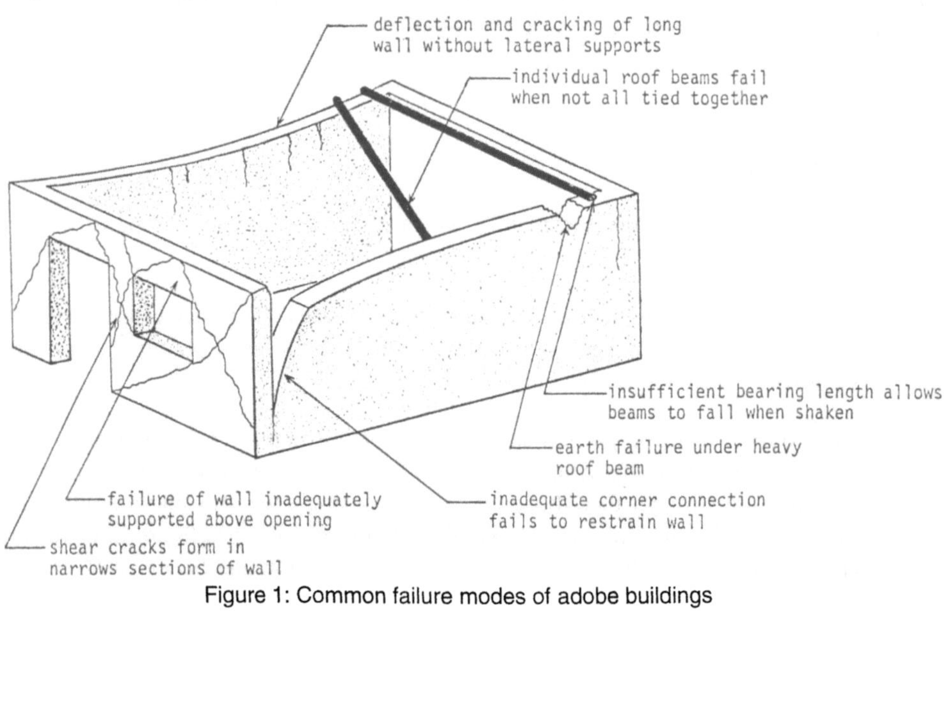

Figure 1: Common failure modes of adobe buildings

roof diaphragm which can provide support at the top of the wall.

iii) The detailing of structural connections is critical, and care needs to be taken in attaching stronger elements to the relatively weak earth material. Poorly designed connections may result in dangerously high local stresses under seismic forces, and may be inadequate to transmit the loads to other parts of the structure. In resisting horizontal forces, it is particularly important that perpendicular walls are well tied together where they meet at corners, and that the roof structure is able to act as a structural diaphragm and is securely connected to the top of the walls (Wasti 1986).

iv) Wall openings such as windows and doorways reduce the overall wall strength. Earthquake forces cause particularly high stresses to develop at the corners of openings, and in narrow sections of wall between adjacent openings. Windows and doorways should therefore be few in number, located well away from each other and from the ends of walls, and kept as small as possible. Strong lintel beams are required where heavy wall material is to be used above an opening (Figure 1).

v) Heavy flat earth roofs greatly improve thermal comfort in extreme climates, and are traditional in many areas, but it is not uncommon for the practice of applying a new mud plaster coating to the roof during regular maintenance to eventually result in a depth of almost 500 mm of earth (Keightley 1975, Wasti 1986). In an earthquake the large mass of a heavy earth roof imposes high forces at the top of the supporting walls and may cause them to fail.

Upper stories constructed with earth floors or earth walls impose similar forces, and it is therefore desirable to limit earth construction to single storey, or to require upper stories and roofs to be of light weight construction.

vi) The planning or layout of earth buildings should be compact and fairly symmetrical. This allows seismic forces to be distributed more uniformly throughout the structure and minimises the development of areas of particularly high stress.

4. CODES OF PRACTICE FOR EARTH CONSTRUCTION

Building codes are used primarily to ensure that buildings are constructed in such a way as to safeguard people from injury or illness, and to safeguard property from damage. For this purpose, codes contain various specifications relating to structural strength and to health requirements. In most countries these provide a limited number of allowable options by specifying the way in which a range of approved materials may be used.

The structural performance of an earth building depends on many factors as briefly indicated in the previous section. In particular, the material properties depend on such variable factors as the soil composition and moisture content, and can exhibit considerable variation even when care is taken to maintain consistency. This variability makes strict quality control difficult, and needs to be taken into consideration in developing codes based upon dependable strength properties.

In most countries there is little, if any, provision in building codes for earth construction, with the result that such buildings are commonly constructed without the normal safety standards being applied, and without the processes of formal approval and inspection. The thoroughness with which building codes are enforced varies greatly between different countries and even different provinces, and is often hampered by a shortage of adequately trained personnel, and by the relatively high costs of the specified reinforcing materials. However the fact remains that often those constructing earth buildings receive little guidance from codes of practice.

The Turkish Earthquake Code (Aytun 1981, Wasti 1986, Yorulmaz1981) is one which does contain a section on adobe buildings. It specifies strength and durability criteria for adobe blocks, and provides maximum and minimum allowable dimensions and spacings for walls and other building elements. A similar approach was taken in developing a proposal for a section on adobe construction in the Peruvian Seismic Resistant Code (Vargas 1981). In the western United States the predominant building code, the Uniform Building Code (UBC), also permits the use of stabilised adobe blocks in a cement mortar, and like the Turkish code, specifies wall dimensions (Barnes 1989). In the 1950's in Australia, the Commonwealth Experimental Building Station (now the National Building Technology Centre) developed specifications for earth construction which were published as Bulletin 5 (Middleton 1987). Unlike the Turkish code and the UBC, this bulletin includes provision for rammed earth as well as adobe and pressed block. Despite the availability of Bulletin 5 there was little activity in earth construction in Australia until the third edition was published in 1981 when a number of Local Governments began recognising it as means of compliance with the building regulations (Schneider 1989).

All of these examples take an empirical approach, defining permissible strengths, techniques and dimensions. The Australian Bulletin 5 provides the most comprehensive guidance for earth construction, but in general the codes which have developed so far are overly restrictive, requiring a sophisticated level of construction, and allowing little scope for alternative approaches to achieving adequate structural performance.

This perhaps reflects the reliance that is placed on tradition and historical data in our understanding of the performance characteristics of earth buildings. Although a considerable amount of modern research has been carried out, particularly on material properties, engineering knowledge of the structural behaviour of earth buildings lags well behind that of concrete, steel or timber structures. However increasing pressure from earth builders who are dissatisfied with the limitations of the current "cookbook" codes is highlighting the need for a more thorough understanding of earth construction.

5. THE NEW ZEALAND BUILDING CODE

In New Zealand any proposal for construction of a building must be approved by the Local Authority before work is commenced. This ensures that the proposed building complies with town planning, sanitation and structural requirements.

Until recently approval was based on compliance with various New Zealand Standards which were adopted by the Local Authorities, and which specified a range of acceptable solutions. In 1991, a new Building Act established a performance-based building control system. The Building Code (NZ Building Industry Authority 1992) now specifies health and safety performance standards which a building must meet, but does not specify what techniques must be used to achieve them. Local Authorities still have the responsibility of ensuring that these are met before issuing permission to build. The existing New Zealand Standards are recognised as specifying acceptable methods of compliance with the Building Code, but there is now greater provision for innovative approaches to construction in that any combination of materials and techniques may be permitted if it is proven to meet the required performance levels. Figure 2 is an extract from the New Zealand Building Code, showing Section B1 which covers structure.

In New Zealand where earth is relatively uncommon as a modern construction material, proposals for earth buildings have usually required detailed engineering calculations before approval has been granted by the Local Authority. This is a time consuming and costly process, and few engineers have either interest or experience in the design of structural earth walls, which has tended to discourage prospective owners from persevering with such a nonstandard material. In contrast, the requirements of light timber construction, the most common form of domestic construction, are well documented in the New Zealand Standard NZS 3604. The majority of New Zealand houses are constructed in accordance with this document, and because it is well understood by builders, designers and Local Authority personnel, building approval is usually granted without difficulty, and a consistent standard is maintained.

6. AN EARTH BUILDING CODE OF PRACTICE FOR NEW ZEALAND

The Earth Builders' Association of New Zealand was established in the late 1980's to disseminate information and provide support for people involved in earth construction, and has been closely linked with the increasing interest in this area in recent years. The Association saw the need to develop documentation to assist owners in gaining Local Authority approval for earth building proposals. With support from the Earth Building Association of Australia, it persuaded Standards New Zealand and the Standards Association of Australia that a code of practice on earth construction should be developed as one of the new set of integrated Australia/New Zealand Standards. A joint technical

Figure 2. The New Zealand Building Code: Section B1 Structure (New Zealand Building Industry Authority 1992)

NEW ZEALAND BUILDING CODE

B STABILITY

B1 STRUCTURE

PROVISIONS

Objective

B.1.1 The objective of this provision is to:

(a) Safeguard people from injury caused by structural failure,
(b) Safeguard people from loss of *amenity* caused by structural behaviour, and
(c) Protect *other property* from physical damage caused by structural failure.

FUNCTIONAL REQUIREMENT

B1.2 *Buildings, building elements* and *sitework* shall have a low probability of rupturing, becoming unstable, losing equilibrium, or collapsing during *construction* or *alteration* and throughout their lives.

PERFORMANCE

B1.3.1 *Buildings, building elements* and *sitework* shall have a low probability of rupturing, becoming unstable, losing equilibrium, or collapsing during *construction* or *alteration* and throughout their lives.

B1.3.2 *Buildings, building elements* and *sitework* shall have a low probability of causing loss of *amenity* through undue deformation, vibratory response, degradation, or other physical characteristics throughout their lives, or during *construction* or *alteration* when the *building* is in use.

B1.3.3 Account shall be taken of all physical conditions likely to affect the stability of *buildings, building elements* and *sitework*, including:

(a) Self-weight,
(b) Imposed gravity loads arising from use,
(c) Temperature,
(d) Earth pressure,

Figure 2: (Continued)

(e) Water and other liquids,
(f) Earthquake,
(g) Snow,
(h) Wind,
(i) *Fire*,
(j) Impact,
(k) Explosion,
(l) Reversing or fluctuating effects,
(m) Differential movement,
(n) Vegetation,
(o) Adverse effects due to insufficient separation from other *buildings*,
(p) Influence of equipment, services, non-structural elements and contents,
(q) Time dependent effects including creep and shrinkage, and
(r) Removal of support.

B1.3.4 Due allowance shall be made for:

(a) The consequences of failure,
(b) The intended use of the *building*.
(c) Effects of uncertainties resulting from *construction* activities, or the sequence in which *construction* activities occur,
(d) Variation in the properties of materials and the characteristics of the site, and
(e) Accuracy limitations inherent in the methods used to predict the stability of *buildings*.

B1.3.5 The demolition of *buildings* shall be carried out in a way that avoids the likelihood of premature collapse.

B1.3.6 *Sitework*, where necessary, shall be carried out to:

(a) Provide stability for *construction* on the site, and
(b) Avoid the likelihood of damage to *other property*.

B1.3.7 Any *sitework* and associated supports shall take account of the effects of:

(a) Changes in ground water level,
(b) Water, weather and vegetation, and
(c) Ground loss and slumping.

Building Industry Authority, July 1992.

committee consisting of earth builders and engineers (including the author) from the two countries is currently working on a set of draft documents.

In examining the needs of earth builders in New Zealand and considering the way in which the various codes of practice relating to construction are currently used, it was found that the most appropriate approach would be to develop a set of three documents. The first will deal with the material strength. It will include specifications on soil type, moisture content, compaction, drying processes and the use of additives for rammed earth, adobe and pressed block construction, and details of appropriate test procedures. The second document, referred to as the Builders' Code, will define maximum and minimum dimensions for buildings which may be approved without the need for specific structural analysis. It will cover only single storey buildings and buildings with an upper level of light weight construction. This will be similar to the approach taken by the Turkish Earthquake Code and the American UBS, and will provide a set of rules which are conservative, but which will allow many earth houses to gain approval without the need for expensive engineering input. It is also essential that a third document, an Engineers' Code, is produced which enables appropriate engineering analysis to be applied to proposed buildings which do not comply with the Builders' Code. This will deal with analytical techniques, and in particular will cover the use of appropriate design strengths and safety factors when dealing with the variability of earth as a structural material. The guidelines set out in the Engineers' Code will allow designers to incorporate the advances in the understanding of earth's structural behaviour as the results of research gradually become available.

These codes of practice incorporate historical knowledge and the results of research carried out in many parts of the world, but are being developed for New Zealand and Australian conditions, and to reflect current practice. In order to gain recognition as an acceptable means of compliance with the New Zealand Building Code, it will have to be demonstrated that the specifications contained in these documents are adequate to achieve the relevant performance criteria. It is hoped that the Earth Building Codes of Practice will be complete and have full recognition within about a year.

7. CONCLUSION

Earth buildings have a history of poor seismic performance, but when well designed and constructed they provide a satisfactory level of resistance. Although soil properties are generally well understood, it is only recently that engineers have begun to examine the seismic response of buildings constructed of earth, and steady but limited work is being carried out on the analysis of structural behaviour. Substantial research in this area requires the financial support of either the construction industry or community interests. Although earth buildings are becoming increasingly popular in Australia, New Zealand and the United States, improved earth construction will neither contribute noticeably to profits in construction materials or services, nor significantly reduce the risk to life or property in developed countries. It is therefore likely to remain difficult to attract substantial funding for this type of research.

In New Zealand at least, it is clear that there is urgent need for some form of code of practice which will avoid the need for individual earth builders to enter into costly and protracted discussions with their Local Authority before gaining permission to build. While engineering knowledge of earth buildings remains so incomplete, any guidelines produced must be conservative, but as greater understanding of the seismic behaviour is gained, more efficient structural design will be possible.

The introduction of a performance-based Building Code in New Zealand has provided an opportunity for the Earth Builders' Association, as this Code makes clear provision for alternative construction materials and techniques to be used provided that the performance criteria are met. The task of developing a set of specifications which will satisfy the requirements of earth builders, and of the Building Code is proving to be a substantial one, but it will greatly assist earth building in New Zealand, and it is hoped that it may prove to be a useful model for elsewhere.

Ultimately however a building code cannot improve construction unless it is successfully applied. If building codes are to be instrumental in reducing the vulnerability of earth buildings in seismically active areas, their requirements must be achievable in terms of the cost and skills involved, and mechanisms must be in place for disseminating information and for ensuring that required standards are met.

REFERENCES

Ambraseys, N.N., Moinfar, A.A., and Tchalenko, J.S., (1972). Gir Earthquake of 10 April 1972, Serial No. 2789/RMO RD/SCE, UNESCO , Paris.

Aytun, A., (1981). Earthen Buildings in Seismic Areas of Turkey. In: *Proceeeings, International Workshop: Earthen Buildings in Seismic Areas* (University of New Mexico, Albuquerque), Gerald W. May, editor, 2, 345-72.

Barnes, R.E., (1989). Current Uses of Earthen Materials in the USA. In: *Proceedings,Self Help Housing* (University of Melbourne, Australia), Dennis Ingemann and Allan Rodger editors, 143-45.

Cano, J.H., (1977). Guatemala Earthquake of Feb 4th 1976: Description and Analysis of Damage caused on Buildings. In: *Proceedings, 6th World Conference on Earthquake Engineering* (New Delhi, India), Sarita Prakashan, editor, Indian Society of Earthquake Technology, Meerut, India, 299-301.

Denker, D., (1989). Integration of the Timeless Principles. In: *Proceedings, Self Help Housing* (University of Melbourne, Australia), Dennis Ingemann and Allan Rodger, editors, 163-68.

Dethier, J., (1982). Down to Earth: Mud Architecture an Old Idea, a New Future, trans. Ruth Eaton, Thames and Hudson, London, 192 pp.

El Fadil, A.A. and Osman, M.A., (1984). Factors Influencing the Use of Earth as a Building Material in Sudan. In: *Proceedings, Earth Construction Technologies Appropriate to Developing Countries* (Brussels, Belgium), Post Graduate Centre for Human Settlement editor, Katholieke Universiteit, Leuven, 111-29.

Etzion, Y. and Saller, M., (1987). Earth Construction: A Review of Needs and Methods. *Architectural Science Review*, 30 (2), 43-8.

Hughes, R., (1983), Material and Structural Behaviour of Soil Constructed Walls. Monumentum (United Kingdom), 26 (3), 175-88.

Keightley, W.O., (1975). Destructive Earthquakes in Burdur and Bingol, Turkey, May 1971: Report and Recommendations, Montana State University and National Academy of Sciences, Washington D.C.

Lunt, M.G., (1980). Stabilised Soil Blocks for Building, Overseas Building Notes, British Research Establishment, no. 184, 15 pp.

Meli, R., Hernandez, O. and Padilla, M., (1980). Strengthening of Adobe Houses for Seismic Actions. In: *Proceedings, 7th World Conference on Earthquake Engineering* (Istanbul, Turkey), Turkish National Committee on Earthquake Engineering, Ankara,465-72.

Middleton, G.F., (1987). Earth Wall Construction, 4th Edition revised by L.M. Schneider. National Building Technology Centre Bulletin 5, 65 pp.

New Zealand Building Industry Authority, (1992). New Zealand Building Code Handbook. Wellington.

Razani, R. and Lee, K.L., (1973). The Engineering Aspects of the Qir Earthquake of April 10 1972 in Southern Iran, Academy of Sciences for National Academy of Engineering, Washington D.C.

Schneider, L.M., (1989). The Formalisation of Earth-Wall Construction in Australia. In: *Proceedings, Self Help Housing* (University of Melbourne, Australia), Dennis Ingemann and Allan Rodger, editors, 147-49.

Sharma, S.K., (1989). Back to Mud: A Step Forward. In *Proceedings, Self Help Housing* (University of Melbourne, Australia), Dennis Ingemann and Allan Rodger, editors, 133-41.

Vargas, J.N., (1981). Adobe Constructions: Basis for a Seismic Resistant Code. In: *Proceedings, International Workshop: Earthen Buildings in Seismic Areas* (University of New Mexico, Albuquerque), Gerald W. May, editor, 457-502.

Wasti, S.T., (1986). Adobe Structures: Current Practice and Research Needs. In: *Proceedings, Middle East and Mediterranean Regional Conference on Earthen and Low-Strength Masonry Buildings in Seismic Areas* (Middle East Technical University, Ankara, Turkey), Mustafa Özder Erdik, editor, 581-88.

Yorulmaz, M., (1981). Turkish Standards and Codes on Adobe and Adobe Construction. In: *Proceedings, International Workshop: Earthen Buildings in Seismic Areas* (University of New Mexico, Albuquerque), Gerald W. May, editor 2, 503-19.

SOME GEOLOGICAL HAZARDS IN EGYPT, THEIR NATURE AND EXTENT

AHMED FAROUK KAMEL
Nuclear Materials Authority, Maadi, P. O. Box. 530, Cairo, Egypt

Abstract: The present paper deals with the geological investigation of the sites susceptible to natural hazards in Egypt. Earthquakes, landslides, and the movement of sand dunes are the most important events causing future disasters. Dealing with the earthquakes, Egypt is a country of low to moderate seismic hazard and earthquakes are active in the northern parts of the country and along the western side of the Red Sea. In the northern parts the dominant rock types are related to the Oligocene, and along the western side of the Red Sea coast the land is covered by the Quaternary deposits. Correlating the seismicity with large - scale tectonic features such as faults, it was found that NNW and E - W faults are responsible for some significant seismic activity. The areas with relatively low seismic activity are covered by thick marine limestone of Cretaceous age lying above the Nubian sandstone in the larger part of the Western Desert. The basement rocks in the Eastern Desert have also low seismic activity The Nile Valley and Delta areas are underlain by thick alluvial sequences that are likely to provide a much less secure foundation during earthquakes than the crystalline bedrock in the Eastern Desert. Damage to buildings is related to clay thickness. Clays swell and slake in water leading to vulnerability of the buildings in areas with considerable clay thickness.

The poor mechanical properties and structural features of most geological formations, the seismic characteristics of the territory, the intense rainfall as well as the human activities are the main causes of landslides. On the other hand, vast areas in the Western Desert is covered by moving sands which can be considered an important event in natural disaster. Sand dunes to the north of the Nile Delta started to encroach the cultivated land. To avoid dune movement over areas of future development, the location should be proportional to the rate of barchan movement.

1. INTRODUCTION

Egypt is bounded to north by the east Mediterranean Sea, to south by Sudan, to the west by Libya and to the east by the Red Sea - Gulf of Aqaba region (Fig. 1). Future disaster is considerably great in the uncontrollably fastest growing cities in Egypt. The present work deals with the study of some geological hazards in the areas which are expected to be subjected to natural disasters. Sedimentary rocks are likely to provide a much less secure foundations during earthquakes than the crystalline bedrocks.

M. I. El-Sabh et al. (eds.), Land-based and Marine Hazards, 49–69.

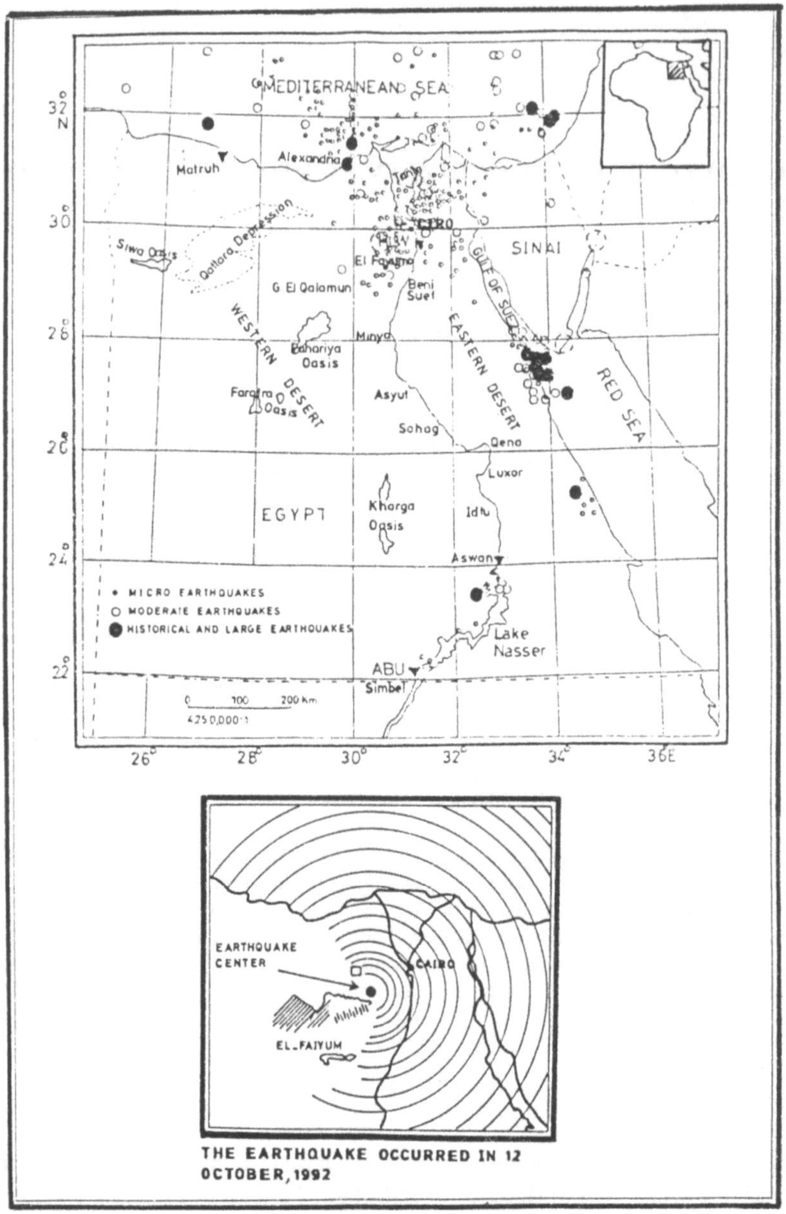

Figure 1: Epicentral distribution of earthquakes which took place in Egypt in the period 2200 B.C. till 1989 (After kebeasy et al., 1982) and the earthquake that occurred on 12 October 1992.

The greatest influence on building vulnerability to seismic shaking is the poor construction standards and thickness of the clay bed in the foundation bedrock. The stability of buildings is variable according to the lithological and structural characteristics of the bedrock. High risk buildings occur in modern cities of Egypt as a result of poor standards of construction as well as the presence of thick layer of clay which causes landslides.

Areas in Western Desert and Northern Egypt are covered with sand dunes. The sand dunes to the north of the Nile Delta and several parts of the Western Desert suffer from hazards due top moving sand. Recently, the high population density constitutes a social and economic problem. The geological study of the developing areas can provide more informations which can help in disaster prevention in these areas.

2. EARTHQUAKES

Egypt is a country of low to moderate seismic hazard (El-Sayed et. al., 1994). It is affected by active faults in its northern part and along the western side of the Red Sea - Gulf of Suez coast. The importance of earthquake activity is rapidly increased remarkably in recent years in Egypt as a result of the industrialization of the country. In November, 1981 an earthquake with total magnitude 5.6 on Richter scale took place in the area of Kalabsha about 70 Km southwest of Aswan. This area is characterized by the presence of basic structural weakness and the pre- existing E - W and NNW faults which have been submerged since the lake Nasser reached its maximum level in 1976. The Aswan earthquake is an extremely important event due to its close location to Aswan Dam which impounds the second largest reservoir (Lake Nasser) in the world.

An earthquake occurred in Egypt on Monday 12 October, 1992 (Badawi and Mourad, 1994). The epicenter is located in Dahshour, about 60 Km to the southwest of Cairo (Fig. 1). The magnitude is 5.9 on Richter scale and the intensity involved damage of poorly constructed buildings and fracturing of other well constructed buildings in an area with radius about 50 Km. In 1847 an earthquake occurred in the same area with the intensity of VIII as given in terms of the Medvedec - Sponheuer - Karnik (MSK) intensity scale of 1964 (Degg and Doornkamp 1990). This earthquake is comparable to the earthquakes occurred in Aswan area in 1981 and the 1992 Dahshour earthquake. The epicenters of the Cairo (Dahshour) and Aswan earthquakes were found in the location of E - W faults which are considered as active faults. According to Said (1962) movements along this direction occurred as early as the middle Cretaceous and seems to have continued more intensively during later times.

Correlating the seismic activity with the geology of the area, Egypt can be divided into areas with high seismic intensity (IX - VI) and areas with relatively low seismic intensity (V to less than IV), as shown in Figure 2.

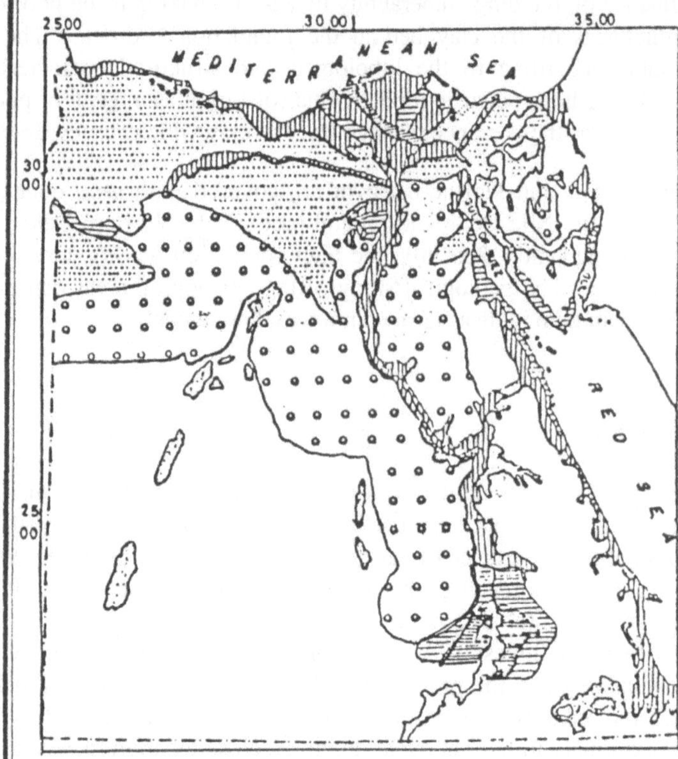

Fig (2) Seismic Regionalization map of Egypt (modified after Kebeasy et al., 1984) and the main geologic units.

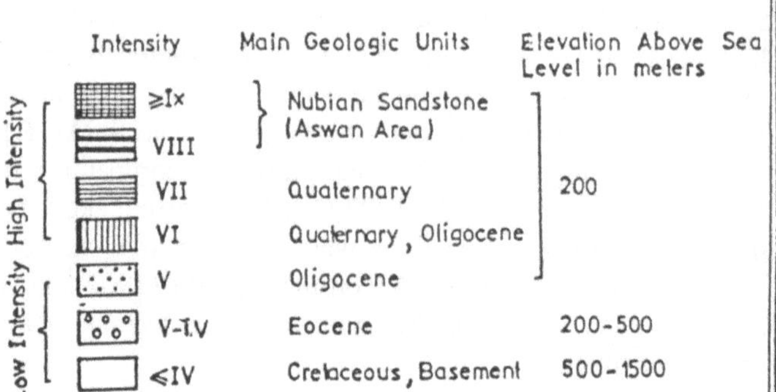

I .The areas with high seismic intensity are in the northern part of Egypt, the Nile Valley and Delta, along the western side of the Red Sea - Gulf of Suez coast and Aswan area. The time sequence of the rock exposure in the northern Egypt is Eocene, Miocene, Pliocene, Pleistocene and Holocene. They are mainly of limestone, marls and loose beach deposits. In September 1955 an earthquake with magnitude 6.7 occurred off the coast of Alexandria, causing catastrophic damage in many localities of Egypt. The principal shock was widely felt in the area represented by a circle with radius more than 300 Km in Egypt and the neighbouring area (Maamoun, et. al. 1978).

According to Said (1981) the beginning of Pleistocene epoch in Egypt was marked by episode of high seismicity. Degg and Doornkamp (1990) stated that earthquake risk is highest in the Delta region due to dense concentration of population and areas of high earthquake hazard are becoming of increasing concern because of the new development that are taking place. The risk is particularly low in the southwest part of Egypt due to a combination of very low population densities and low hazard status.

The eastern Aswan area is covered by well exposed basement igneous- metamorphic rocks of Precambrian age which are overlain by Nubian sandstone sediments and Wadi Natash volcanics of Cretaceous age (El-Shazly et al., 1974). The geology of the western Aswan is dominated by sedimentary succession varying from Cretaceous to Quanternary and basement rocks represented by small outcrops of Aswan monumental granite and associated rocks. Due to the fluctuation of water level in Lake Nasser, the 1981 earthquake still continue to occur around the northern part of the Lake but with low magnitude. Toppozada et al., (1984) concluded that Aswan earthquake makes the intersection of two major of pre - existing E - W and N - S faults which have been submerged since the Lake Nasser reached its maximum level in 1976. Kamel (1994) found that the E - W and NNW fractures have high density in the area around Aswan where seismic intensity is high.

Degg and Doornkamp (1990) stated that increase in the earthquake activity is largely man induced and related to the impounding of large volume of water behind the Aswan High Dam to create Lake Nasser. A set of fractures produced by the earthquake were observed in the recent deposits of Wadi Kalabsha to south of Aswan town. The ground cracks are vertical, some are straight while the others are sinuous. The main fractures trend N 60° E in clay and sandstone bedrock with length of 100m, width 30 cm and depth 1.5m (Fig. 3). The other fractures are bifurcating with the trends S 35° E, S 60° E, and N 80° E showing discontinuity and notable variation in thickness. The length is 5 - 15 m, width is 5 - 20 cm and depth is 60 - 75 cm (Fig. 4) .The main trends of fault in Aswan area are E - W and NNW.

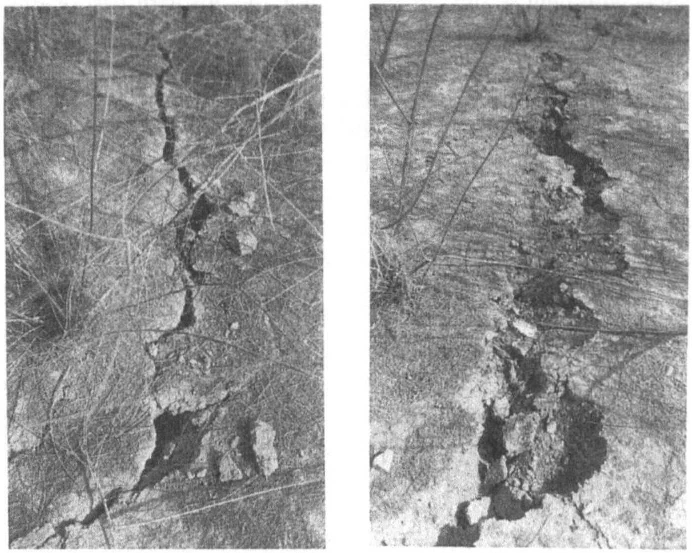

Figure 3: main fracturestrending S 60oE in clay and sandstone bedrock with length 100 m, width 30 cm and depth 1.5 m in kalbasha area to the southwest of Aswan town by 80 km.

Figure 4: Bifurcating fractures trending S 35 E, S 60 E and N 80 E in the alluvium/sand showing discontinuity and notable variation in thickness in Wadi Kalabsha southwest of Aswan. The length is 5-15 m, width is 5-20 cm and depth is 60-75 cm.

The Eastern Desert and Sinai form one unit whose geomorphology is closely connected with its geological structures.

The Arabo - Nubian massif made of igneous and metamorphic rocks of the basement complex occupies the eastern side of the Eastern Desert and extends to the southern part of Sinai across the Gulf of Suez. Along the western side of the Red Sea Coast, the area is covered by Quaternary deposits represented by coral reefs and raised beaches. The Red Sea earthquake of March 1969 with magnitude 6.3 is one of the largest events in recent history which has shaken the whole country and caused damage in an abnormally wide area in Cairo (Kebeasy, 1988). Kamel (1994) found that there are two broad zones of high density of ENE and WNW fractures in the structural map which are correlated with the same zone of high seismic activity on the western side of the Red as shown in the seismic map constructed by Kebeasy et al., (1982;1984). According to Kulhanix et al. (1992), the distribution of earthquake epicenters in the Red Sea shows that there is a seismic gap in the central part and a seismic connection between Egypt and the Red Sea.

II .The areas with low seismic intensity are covered with Oligocene to the north, Ecocene in the central part and Cretaceous to the south of Egypt, in addition to the basement rocks of Precambrian age in the Eastern Desert and southern Sinai. The Oligocene is continental to fluviatile sands, flint gravel with silicified wood and vertebrate remains, volcanic rocks are associated (Said, 1962). The Eocene rocks are thick marine limestone beds with chert and minor clay beds forming high cliffs and plateaux overlooking the Nile and covers most of the central part of the Western Desert, north Eastern Desert and central Sinai (The Egyptian Geological Survey, 1981). The Cretaceous is represented by grey shale overlain by phosphate rocks. Oyster bank is found at the base in southern Egypt and replaced by carbonate rocks in the north. Quartzo - sandstone, shale, marl, variegated sandstone and limestone are of wide distribution in Sinai and north Eastern Desert (Said, 1962).

The intensity of the earthquake is related to the distribution of lithofacies, structures and geomorphological setting. The calcareous rocks, marl and sandstone can be affected by earthquake more than the basement rocks. Earthquake damage is generally larger over soft surface deposits and Quternary pyroclastic layers. Hilly areas are affected by various intensities of earthquakes.

According to Ezzat (1965), The complicated seismic - geological conditions can be divided into near surface and deep conditions. The near surface conditions originate intense surface waves with big angle of incidence while the deep seismic geological condition are attributed to structural and stratigraphic reasons.

Dealing with the tectonic setting of Egypt, Degg and Doornkamp (1990), stated that the African plate is drifting in a northerly direction to collide with plates of the European land mass. The areas are associated with seismic activity of shallow and intermediate depth. To the east of Egypt lies the Red Sea rift. This is a zone of plate separation along which Sea - floor separation forces the African and Arabian plates apart. The zone is associated with shollow seismicity. The northern part of the Red Sea rift is particularly significant in terms of Egyptian seismic hazard. There are no major tectonic structures lying to the immediate west and south of Egypt in Libya and Sudan respectively. Both areas are situated within the African plate and are relatively stable. A similar situation exists within Egypt itself away from the margins of the major tectonic structures. Seismic activity in all of the continental (interplate) areas is of shallow focal depth.

Some active faults in Egypt, such as the E - W faults, are related to the continued movement of the Arabian and African plates. These active fault have their influence on the intensity of earthquakes.

Earthquake activity can be correlated with large scale tectonic features and landsat imagery interpretations have contributed effectively to the delineation of the major fractures. In turn this leads to the understanding of the structural setting of Egypt as an essential element in judging the cause of seismic activity and assessing its potential in the future. Basic structural weakness are inherent in the areas with abundant structural elements. Some of the fractures are proved to go deep in the earth crust for several kilometers. The epicenters of earthquakes are in the environs of important faults or fractures and their intersection. Maamoun and Ibrahim (1978) correlated the seismicity in Egypt with large - scale tectonic features and recognized two main seismic dislocation zones, the Red Sea - Gulf of Suez due to NNW faults and Egypt - Mediterrenean coastal dislocation zone due to deep E - W faults. These faults are responsible for some significant seismic activity. The determination of the main seismic zones in Egypt has been made by representing the geological structures in relation to the distribution of historical and recent earthquakes. Ibrahim (1988) stated that the size of the earthquake that takes place inside any region is a measure of the tectonic activity of this region. The Mesozoic and Cenozoic tectonic epochs in the earth's history are marked by rather pronounced processes of seismic activation.

3. LANDSLIDES

Landslides incidence correlates fairly well with earthquake magnitude. In Egypt landslides are often located in limited areas with some building or civil work. Therefore, the geological study is needed in the sites of land development and construction.

The mountainous backbone of Egypt is exposed in the Eastern Desert, running parallel to the Red Sea, in the southern part of Sinai and southern end of the Western Desert. These mountains have high relief and the highest peak is that of Gebel Katherina in Sinai.

Landslides and mass movement occur in limited areas. El-Etr et al. (1989) have shown that mass movement on slopes is greatest in mountains and hills which border the Red Sea tectonic zone including the Gulfs of Suez and Aqaba. These areas of high relief are characterized by a relatively rainy climate and steep, incised terrain susceptible to rockfall and landsliding, and are occasionally subject to high magnitude earthquakes.

Degg and Doornkamp (1990), stated that mass movements on slopes can be triggered by earthquakes .They added that although movements can occur on the most gentle of slopes, the potential for earthquake induced failure is greatest on unstable steep slopes and cliff lines.

The northern part of Egypt and the Nile Valley as well as Aswan area are used for commercial activities and habitation and the greatest influence on building vulnerability was the nature of the underlying subsoil. It is therefore of concern to note that high risk buildings are becoming increasingly prevalent in many of the modern cities and damage of buildings can be explained in terms of poor standards on construction.

The floor of the Nile Valley and most of the Delta surface is covered by an average of 9m - thick clay, bordering this floor are a series of gravel and sands of late Pliocene and Recent times (El-Nakkady, 1958). The Egyptian Nubian sandstone is characterized by the presence of clay layers and according to Basta et al., (1979), most of the clays swell and slakes in water, therefore in areas where the clay layers have considerable thickness, the building suffer vulnerability. This can be seen in Nasser city in Aswan area where the buildings constructed in 1967 have been subjected to fracturing. The detailed geological study of the area by the examination of the rock type through a lithostratigraphic section indicated that the underlying rocks belong to the lower unit of the Nubian sandstone which overlies uncomfortably the granite rocks and dip to the northwest by an angle ranging from 2°- 5°, (Fig. 5). The site of the fractured buildings is composed of sandstone and clay beds and when water reached the clay layers, the rocks swell and cause a type of sliding over the hard granite rocks leading to instability of the buildings. Therefore, the high clay content explains the instability of foundations. Quaternary pyroclastic layers are also affected by dangerous landslide especially where they overlay clayey materials.

The stability degree of limestone is variable according to lithological characteristics and structure as well as occurrence of clayey interlayers.

The area of Gebel (Mount) El-Mokattam district lies to the east of Cairo, where the youngest Eocene Formation is represented by foossiliferous marly limestone beds. The basal Pliocene conglomerate 7.5m thick is found in several localities alternating with gritty and clayey layers. Recently the new high building constructed over these sediments suffered cracking due to its bad material character and presence of clayey layered as well as a thick layer of conglomerate (Awad et al., 1953).

Moustafa et al., (1991) found that the slopes on the southern and southwestern sides of the upper plateau of Gebel El-Mokattam where the city was built, have been affected by slope failure and there is a retreat 1.7m/ year of the edge of the plateau where several buildings and road have been damaged. This is attributed to the fact that foundation bedrock is highly jointed and vertical joints act as potential failure surface. The limestone of the middle plateau are more suitable foundation bedrocks since they are not interbedded with expansive clays and the slopes are stable.

Recent studies carried out by Abdel-Tawab and Ibrahim (1991) on the New Minia City located on the eastern side of the Nile Valley in Upper Egypt, showed that the area is covered by inhomogeneous fossiliferous limestone which is considered by Said (1962) to belong to the Middle Eocene. The rocks are highly cavernous and dissected by large number of minor faults and joints which cause instability of the foundation bedrock. The detailed geological study revealed the presence of deep caves in the building site. These caves may be formed due to rain water or ground water passing through the fractures and the dissolution of soluble materials in the limestone.

It was found that similar caves occur in other cities in Upper Egypt such as Beni Suef where the bedrock consists of limestone. The future activity of urbanization of these cities may cause an increase of the ground water level which contribute to the unfavourable geologic conditions of the foundation bedrock.

Disaster prevention is closely linked to the geological and soil mechanic studies. Advance in geologic mapping of landslide terrain has emphasized the possibility to identify areas of high landslide risk. Core drilling is the most conventional means of investigation of the bedrock foundation and can be carried out before a definite study.

4. SAND DUNES

Sand dunes are important topographic features prevailing in the Western Desert and Northern part of Egypt (Fig. 6). El-Baz, (1979) revealed two major regimes, a narrow land along the Mediterranean Sea coast and the north - northwest Great Sand Sea in the Western Desert. Sand dunes movement is controlled by the effective wind and the size and length of the dune. In the present study we can consider the temperature variation and humidity in the different seasons as additional factors having their influence on the movement of sand dunes. The sand dunes in the Western Desert and northern Egypt can be dealt with as follows :

4.1 Sand Dunes in the Western Desert :

The western Desert is a rocky plain which slopes gently to the north and it is broken by the Qattara Depression and the Oases of Siwa, El-Faiyum, El-Bahriya, El-Farafra, El-

A

B

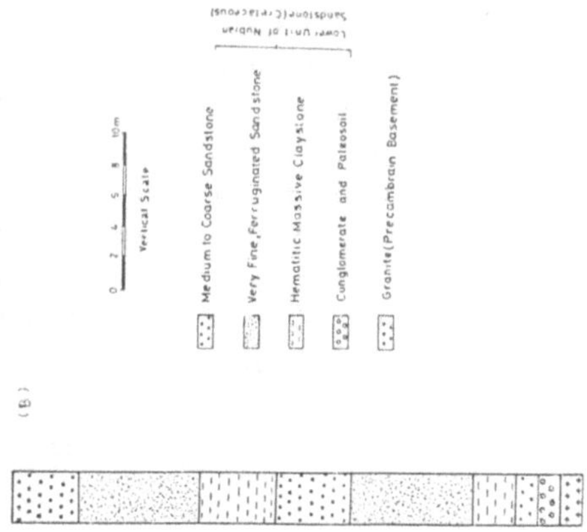

Figure 5: The Site Where the Buildings Suffered Cracking in Nasser City, Aswan Area (A) and a lithosgratigrahic Section in the Nubian Sandstone (B).

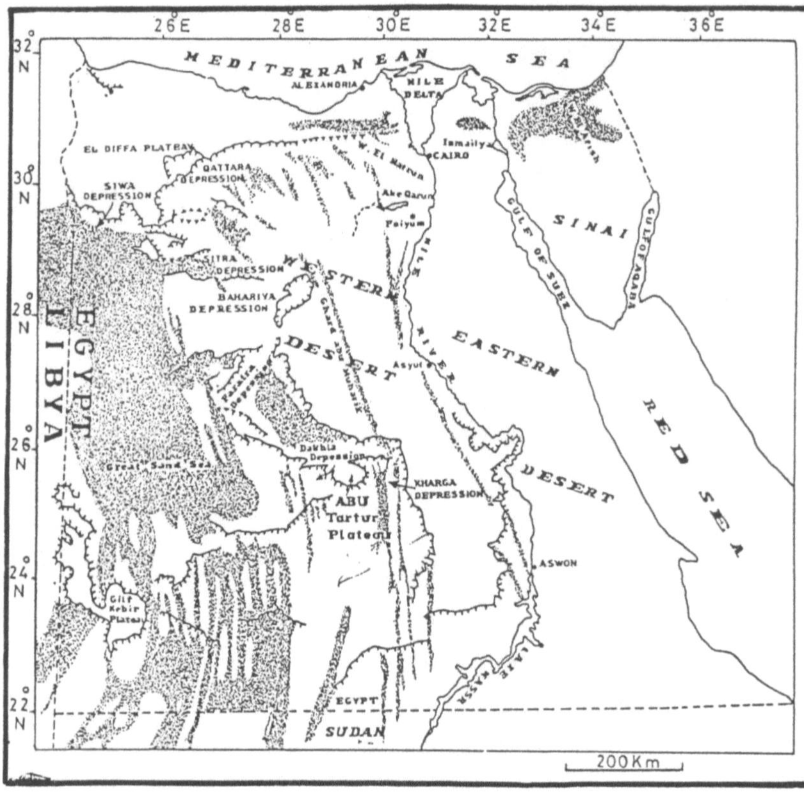

Figure 6: Sand dune belts in Egypt as mapped from Apollo Soyuz photographs and Landsat images (modified from Gifford et al.(1979) and El baz and Walfe (1981)).

Dakhla and El-Kharga. The drop in the topography in the depression is a factor affecting the change of the Sand Sea from one mass of sand extending for about 330 Km (Ashri, 1973), into separated brachans whose directions are affected by the surrounding escarpments. Gifford et al. (1979) stated that the dune orientation changes from north - northwesterly on the northern part to north - northeasterly in its southern part. The temperature recorded by the meteorological station in El-Kharga and Siwa Oases and El-Naturun Depression shows marked variation. It is greater in El-Kharga Oasis (13.9 - 32.2°C) and relatively lower in Wadi El-Naturun (13.3 - 27.6°C) reaching its maximum in June, July and August, (Fig. 7). The relative humidity is greater in Wadi El-Naturun (44 - 63%) and lower in El-Kharga Oasis (21.4 - 46%) increasing in November - January, and decreasing in May and June (Fig. 8). The relative speed of the wind is 1.8 km/hour, and in summer the winds are stronger and less variable (El Baz and Walfe, 1981).

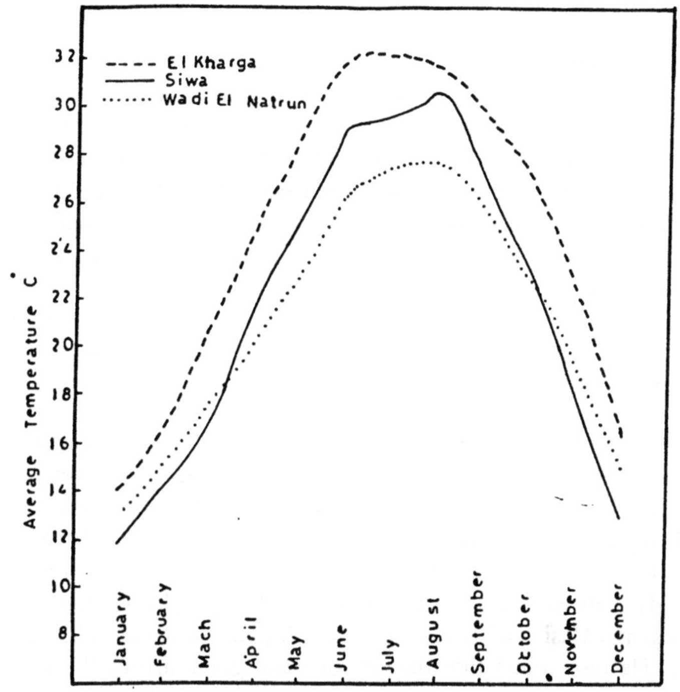

Figure 7: Average temperature in the depressions, western desert of Egypt.

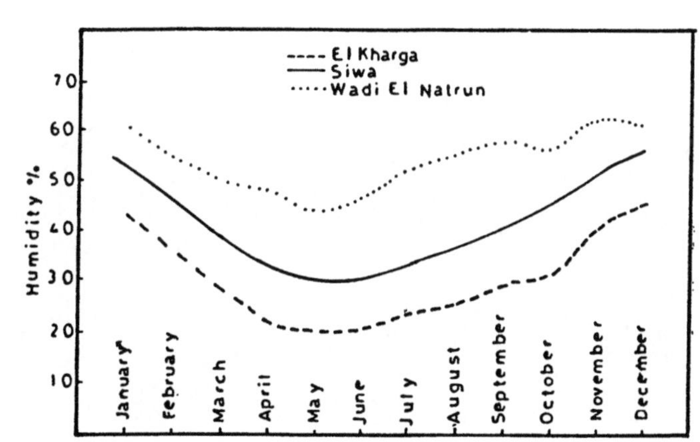

Figure 8: Relative humidity in the depressions in western desert of Egypt.

In El-Faiyum Depression, the direction of the sand dunes becomes towards the south, some have reached the borders of the fertile land and started to encroach the agricultural land. In Siwa Oasis, sand dunes are oriented in an east direction which is the main trend of the depression. The hills in the El-Bahariya Oasis form discontinuous lines in its center and appear to affect the local wind regime. The eastern half of the depression is covered by sand dunes of longitudinal type (Fig. 9A) which arrange to form a belt with sharp crest in the outer part of the depression (Fig. 9B). In El-Kharga Oasis, the dune of simple and compound type are oriented north - south on its eastern side and northeast in the northern part of the Oasis. The barchan vary in diameter from less than 40m to more than 500m (Ashri, 1973). Sand dunes in El-Kharga Oasis attack the buildings, roads and areas planned for future developments.

Generally, when the temperature reaches its maximum and humidity reaches its minimum, wind will be more effective in eroding the barren rocks and the movement of the sand dunes will be considerable. Field measurements carried out by Beadnell (1910) indicated that the annual displacement of the dunes in Western Desert varies between 10.8m and 18.8m. The same findings were reached by Ashri (1973) in his work based on a comparative study between two series of aerial photographs of brachans at various localities in El-Kharga Oasis and found that dune displacement varies between 20.8m and 100m per year. He also found that the predominant grain size of the dune sand range between 0.1mm and 0.4mm and the roundness varies with the grain size. The early Tertiary limestone to the north and the Nubian sandstone to the south can be considered as the possible source of the vast amount of sand (El-Baz, 1992).

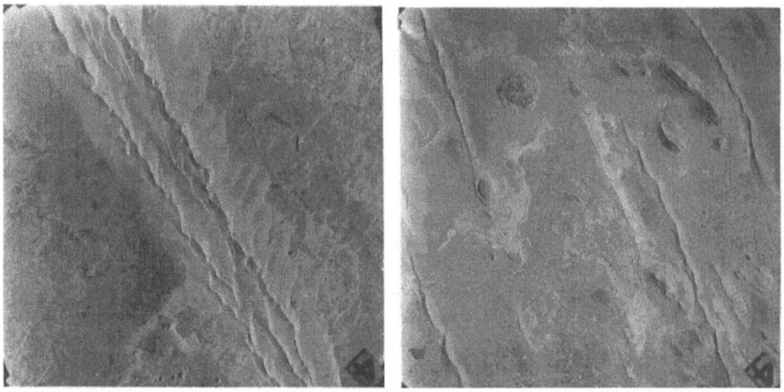

Figure 9: A bundle of closely-spaced parallel sand dunes (A) to the south of El bahariya oasis, and separate widely-spaced sand dunes in the oasis (B) where the trend of the dunes is controlled by the outcrops.

4.2 Sand Dunes in the Northern Egypt :

Complex sand dune belts dominate in the northern part of Egypt along the Mediterranean Sea shore. These dunes are of active type reflecting the convergence of several wind trends. The backshore plain is marked by narrow zone of low dunes. The moving sand dunes have their hazards on the areas covered by cultivation to the north of the Nile Delta.

The sand dunes in the northern Sinai are represented by a complex of crescentic and transverse dunes (Fig. 10A) and linear feathered dunes to the south (Fig. 10B). Kamel et al., (1982) found that the prevailing wind responsible for the formation of most sand dunes is directed from the northwest to the northeast, where many sand dunes attain the same trend. They found that to the south the dunes are generally extending parallel to the regional structure where the mountains control the direction of wind but further northwards along the Mediterranean Sea, the lack of relief results in a homogenous type of transverse sand dunes. The brachans join together to form longitudinal lines extending for few kilometers.

Sand dunes cover large area in Gamasa to the north of the Nile Delta. They are well developed to the west and are found in the form of feathered and barchan types (Fig. 11A) while in the eastern side of Gamasa the dunes are present in linear arrangement and as sand hillocks (Fig. 11B). All the sand dunes in Gamasa area have started to encroach the cultivated land in the southern part.

The average temperature recorded by the meteorological station at El-Sallum, Mersa-Matruh and Port-Said shows that there is general variation from east to west. Temperature increases in Port-Said (14.2 - 27.3 ° C) and decreases in Mersa-Matruh (12.8 - 25.5 ° C) while humidity increases in Mersa-Matruh (61 - 73%) and decreases in El-Sallum (54 - 66%) as shown in figures 12 and 13.

The sand dunes at some localities in north Sinai at El-Arish and to the north of the Nile Delta between Rosetta and Damietta have been recently the subject for field and laboratory studies due to presence of heavy minerals of economic importance. According to Misak and El-Shazly (1982), the source of the blown sand in northern Sinai may be related to the Miocene rocks and old terrain but later on El-Shazly et al., (1986) believed that the genesis of the sand dunes in northern Sinai is of a complex nature being derived from the River Nile and from the erosion of various rocks and sediments along Wadi El-Arish drainage system.

Several attempts have been made to reduce the movement of sand dunes over the cultivated land and areas of future development. If there is necessity to utilize areas lying along the paths of dune belts, the location should be proportional to the rate of barchan movement in the lower part of the belt in order to avoid dune encroachment over a relatively long period.

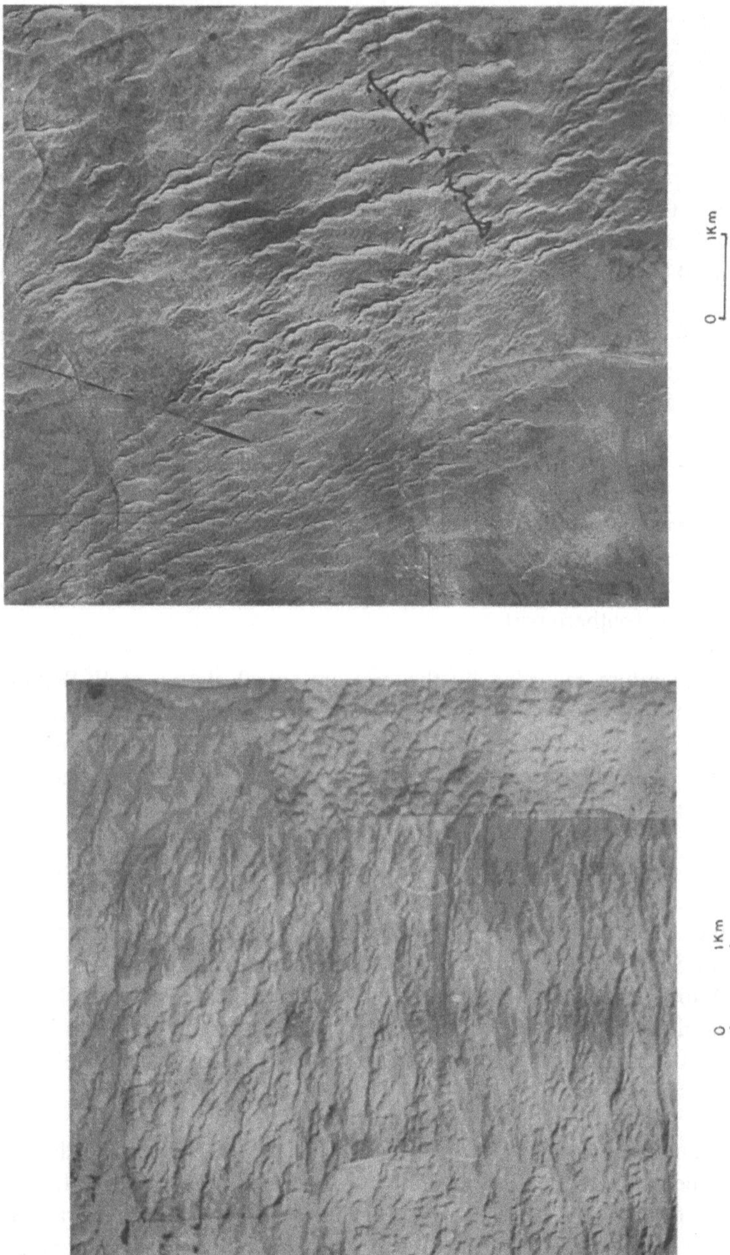

Figure 10: Photomosaic showing a complex pattern of concentric and transverse sand dunes (A) and linear sand dunes in a feathered pattern (B) in the Northern Sinai.

A

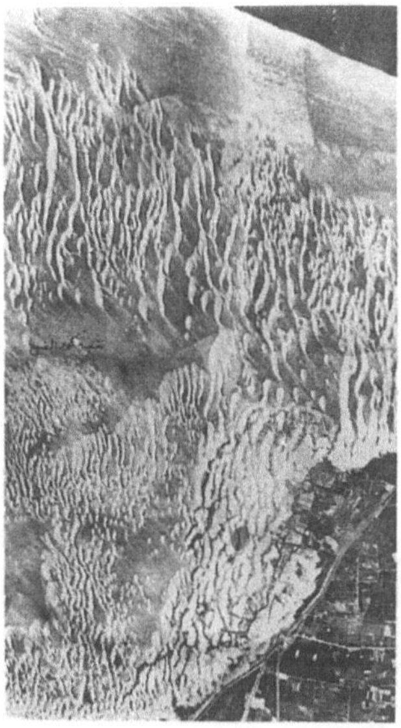

B

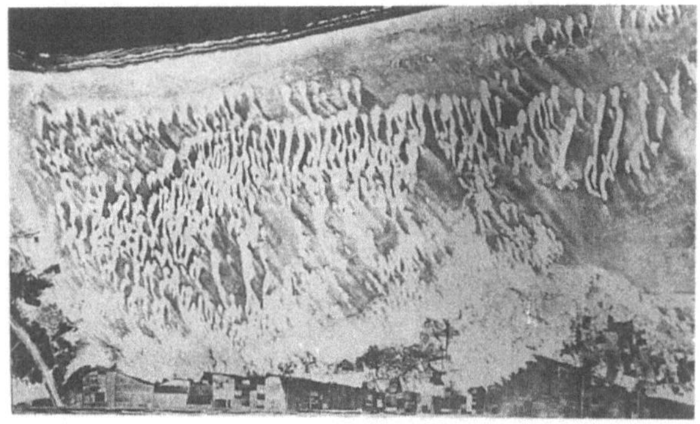

Figure 11: Photomosaic Showing Linear Dunes, Barachans and Sand Sheets to the West (A) and East (B) of Gamasa. The Dunes are Encroaching the Cultivated Lands in the Southern Parts.

5. CONCLUSION

The northern part of Egypt, The Nile Valley, the area along the western side of the Red Sea coast and the southern part of Egypt around Aswan area are considered to be seismically active. These areas differ greatly from other areas with low seismic activity in geology, geomorphology and topography. Sufficient attention should be paid to the E - W faults in planning constructions and selecting their sites in order to avoid potential disasters since the E - W faults are active and dominate in areas subjected to earthquakes. Aerial photography and landsat imagery interpretation contribute effectively to the understanding of the geologic - structural setting of areas affected by earthquakes. The time sequence of the rocks subjected to earthquakes with high intensity is Miocene, Pliocene, Pleistocene, and Holocene. The areas with low seismicity are covered by Cretaceous, Eocene and Oligocene in addition to the basement rocks.

Large areas in the Delta are underlain by thick alluvial sequence that are for the most part unconsolidated and occasionally water saturated. Earthquake damage is larger over these areas than the crystalline bedrocks. Recently, high earthquake hazard are becoming of increasing concern because of new developments that are taken place in the areas growing uncontrollably. Aswan areas is of particular concern due to recent development including the construction of the Aswan high Dam. The earthquake occurred in 1981 is attributed to the creation of Lake Nasser behind the Dam.

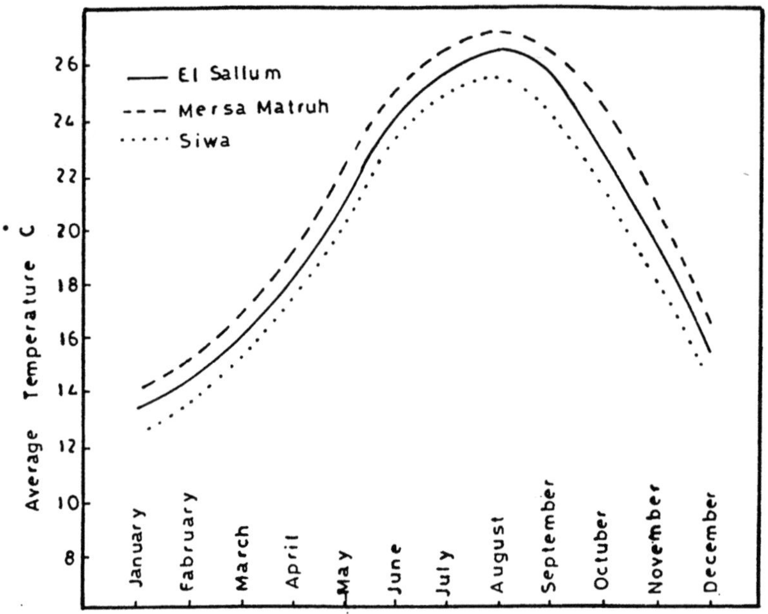

Figure 12: Average temperature in Northern Egypt.

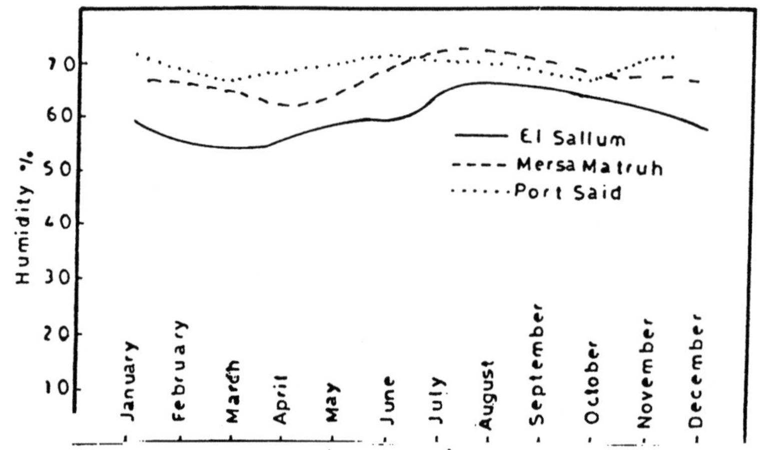

Fig 13: Relative humidity in Northern Egypt.

The middle and Upper Eocene limestone in Egypt is characterized by the presence of deep caves which cause instability of the foundation bedrocks in the sites of the new cities. This can be found in Gebel El-Mokattam area located to the east of the Greater Cairo.

Further hazards in Egypt include the encroachment of sand dunes over large part in the Nile Valley and Delta, lowering of the Nile level, long term decreased flood plain erosion of the Nile Delta front, saline groundwater attack on buildings and monuments, intense rainfall in Upper Egypt, rockfall in the mountainous areas e.g. Gebel El-Mokattam in the Greater Cairo and increasing in the density of population in limited areas. The application of shared knowledge among experts in many nations, the incorporation of geological information principles in the planned areas, land use regulation, the application of new techniques in the Satellites Remote Sensing can help in hazard mitigation.

It should be recommended the necessity of the presence of meteorological stations in the Oases and in the southwestern parts of the Western Desert and seismological stations in the fast developing parts of the country. We must deal with the sand dunes expected to hinder cultivation or housing in the near future especially in the northern part of Egypt.

Data obtained from the detailed analysis of the geological conditions to identify subsoil units should be used as a basis for implementing measures aimed at reducing the size of future hazard losses. The type of construction and building's height must be considered in the future.

REFERENCES

Abdel-Twab, S. and Ibrahim, H. A. A., (1991) : Unfavorable geologic setting of the New Minya City, Upper Egypt. Faculty of Science, Ain Shams Univ., pp. 422 - 429.

Ashri, A. H. (1973) : The movement of sand dunes at Kharga Oasis, *Egypt. J. Geol.*, 17, No. 1, pp. 37 - 46.

Awad, G. H., M.I. Farris ,.and H. L. Abbas, (1953) : Contribution to the stratigraphy of the Mokattam area, east of Cairo. extrait du Bulletin de l' institute de Desert d'Egypt - Tomo III - No. 2, Juillet 1953, pp. 105 - 107.

Badawi, H.S. and S.A. Mourad (1994). Observations of the 12 October, 1992 Dahshour Earthquake in Egypt. *Natural Hazards*, 10: 261-274

Basta, E. Z., M. Abdel-Maksoud, and A.T. Abdel-Aziz, (1979) : Properties and Mineralogical constitution of swelling clays from two correlative sections in the Nile Valley and Kharga Oasis, *Egypt. J. Geol.*, Vol. 23, No. 1 - 2, pp. 27 - 40.

Beadnell, H. J. L., (1910) : The sand dunes of the Libyan Desert : *Geogr. Jour.*, V. 35, pp. 379 - 394.

Degg, M. R. and A. Doornkamp, (1990) : Earthquake hazards atlas, 2 - Egypt based on the R. O. A. Earthquake hazard zonation scheme R. O. A. Reinsurance Office Association Aldernary House, Queen Street, London EC 4N IST, United Kingdom, P- CI. 2.1. - c1. 2.24.

El-Baz, F., (1979) : Egypt as seen by landsat : Cairo, Egypt Dar Al Maaref, Egypt.

El-Baz, F., (1992): Origin and evolution of Sand Sea in the Great Sahara, The first International Conference on Geology of the Arab World, Cairo Univ., Egypt (Abstracts)

El-Baz, F. and B. Walfe, (1981) : Wind pattern in the Western Desert. Annals of the *Geological Survey of Egypt.*, Vol. XI, pp. 119 - 139.

El-Etr, H. A., M. S. Youssef, M. M. Hamza, and A. A. Hussein, (1989) : Landslide vulnerability in the Arabian Peninsula and Egypt, In Brabl E. E. and Harros, B. L. landslides Extent and economic significance, Balkama (Rotterdam), pp. 233 - 237.

El-Nakkady, S. E., (1958) : Stratigraphic and petroleum geology of Egypt, Univ. of Assiut, Monograph series No. 1, p. 215.

El-Sayed, A.R. Wahlstrom and O. Kulhanek. (1994) Seismec hazard of Egypt. *Natural Hazards*, 10: 247-260

El-Shazly, E. M., M.A. Abdel-Hady, M.A. Ghawaby, I.A. El-Kassas, (1974) : Geological interpretation of ERTS - i Satellite images for east Aswan area, Egypt, Proc. 9th intern. symp. Remote Sensing of Environment, 15 - 19 April 1974, Ann. Arbor, Michigan, pp. 105 - 117.

El-Shazly, E. M., A. F. Kamel, M.A. Morsy, and S. H. Atia, (1986) : Studies on beach and dune sands on the Mediterrnean Sea coast at El-Masaid, West of El-Arish, Sinai, *Egypt J. Geol.*, 30, No. 1 - 2, pp. 1 - 8.

Ezzat, R. H. (1965) : The use of regulated directional reception (RDR) in areas of complicated seismic geological conditions, *J. Geol. U. A. R.* vol. 9. No. 1, pp. 33 - 44.

Gifford, A. W., D. M. Worner ,and F. El-Baz, (1979) : Orbital observations of sand distribution in the Western Desert of Egypt. In El-Baz, F. and Warner, D. M. ed. Apollo - Sayuz Test Project Summary Science Report, Vol. II : Earth observations and photography : NASA SP - 412, Washington, D. C., pp. 219 - 236.

Ibrahim, E. M., (1988) : tectonic activity in Egypt and its near vicinity and their estimation from the study of seismomagmatic activity and technology. National Research Institute of Astronomy and Geophysics, Egypt.

Kamel, A. F., (1994) : Regional fracture analysis south latitude 29 ° N of Egypt and their influence on earthquakes. *Natural Hazards* 9: 235-245

Kamel, A.F., M.M. El-Rakaiby, and I.A. El-Kassas (1982): Photo-interpretation of sand dune belts in Northwestern Sinai, Egypt. *Journal of Geology* Special Issue, Part I, pp. 57-70

Kebeasy, R. M., (1988) : Project of site selection of seismological stations of the Egyptian National network. Academy of Scientific Research and technology. Helwan Institute of Astronomy and Geophysics, p. 199.

Kebeasy, R. M., M. Maamoun and E.M. Ibrahim, (1982) : Aswan Lake induced earthquake, Prelimanary investigation, Helwan Institute of Astronomy and Geophysics, Helwan, Egypt, p. 8.

Kebeasy, R. M., M. Maamoun, and E.M. Ibrahim, and A.M. Megahed, , (1984) : A report of regional historical seismicity submitted to the Advisory Panel of seismic Activity and Dam Stability Studies. Academy of Scientific Research and Technology, Egypt, p. 17.

Kulhanex, O., I. Korrat, and A. El-Sayed (1992) : Recent seismic activity and general tectonics of the Red Sea, Egyptian Geophysical Society, the 10th Annual meeting (Abstract).

Maamoun, M. E., and E.M. Ibrahim ,, (1978) : Tectonic activity in Egypt as indicated by earthquakes. Acadmy of Scientific Research and Technology, *Helwan Institute of Astronomy and Geophysics*, Bull No. 170, p. 20.

Misak, R. F. and M. El-Shazly, (1982) : Studies on the blown sands at some localities in Sinai and the northern Western Desert, Egypt. Egypt. *J. Geol. Special* Issue, Part 1, pp. 47 - 56.

Moustafa, A. R., F. El-Nahas, and S. Abdel-Tawab ,(1991) : Engineering geology of Mokattam city and vicinity eastern Greater Cairo, Egypt. *Engineering Geology*, 31, pp. 327 - 334

Said, R. (1962) : The geology of Egypt : Amsterdam, Elsevier Publ. Co., p. 377.

Said, R. (1981) : Geological evolution of the River Nile, Springer Verlag, New York, p. 151.

The Egyptian Geological Survey (1981) : Geological map of Egypt scale 1:2000,000. The Egyptian Geological Survey and Mining Authority, 3 Salah Salem Strret, Abbasiya, Cairo, Egypt.

Toppozada, T. R., F.K. Boulos, S.F. Helnin, , A.A. El-Sherif, A.A. El-Sayed, N.A. Basta, F.A. Shatiya, Y.S. Melik, C.H. Cromer, and D. L. Parke, (1984) : Seismicity near Aswan High Dam, Egypt. Following the November earthquake. *Annals of the Geol. Surv. of Egypt*, Vol. XIV, pp. 107 - 126.

THE ECOLOGICAL IMPACT OF FIRE ON THE PEATLAND FOREST IN THE DA XINGAN MOUNTAIN, CHINA

ZHAO KUIYI
Changchun Institute of Geography, Chinese Academy of Sciences, 10 Gongnong Road, Changchun 130021, P.R. China

Abstract: A major wild fire swept through the northern part of Da Xingan mountains of China between May 6 and June 2, 1987. After the fire the burned and unburned areas of peatland forest were compared for differences in their ground vegetations and microclimates, etc. There are changes of composition in ground vegetation. The burnt peatland forests in the valley become shrub swamps without trees and young birch swamps. If surface water decreases, it can return to the peatland forest where large (*Larix gmelini*) are dominant. However, since the water increase naturally, tussocks develop and fens will substitute shrub swamps.

1. INTRODUCTION

Between May 6 and June 2, 1987 a major wild fire swept through the northern part of Da Xingan mountains of China. The burned area is situated in the far north of China from 52°29' to 53°33' N, and 121° 45' to 125° E, an area of about 1.14 million ha. It is in the frigid-temperature zone and coldest part of china (Ding Dewen, 1987). The annual mean temperature is about -4° C. The lowest temperature recorded in China is -53°C at Mohe. Snow covers the forest floor in Da Xingan mountains at least six months of the year between late October and Early May. The rainfall in the region is about 350 to 500mm, 61 % of which falls during June to August. The average relative humidity in the region is 70-75%.

The landscape in this region consists of low mountains and hills with an elevation of 600-1000m. They are covered by the brown coniferous forest soil. The peat soils and swamp soils occur in the valley or on terraces.

The forest of Da Xingan mountain is dominated by larch (Larix gmelini) and mixed with pine (Pinus silvestris var. mongolica). White birch (Betula platyphylla) is associated with the conifers (Zhou Yiliang, 1991).

M. I. El-Sabh et al. (eds.), Land-based and Marine Hazards, 71–78.

2. THE STUDY AREA IN THE DA XINGAN MOUNTAIN AND METHODS

2.1 Damaged Area

The burned area was 220km west to east, and 108km north to south. Total area burned was 1.14 x 10^6 ha, which can be divided into a severely burned area (350110ha, 31% of the total burned area), a moderately burned area (314300ha, 27% of the total burned area) and a lightly burned area (275600 ha. 24% of the total area). The total volume of the burned wood was 95.23 million m^3, approximately equal to half of the total wood accumulation of the Forestry Service.

Table 1: The environmental conditions microclimate observation

Obsevational Place	***Mire types***	**Position**	**Landforms**	**Water Supply**	**Soils**	**Crown Density**	**Tree Death Rate**
I Changing Forestry Centre	*-Larix gmelini -Ledum palustre -Var. angustum -Sphagnum spp*	52°56' N 123°19'18" E	Valley Altitude 610m Slope 2-3°	Runoff Precipitation Fluvial	Humus Swamp Soil	5-6	0
II Changing Forestry Centre	*-Betula fruticosa -Sphagnum spp*	52°56'0.2" N 123°19'10" E	Valley Altitude 610m Slope 2-3°	Runoff Precipitation Fluvial	Humus Swamp Soil	0	99.75
III Changing Forestry Centre	*-Betula fruticosa -Sphagnum spp*	52°52'40" N 123°9'32" E	Valley Altitude 551m Slope 3-4°	Runoff Precipitation Fluvial	Peaty Soil	0	99.70

2.2 Methods

In order to observe the effect of environment change on forest ecology after the big fire, we chose three sites in peatland forest to make comparative observation (Table 1). The first site was forested with Larix gmelini, led Palustre var. angustum and Sphagnum spp, where fire did not damage the wood. The second site killed over 99% of al Larix gmelini. Dead trees were not removed. Site 2 was 300m away from the first site. The third site was 15km away from the first one which was severely burned. Death rate of trees was 99.7%. Using quadrats of 10x10m, 5x5m , and 1x1m for timber, shrub and herb respectively, species cover, density, frequency, dominance, number of plant and biomass, etc., comparative biomasses of observations of height growth of some mosses were taken between burned and unburned area.

3. THE EXTREME FIRE IMPACT ON THE MICROCLIMATE OF PEATLAND FOREST

The burned area showed an increase of area surface temperature. Daily average soil temperatures (0-20cm) were all higher. The daily range of temperature of the earth's surface also increased. The highest temperature of the earth's surface in the severely burned area was 10°C higher than that of non-burned area (Table 2).

Table 2: Comparison of daily mean temperature at different soil depths between burned and unburned swamp

Observational place		Depth (cm)				
		0	5	10	15	20
		Daily mean temperature (°C)				
I	Burned	7.4	2.4	4.6	3.9	3.4
II	Unburned	0.4	1.9	1.4	1.6	1.6
	Difference	7	0.5	3.2	2.3	1.8

(Changying Forestry Centre; July 26, 1990)

There was higher evaporation in severely burned areas. The degree of absolute atmospheric humidity was higher than that in unburned areas, e.g. the moisture between heights of 150cm and 20cm at 11 am on July 28, 1990, were 2.7 hPa in the burned area and 1.1hPa in the unburned area, while at 2 pm they were 1.7 hPa and 0.5 hPa respectively.

There was high radiation during the daytime and low refractive index in the severely burnt area. Because there were no trees, total solar radiation reaching the peatland was greater than at the unburned site. A large amount of short-wave radiation is absorbed, with low refraction. The surface refraction index of the burned and unburned areas on September 13 are 0.15 and 0.18 respectively.

There are reductions of total evapotranspiration in the severely burned areas. The radiation balance of daytime in the severe one on September 13, 1990, is 1492.77 J/cm^2 day (equal to 5.99 mm/day of evaporation). That in the unburned one is 1474.14 J/cm^2 day. In warm days 5 mm/day is evaporated by trees. If this evaporation from trees is added, The evaporation from the unburnt area is 10192 mm/day which greatly exceeds that of the severely burned areas.

4. CHANGES OF THE VEGETATION OF THE PEATLAND

4.1 The Changes of Dominant Plants Under the Trees

Through the analysis of the importance of values of the quadrat plants there are changes of the ground vegetation in the severe burned and unburned area. There are 18 plant species (Vascular plant) under trees in the first experimental area while in the second area there are 22. The number of species common to both burned and unburned is 9, which is 50% and 40% of total species respectively. In order of importance the first 5 in the unburned area are: ledum palustre var. angustum, Vaccinium vitis-idaea, Eriophorum vaginatum, Chamaedaphne calyculata, Care globularis (Table 3). In the severely burned area the important species are: Betula Fruticosa, *Vaccinium uliginosum, Eriophorum vaginatum, Ledum palustre* var. *angustum,* Vacciunium vitis-idaeain (Table 4). After the fire, the greatest shift in ranking occurred for *Betula fruticosa* and *Vaccinium uliginosu.* These species, while not being ranked in the first five in the unburnt area, were ranked first and second on the list of the ground vegetation in the burned areas. The height and cover of ground vegetation are obviously increased in the burnt area. e.g., *Betula fruticosa* in the severely burned area is on average 20-40cm higher than in the unburned area. The plant numbers of every bundle in burned area are doubled or even tripled.

4.2 Over-Growth of the Peatland Vegetation

After the fire, peatland vegetation lost their cover of trees, thus enabling the solar radiation to reach the earth surface directly. Because of the change of the swamp's microclimate, the environment is favourable to the growth of the peatland vegetation, except for the change in the dominant species as mentioned above. The growth amount of under the tree plants is changed obviously. From a survey carried out in June and September, 1991, the plant output was found to be 466 g/m^2 in the unburned area (including shrub) and 1315 g/m^2 in burned areas.

4.3 The Change of Moss Growth

Two groups of mosses of *Sphagnum fuscum, S. magellanicum, S. girgensohnii, S. cuspidatum* were taken from burned and unburned areas to make a comparison for growth. In the burned area the growth of four kinds of sphagnum mosses was as high as 1.2cm in a summer. However, their growth in unburned area is about 1.85cm, which means 0.65 cm higher than the mosses in burned area (Figure 1). After the fire the pH value of peat soil water is 5.6 in burned area, a value higher than the pH of 4.5 in the unburned area. The higher pH results from the higher ash content in the surface of burned area.

Table 3: The importance value of ground vegetation in swamp (non-fire)

Name of Plant	Density	Relative Density	Frequency	Relative Frequency	Dominance	Relative Dominance	Importance Value
Ledum palustre var. angustum	75.4	27	1	9.76	0.35	0.67	37.43
Vaccinium vitis-idaea	75.6	27.2	1	9.76	0.09	0.06	37.02
Eriophorum vaginatum	34.75	12.5	0.5	4.88	0.01	0.0067	17.39
Chamaedaphne calyculata	24.75	8.9	0.5	4.88	0.08	0.054	13.83
Carex globularis	30	10.8	0.25	2.44	0.01	0.0067	13.25
Vaccinium uliginosum	8.3	3.2	1	9.76	0.12	0.0805	13.04
Betula fruticosa	6.6	2.4	1	9.76	0.13	0.087	12.25
Oxycoccus microcarpus	7.25	2.6	0.5	4.88	0.01	0.0067	7.49
Salix vyrtilloides	5.6	2.0	0.25	2.44	0.06	0.04	4.48
Salix sp	5.25	1.86	0.5	2.44	0.03	0.02	4.32
Rubus arcticus	0.25	0.09	0.25	2.44	0.01	0.0067	2.54
Larix gmelini	0.25	0.09	0.25	2.44	0.01	0.0067	2.54
Eriophorum angustifoiium	0.25	0.09	0.25	2.44	0.01	0.0067	2.54
Alnus sibirica	0.75	0.27	0.25	2.44	0.03	0.02	2.54
Smilacina trifoila	1.25	0.45	0.25	2.44	0.02	0.013	2.54
Calamagrostis angustilolia	0.25	0.09	0.25	2.44	0.01	0.0067	2.54
Carex rychophysa	0.25	0.09	0.25	2.44	0.01	0.0067	2.54
Malanthenum bifolium	0.25	0.09	0.25	2.44	0.01	0.0067	2.54

Table 4: The importance value of ground vegetation in swamp (fire)

Name of Plant	Density	Relative Density	Frequency	Relative Frequency	Dominance	Relative Dominance	Importace Value
Betula fruticosa	42.33	12.29	1	11.39	0.5	11.25	34.94
Vaccinium niiginosum	32.25	12.99	1	11.39	0.35	7.8	31.03
Eriophorum vagenatum	60.1	17.45	1	9.97	0.05	1.07	28.49
Ledum palustre var. angusium	26.35	7.36	1	11.39	0.4	9.05	28.07
Vaccinium vitis-idaea	47.19	13.72	0.84	9.8	0.065	1.46	24.98
Rhododendron carvifoiium	16.54	4.8	1	11.39	0.25	5.64	17.33
Carex globularis	16.5	4.8	0.59	6.6	0.2	4.49	15.89
C. schmidtii	23.5	6.8	0.42	4.85	0.095	2.17	13.82
Salix brachypoda	19.44	5.6	0.67	6.79	0.05	1.07	13.46
Chamaedaphne calyculata	9.68	2.8	0.67	8.05	0.4	0.85	11.7
Salix rosmarinifolia	4	1.6	12	0.5	6.4	1.2	8.8
Ainus sibirica	0.2	0.8	0.33	2.99	0.17	3.9	7.69
Calamagrostis angustifolia	4.5	1.35	0.33	1.75	0.15	3.4	6.5
Poa sp	3.6	1.0	0.33	3.29	0.002	0.042	4.33
Rubus chamaenorus	0.51	0.15	0.2	0.33	0.1	2.4	2.88
Sangutisorba officinalis	2	0.58	0.2	1.99	0.01	0.21	2.78
Potentilia Yokussaiana	2.5	0.73	0.2	1.99	0.004	0.086	2.72
Corydalls pallida	1.5	0.44	0.33	1.99	0.005	0.107	2.54
Ribes repens	0.65	0.17	0.2	0.37	0.15	1.34	1.88
Saussurea amurensis	0.25	0.07	0.33	1.99	0.01	0.21	2.54
Rosa davurica	2.36	0.69	0.33	0.33	0.01	0.21	1.23
Dasiphora parvifolia	1.24	0.36	0.33	0.33	0.02	0.43	1.12

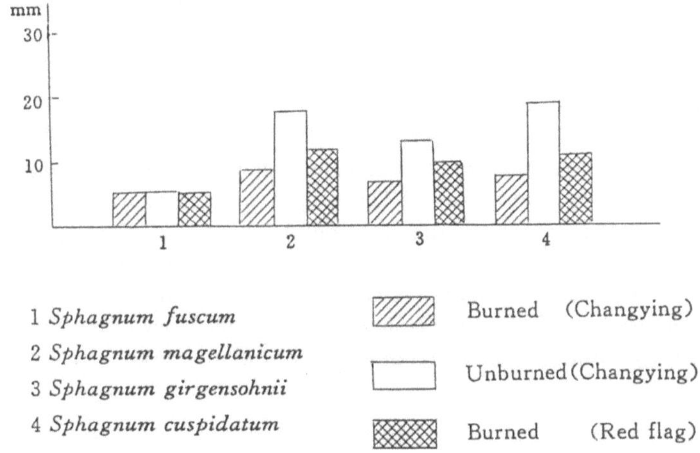

Figure 1: Comparison of growth of mosses between burned and unburned areas.

4.4 The Succession Tendency of Peatland Vegetation After the Fire

Larix gmelinii, Ledum palustre var. angustum, Sphagnum spp are used as examples to indicate the succession tendency of the swamp vegetation after the big fire. *Larix gmelinii, ledum palustre var. angustum, Sphagnum* spp are slow to recover. Meanwhile *Betula platyphylla* and *Populus davidianna* have a stronger renewing ability. After the fire, the earth's surface temperature increases and is favourable for growth of new branch. For several years the severely burned areas are occupied by birch and poplars, which can continue growing for 70-80 years. Under suitable conditions the larch will reestablish itself eventually leading to the recovery of *Lirix gmelinii, ledum palustre* var. *angustum, Sphagnum* spp swamp.

After burned the plants in the valley such *Larix gmelinii, Ledum palustre* var. *angustum, Sphagnum* spp swamp, under tree shrubs *Betula fruticosa, Vaccinium uliginosum*,etc. became the superior plants rapidly. Ground layer are still covered with mosses. Forming *Betula fruticosa, Sphagnum* spp *swamp*, which is a transition type. If water decreases, it can return to the swamp which *Larix gmelinii* is superior. If it is not given a mankind intervention, the water is increased naturally in the peatlands. The tussocks will be developed. The sedge fen will substitute shrub swamp.

At the valley with slope, after *Larix gmelinni, Ledum palustre* var. *angustum, Sphagnum* spp swamp are burned, it was substituted by shrub swamp which are *Betula fruticosa* and *Vaccinium uliginosum* without larch. Partial moss-hummocks have been destroyed. The gravels are seen on the earth surface after the ice core melted. If we let it renew naturally, this type could slowly recover to larch through forest of birch and poplar. If the environment is destroyed continuously, it may turn to crushed stone slope.

5. CONCLUSIONS

This paper deals with the relationship between fire and vegetations following the major wild fire that swept the northern part of Da Xingan mountains in China. The study describes the burned and unburned areas in order to study the plant species changes, density, frequency, dominance and biomass. We also discuss the different environmental factors such as temperature, moisture, evaporation, radiation, heat energy and soil pH. The most important finding of our study is the progressive recovery of the forest site, after a major fire, either by natural regeneration or by reforestation or both.

REFERENCES

Ding, Dewen, 1987: Study of fire in the Da Xingan Mountain on permafrost environment and forest succession. Lazhou Institute of Glaciology and Gryopedology, Academia Sinica.

Zhou, Yiliang, 1991: Vegetation of the Da Xingan Mountain in China. Publishing House of science.

EL NIÑO/SOUTHERN OSCILLATION, INDIAN MONSOON AND WORLD GRAIN YIELDS - A SYNTHESIS

MADHAV L. KHANDEKAR
Atmospheric Environment Service, Downsview, Ontario, CANADA, M3H 5T4

Abstract: Large-scale atmospheric circulations and anomalies have been shown to have a significant impact on seasonal weather over many parts of the world. In this paper, the impact of large-scale atmospheric and oceanic anomalies on the Indian monsoon are analyzed with particular reference to El Niño/Southern oscillation (ENSO) events of the last hundred years.

A possible link between ENSO, Indian monsoon and grain yields is investigated by analyzing world-wide grain yield data in conjunction with ENSO event and floods and droughts in the Indian monsoon. It is found that an ENSO event is generally associated with a drought in the Indian monsoon followed by a low grain yield over South Asia and Australia and high grain yield over the north American prairies. Implications of this analysis for estimating world grain yields in advance are considered.

1. INTRODUCTION

The Indian monsoon is a vigorous atmospheric system which manifests itself every year during the northern hemisphere summer and forms an important component of the large-scale tropical monsoon system extending from West Africa to Northeast China. The phrase "Indian Monsoon" is defined here to mean the monsoon flow and associated rainfall prevailing during June to September over the Indian subcontinent consisting of India, Pakistan, Bangladesh and Sri Lanka. Since the pioneering studies of Sir Gilbert Walker and his associates (Walker, 1923, 1924; Walker and Bliss, 1932), the Indian monsoon has been shown to be influenced by large-scale atmospheric anomalies like the Southern Oscillation and El Niño conditions in the equatorial eastern Pacific, in several recent studies (Shukla and Paolino, 1983; Mooley and Parthasarathy, 1984; Khandekar and Neralla, 1984; Elliott and Angell, 1987).

In a couple of landmark papers, Bjerknes (1966, 1969) analyzed the impact of the sea-surface temperature (SST) anomalies in the equatorial eastern Pacific on large-scale atmospheric flow patterns and proposed a thermally direct circulation in the equatorial plane which he called the Walker circulation in honour of Sir Gilbert Walker and his pioneering studies. The Walker circulation is now recognized as the dynamical link between the ENSO and weather and climate over the entire Indo-Pacific region. Since the

M. I. El-Sabh et al. (eds.), Land-based and Marine Hazards, 79–95.

publication of Bjerknes' papers, several hundred studies have been reported on El Niño and its possible connection to global weather and climate anomalies. Several of these studies (ex. Rasmusson and Carpenter, 1983; Khandekar and Neralla, 1984; Mooley and Paolino, 1989) have investigated the impact of El Niño and La Niña (inverse El Niño) events on the Indian monsoon droughts and floods.

In this paper, the historical development of the connection between El Niño, La Niña and the Indian monsoon is briefly reviewed and important result from some of the recent studies are presented in Section 2; the paper further examines the impact of other large-scale atmospheric features namely Quasi-Biennial Wind Oscillation (QBO) of the equatorial stratosphere and Eurasian snow cover on the Indian monsoon. Based on this analysis, a conceptual model of the large-scale atmospheric flow for the drought and flood conditions of the Indian monsoon is presented in Section 4. In Section 5, the impact of El Niño/La Niña on the Indian monsoon and on the world grain yield is analyzed. The implications of this study for foreshadowing Indian monsoon and for estimating world grain yields in advance are finally considered.

2. INDIAN MONSOON AND ENSO

The ENSO and Indian monsoon connection goes back to the pioneering studies of Walker (1924) on correlation in seasonal variation of weather. From a large number of correlations between Indian monsoon and worldwide weather elements, Walker identified a parameter "South American Pressure" (defined as mean of surface pressure over Buenos Aires, Santiago and Cordoba during April and May) which emerged as one of the most reliable predictors for estimating June to September rainfall over two main subdivisions of India namely northwest India and the peninsular India. Besides the parameter South American pressure, Walker defined other parameters like Java (Indonesia) rain, equatorial pressure, etc. which had significant correlations with Indian monsoon. These parameters represented atmospheric conditions during the pre-monsoon months over the Indo-Pacific region which influence the subsequent monsoon rainfall over India and vicinity. An incisive analysis of the correlation structure between the Indian monsoon and world-wide weather elements led Walker and Bliss (1932) to define the Southern Oscillations (SO) as a standing fluctuation over the Indo-Pacific region which provide an exchange of air between eastern and western hemispheres. In a comprehensive study, Berlage (1966) presented a schematic of the SO in terms of isopleths of correlation of monthly mean station level pressure at world-wide locations with that at Djakarta, Indonesia. Studies by Troup (1965) and Wright (1977) have confirmed the essential features of the SO as defined by Walker and Bliss. Other studies (ex. Khandekar, 1979; Angell, 1981; Pant and Parthasarathy, 1981; Shukla and Paolino, 1983) have demonstrated a link between the SO and the Indian monsoon and the utility of a SO index for prediction of summer monsoon rainfall over India. More recent studies (Rasmusson and Carpenter, 1983; Mooley and

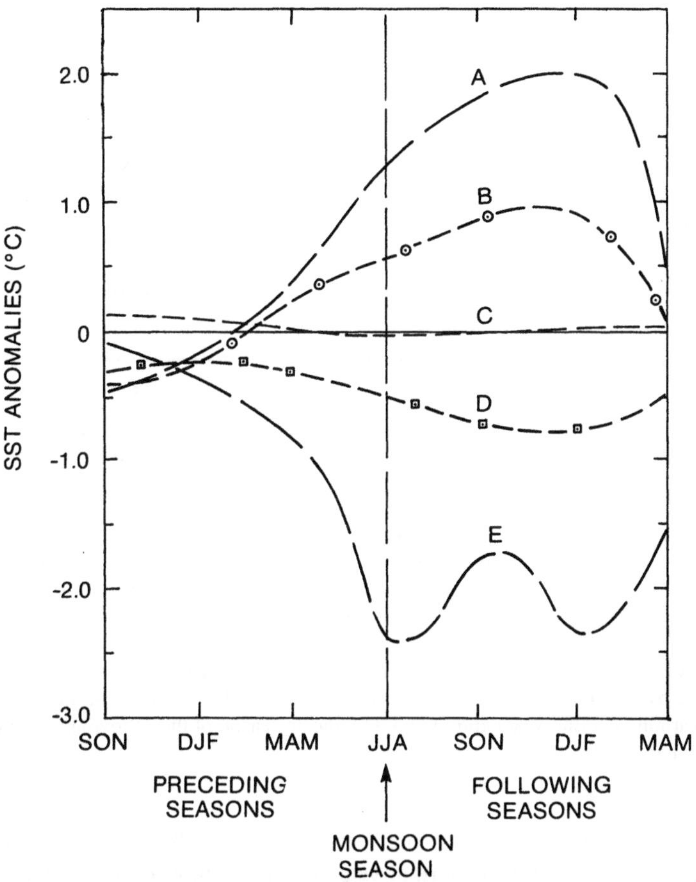

Figure 1: SST anomaly composites for drought, flood and normal years of Indian monsoon. Curve A: SST anomaly for an extreme drought year of 1972. Curve B: SST anomaly composite for 10 severest drought years. Curve C: SST anomaly composites for 30 normal years. Curve D: SST anomaly composites for 10 severest flood years. Curve E: SST anomaly for an extreme flood year of 1916. Years of data: 1901-1980 (from Khandekar and Neralla, 1984).

Parathasarathy, 1984) have demonstrated the impact of sea surface temperature (SST) anomalies in the equatorial eastern Pacific associated with El Niño events on the Indian monsoon. These and many other studies have firmly established the impact of ENSO events on the Indian monsoon. Based on 80 years (1901-1980) of data, Khandekar and Neralla (1984) have demonstrated that a deficit (surplus) Indian monsoon is almost always associated with positive (negative) SST anomalies in the equatorial eastern Pacific before, during and after the monsoon season of June to September. In Figure 1 are shown composite SST profiles for 10 drought, 10 flood and 30 normal years of Indian monsoon. Also shown in the Figure are SST profiles for an extreme drought and an extreme flood year. These profiles extend over seven seasons in all, three pre-monsoon, three post-

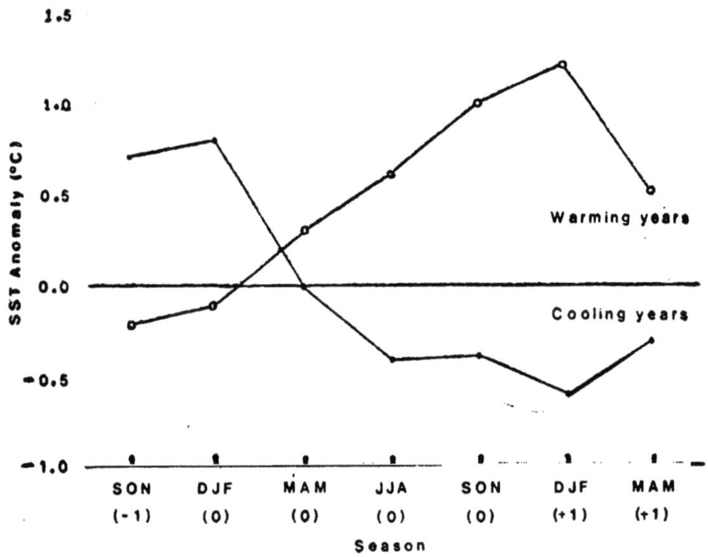

Figure 2: Mean SST anomaly during a warming and a cooling event over the eastern south equatorial Pacific extending over seven seasons with monsoon season (June, July, August) at the centre. (0) indicates season of the warming/cooling years and (+1), (-1) indicate season of the following/preceding year (from Mooley and Paolino, 1989).

monsoon and the monsoon season in between. A noteworthy feature of Figure 1 is that the composite SST profile for 30 normal years of Indian monsoon is remarkably close to zero before, during and after the monsoon season.

In another recent study, Mooley and Paolino (1989) analyzed SST data of eastern south equatorial Pacific (0°-10°S, 80°-180°W) from 1871 to 1984 and identified 27 years of warming events and 23 years of cooling events in the equatorial Pacific. The warming and the cooling events were identified by the SST profile extending over seven seasons and centred around the three months of the Indian monsoon, namely June, July and August. The mean SST profile for the warming and the cooling event is shown in Figure 2. Mooley and Paolino further obtained normalized anomalies of Indian monsoon rainfall corresponding to each of the 27 warming events and 23 cooling events and these values are presented in Table 1. The values of Table 1 clearly shows that the warming (cooling) event in the eastern south equatorial Pacific is almost always associated with deficit (surplus) rainfall for the Indian monsoon.

Table 1: Normalized anomaly of Indian monsoon rainfall during years of warming and cooling events over equatoria eastern Pacific (from Mooley and Paolino, 1989)

No.	Years of warming events	Normalized rainfall anomaly (cm)	Years of cooling events	Normalized rainfall anomaly (cm)
1	1876	-0.92	1872	0.70
2	1877	-2.99	1874	1.43
3	1883	-0.04	1878	1.47
4	1888	-0.51	1885	-0.12
5	1896	-0.34	1889	0.90
6	1899	-2.70	1892	1.66
7	1902	-0.73	1897	0.46
8	1904	-1.24	1898	0.34
9	1905	-1.65	1900	0.40
10	1911	-1.43	1903	0.07
11	1913	-0.84	1906	0.37
12	1918	-2.46	1908	0.52
13	1925	-0.59	1916	1.18
14	1930	-0.63	1938	0.67
15	1939	-0.76	1942	1.28
16	1940	-0.02	1946	0.59
17	1941	-1.48	1954	0.40
18	1948	+0.24	1955	0.94
19	1951	-1.39	1964	0.82
20	1957	-0.82	1967	0.08
21	1963	+0.04	1970	1.05
22	1965	-1.75	1973	0.72
23	1968	-1.18	1983	1.29
24	1969	-0.28		
25	1972	-2.40		
26	1976	+0.04		
27	1982	-1.40		
Mean		-1.05		+0.75

Long period (1887-1984) normal rainfall is 85.2 cm and standard deviation is 8.3cm

3. INDIAN MONSOON AND OTHER LARGE-SCALE ATMOSPHERIC FEATURES

Many of the monsoon predictors discovered by Walker were directly or indirectly related to the SO or its derivatives. However, there were a few which were outside the domain of the SO and which were attempted by Walker as possible predictors for Indian monsoon. One such parameter was the Himalayan snow cover during winter which was in fact identified and used as a predictor by H.F. Blanford, Walker's predecessor. Blanford recognized the importance of Himalayan winter snow cover on the subsequent monsoon rainfall over parts of the Indian subcontinent and made the first monsoon forecast in 1881 on the hypothesis that the "varying extent and thickness of the Himalayan snows exercise a great and prolonged influence on the climatic conditions and weather of the plains of northwest India" (Blanford, 1884). Walker attempted to quantify the available snowfall reports from various locations in the Himalayas by developing an index of snow cover. Using a snow cover index, Walker obtained a correlation of -0.40 between the winter snow cover over the Himalayas and the subsequent (June-September) monsoon rainfall over northwest India for the period 1876 to 1910. The Himalayan snow cover was considered an important predictor in the multiple regression equations developed by Walker (around 1906) for foreshadowing of Indian monsoon. In later years, while upgrading the multiple regression equation for forecasting northwest Indian monsoon, the predictor snowfall accumulation was dropped since the quality of the snowfall reports became less reliable and the snowfall accumulation as a predictor appeared to lose its statistical significance.

With the advent of satellite technology, reliable estimates of the snow cover over Himalayas and over Eurasian highland region have become available for the last twenty years or so and this has prompted Hahn and Shukla (1976), Dey and Bhanu Kumar (1983) and Dickson (1984) among others to examine Blanford's hypothesis and to show that the Indian monsoon is indeed inversely related to the extent of the Eurasian snow cover during the preceding winter (December-February). The inverse relationship between the Eurasian snow cover and the Indian monsoon has also been numerically simulated by Barnett et al. (1989) and Yasunari et al (1991) using a coupled atmosphere-ocean model. An analysis of the snow cover data and Indian monsoon of recent 24 years (Khandekar, 1991) suggest that in general, heavier (lighter) than normal Eurasian snow cover in winter (December-February) is followed by deficit (surplus) monsoon rainfall during the following summer (June-September) months. The present operational model of the India Meteorological Department (see Gowarikar et al., 1989) includes Eurasian as well as Himalayan snow cover as important predictors for seasonal forecasting of Indian monsoon.

In recent years, the equatorial stratospheric wind oscillation, commonly known as QBO (Quasi Biennial Oscillation) has been recognized as an important factor in foreshadowing Indian monsoon. Mukharjee et al. (1979) found that a strong easterly phase of the QBO was associated with a deficit monsoon, while Bhalme et at (1987) have shown that large-scale floods never occur with easterly phase of the QBO while droughts almost never occur with a westerly phase of the QBO. The impact of QBO on the large-scale behaviour of the Indian monsoon is now well recognized and a suitable index of QBO during the pre-monsoon months is now considered as an important predictor for foreshadowing of the Indian monsoon.

Besides the parameters discussed above, there are several other parameters used for foreshadowing the Indian monsoon (see Gowarikar et al., 1989; Thapliyal and Kulshrestha, 1992); however, most of these other parameters can be shown to be directly or indirectly related to the three large-scale features namely ENSO, Eurasian snow cover and QBO as discussed earlier. For example, the monthly mean latitudinal location along 75°E of the axis of 500-hPa ridge during April has been found to be highly correlated with the subsequent all-India monsoon rainfall and has been included as an important parameter in the operational seasonal forecasting of Indian monsoon. The relationship between the April ridge location and Indian monsoon was first discovered by Banerjee et al. (1978)and later examined in detail by Mooley et al. (1986) based on data from 1939 to 1984. A recent study by Krishna Kumar et al. (1992) shows a steady south-to-north progression of the ridge location during the pre-monsoon months of March to May. The study (by Krishna Kumar et al.) further obtains an interesting reversal of correlation between the ridge location from march to May and all-India monsoon (June-September) rainfall, the ridge location for March having a correlation of -0.47 while the ridge location for April having a correlation of +0.63 with the subsequent monsoon rainfall, both the correlation values being statistically significant. The ridge location and its progression from south to north is considered to be an indicator of the evolution of the mid-tropospheric circulation over India during the pre-monsoon months. However, a study by Bhanu Kumar (1988) analyzes the relationship between the location of the ridge and the Eurasian winter snow cover and demonstrates an intimate link between the extent of the Eurasian snow cover and the south to north progression of the 500-hPa ridge location during the pre-monsoon months. Thus, the Eurasian winter snow cover significantly influences the pre-monsoon atmospheric circulation at the mid-tropospheric and presumably at the surface level over the Indian subcontinent. Accordingly, predictors, which are based on pre-monsoon circulation patterns over the Indian subcontinent can be shown to be related to the Eurasian winter snow cover. Another predictor which is presently used in the operational monsoon forecasting is the Indian ocean equatorial pressure during the pre-monsoon months; this predictor can be shown to be related to an index of the SO as defined by Walker and Bliss.

4. A CONCEPTUAL MODEL FOR DROUGHT AND FLOOD IN THE INDIAN MONSOON

Based on the earlier discussion, it can be concluded that the broad characteristics of the Indian monsoon appear to be governed by three (seemingly) independent features of the large-scale atmospheric circulation namely ENSO, Eurasian winter snow cover and the stratospheric quasi-biennial wind oscillation known as the QBO. Local and regional factor like topography and passage of monsoon depressions along the axis of the monsoon trough can produce significant variations in the rainfall distribution in a given year; nevertheless, the broad features of the Indian monsoon can be considered to be governed by the three large-scale features mentioned above. These large-scale features exhibit interannual variation and can therefore produce different impact on the Indian monsoon in different years. Depending on the phase of the ENSO and the QBO and further depending on the extent and the amount of Eurasian winter snow cover, there could be a drought or a flood in the Indian monsoon in a given year. An El Niño condition in the equatorial eastern Pacific associated with an easterly phase of the QBO and a heavier Eurasian winter snow cover would lead to a drought in the Indian monsoon, while a La Niña (inverse El Niño) condition associated with a westerly phase of the QBO and a lighter Eurasian winter snow cover would lead to a flood in the Indian monsoon. The large-scale atmospheric flow patterns that would prevail during a drought or a flood in the Indian monsoon are shown in Figure 3(a,b) in a conceptual large-scale flow model. The mechanics and dynamics of the flow patterns leading to a drought or a flood in the Indian monsoon are discussed in the following:

Figure 3a shows the large-scale circulation associated with an El Niño condition in the equatorial eastern Pacific, an easterly phase of the QBO with moderately strong winds (-25 to -30 ms-1) at about 50mb level and a weak tropical easterly jet stream (wind speed -15 to -20 ms-1) at about 150mb level and centred roughly around 12°N. The Walker circulation along the equatorial vertical plane shows an ascending motion off the coast of South America and a descending motion over the maritime continent of Indonesia. The easterly phase of the QBO is assumed to be moderately strong while the tropical easterly jet at 150mb is depicted to be weaker in response to heavier Eurasian winter snow cover (see Barnett et al., 1989). The vertical easterly shear from upper troposphere (150mb) to lower stratosphere (50mb) will produce positive upper troposphere geopotential height anomalies (Reid and Gage, 1985) in the equatorial region. The impact of these positive anomalies, according to Gray et al (1992) will be to strengthen upper level divergence along the equator and inhibit the overall off-equator upper level divergent monsoon circulation and the associated convective activity. Thus, a weaker monsoon circulation will be established over peninsular and central India. The easterly phase of the QBO also helps suppress monsoon storm and depression activity in the Bay of Bengal and along the axis of monsoon trough as documented in a recent study by Sikdar et al (1993). An

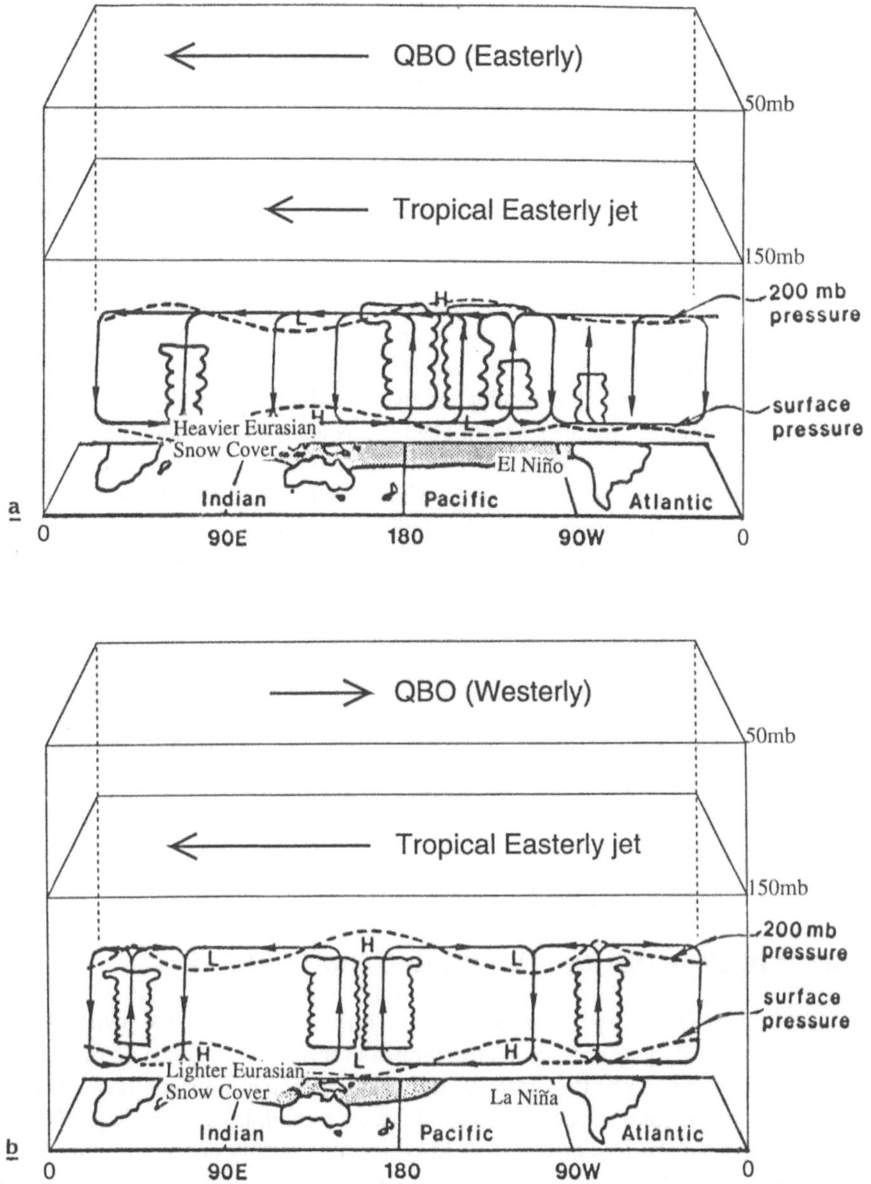

Figure 3 (a,b): A conceptual large-scale atmospheric flow model for a. drought and b. flood in the Indian monsoon. The Figure shows a schematic of the Walker circulation in the equatorial zonal plane associated with an El Niño (La Niña) event, the tropical easterly jet stream at 150 mb and centred around 12°N, easterly/westerly phases of the quasi-biennial oscillation (QBO) at 50 mb and heavy/light Eurasian winter snow cover. See text for details.

overall weak or a drought monsoon will thus be established over India and vicinity.

Figure 3b shows the large-scale circulation associated with a La Niña (or a cooling) event, a westerly phase of the QBO (with wind speed about 15 ms-1) and a moderately strong tropical easterly jet (with wind speed -25 to -30 ms-1) at 150mb in response to lighter Eurasian winter snow cover. The westerly phase of the QBO, according to Gray et al (1992) will produce enhanced off-equator convection associated with the monsoon circulation and this together with the ascending branch of the Walker circulation over Indonesia and vicinity will lead to an active or a flood monsoon over India and vicinity.

The conceptual model of Figure 3(a,b) depicts large-scale atmospheric flow patterns associated with two extreme phases of the Indian monsoon; these extreme phases of the Indian monsoon are associated with the two phases of the SO as defined by Philander (1985). Using satellite data of outgoing longwave radiation as well as station data of precipitation and sea level pressure, Meehl (1987) has analyzed the interannual variability of the tropical Pacific and the Indian ocean regions and has hypothesized a sequence of processes which evolve during a strong and a weak annual cycle, the strength of the cycle being identified with the strength of the Indian monsoon. The conceptual model developed here extends Meehl's analysis to include the impact of the upper tropospheric and lower stratospheric circulation on the Indian monsoon region. No attempt is made in this study to hypothesize a sequence of events that would help evolve one phase of the monsoon from the other. However, a recent study by Khandekar (1991) hypothesizes a sequence of events that would initiate an El Niño (La Niña) event following a lighter (heavier) Eurasian snow cover and a surplus (deficit) Indian monsoon. This sequence of events can be applied in the present study to hypothesize an evolution of a drought monsoon from a flood monsoon and vice-versa. The time-scale of this evolution would be roughly comparable to the time-scale between an El Niño and a La Niña event in the equatorial eastern Pacific.

5. ENSO, INDIAN MONSOON AND WORLD GRAIN YIELDS

The impact of ENSO events on grain yields in different parts of the world has been recognized by climatologists and agrometeorologists for the last 25 years or more. The El Niño of 1972 triggered one of the worst droughts in the Indian monsoon (see Bhalme and Mooley, 1980) and created a serious food shortage in India. The 1972 El Niño also adversely affected the Australian wheat yield while significantly enhancing the U.S. corn yield. During the last ten years or so, several studies on the impact of ENSO on grain yields have been reported (ex. Handler, 1984; Nicholls, 1985; Garnett and Khandekar, 1992). Handler has analyzed the impact of SST anomalies in the equatorial eastern Pacific on the U.S. corn yield and has found that in general above (below) average corn yield is associated with warmer (cooler) SST values in the equatorial eastern Pacific. Nicholls has found a significant negative correlation between Australian wheat and barley yield and an index of SO namely Darwin sea-level pressure during June-August. The recent study by Garnett and Khandekar (1992) reveals that the Indian monsoon droughts (floods) are in general associated with a deficit (surplus) yield for Indian rice as well as for Australian wheat and with a surplus (deficit) yield for U.S. corn. The interannual variation of seven major grain yields of the world together with variation in Indian monsoon is shown in Figure 4. The Figure also shows large-scale droughts and floods in the Indian monsoon and mean yield for the seven major world grains based on data from 1967 to 1990. The various curves of Figure 4 reveal a direct relationship between Indian monsoon and rice yield in India and wheat yield in Australia. The Figure further reveals an interesting inverse relationship between the Indian monsoon and the U.S. corn yield. In order to get a quantitative assessment of these apparent relationships, linear correlation coefficients between the Indian monsoon and the various grain yields as well as between grain yields and various parameters like SST in the equatorial eastern Pacific, SO, Eurasian snow cover and equatorial stratospheric QBO, were calculated and these correlations are presented in Table 2. These correlations are based on 24 years of data (1967-1990) and those values which are statistically significant (at 5% and 1% level) are shown with asterisks. There are several positive and negative correlation coefficients which are highly significant and reveal following interesting relationships between the grain yields and various parameters:

1. The Indian monsoon is strongly and positively correlated with the Indian rice yield while strongly and negatively correlated with the U.S. corn yield. There is only a weak correlation between Indian monsoon and U.S.S.R. wheat yield or Australian wheat yield.

2. The SST anomalies in the equatorial eastern Pacific are positively correlated with Canadian wheat yield and also with U.S. corn yield, while negatively correlated with Indian rice yield.

3. The SO index is in general negatively correlated with U.S. and Canadian grain yields as well as with Argentinean wheat yield, while strongly and positively correlated with Australian wheat yield and only weakly correlated with Indian rice yield.

4. The Eurasian snow cover is strongly but negatively correlated with Indian rice yield and positively correlated with U.S.S.R. wheat yield.

5. The equatorial stratospheric wind oscillation appears to have a positive influence on Australian wheat yield and a negative influence on Argentinean wheat yield.

Most of the above relationships can be interpreted in the light of earlier discussion. For example, warmer SST anomalies in the equatorial eastern Pacific in general favour U.S. and Canadian grain yields because of the possibility of increased precipitation over the

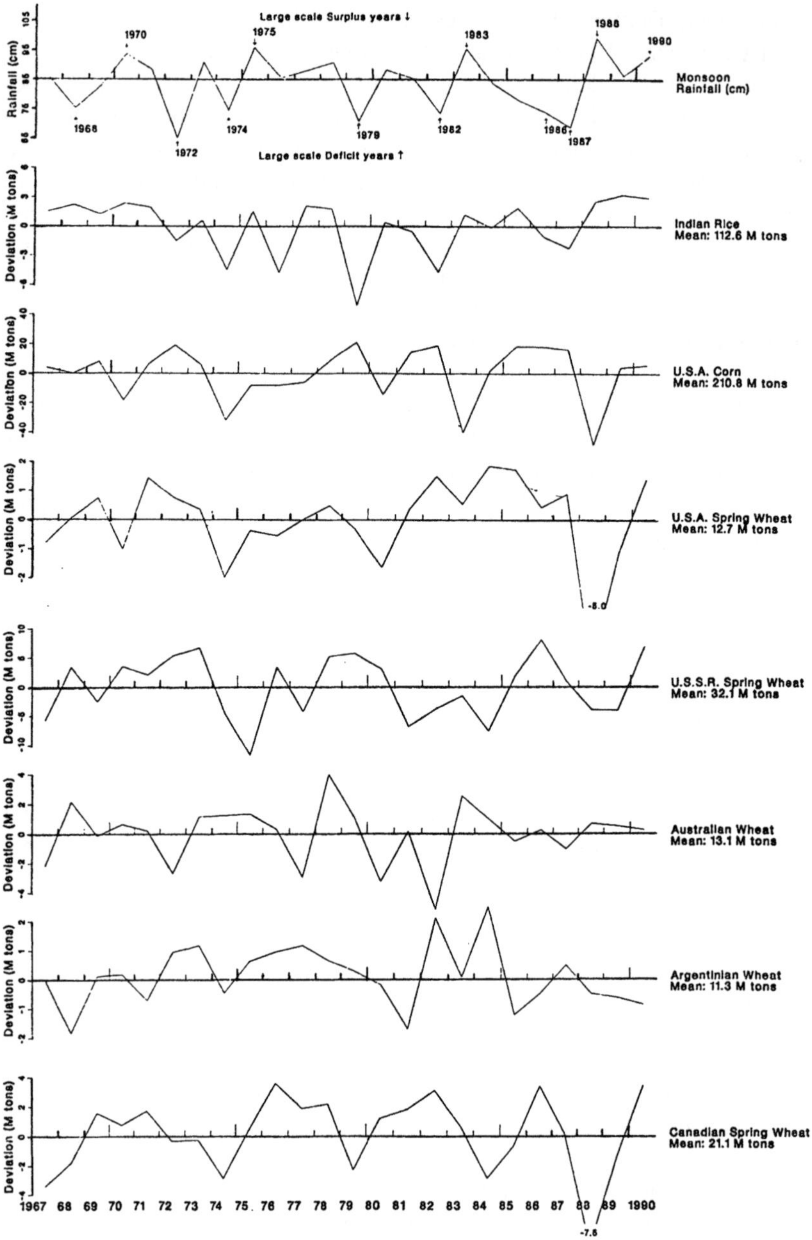

Figure 4: Interannual variation of Indian monsoon and seven major grain yields of the world. Large-scale droughts and floods in the Indian monsoon are shown by years and mean yield for various world grains are shown in millions of tons.

north American prairies (Handler, 1984), while the SO has in general a very strong positive impact on Australian wheat yield as documented by Nicholls (1985). The adverse impact of the Eurasian snow cover on Indian monsoon (as discussed earlier) and consequently on Indian rice yield is reflected in the strong negative correlation, while the positive correlation between Eurasian snow cover and U.S.S.R. wheat yield reflects the usefulness of snow cover in providing adequate soil moisture for increased wheat yield. The positive influence of the stratospheric QBO on Australian wheat yield appears to be due to the possibility of increased precipitation over northern Australia in response to the large-scale flow patterns of Figure 3B which are conducive to flood conditions in the Indian monsoon. Similar flow patterns over northern Australia associated with a westerly phase of QBO and a La Niña condition could produce increased convective activity and thus increased precipitation. Finally, the SO is strongly but negatively correlated with Argentinean wheat yield reflecting the fact that ENSO events influence precipitation patterns in northeastern Argentina (see Ropelewski and Halpert, 1987).

Table 2: Correlation coefficients of various parameters vs. grain yield. (from Garnett and Khandekar, 1992)

Parameter	Months	Cdn spring wheat	US corn	Indian rice	USSR spring wheat	US spring wheat	Australian wheat	Argentinian wheat	World grain yield
Monsoon rainfall		-0.029	-0.597**	0.628**	-0.237	-0.337	0.339	-0.061	-0.302
SST	DJF	-0.031	-0.290	0.166	0.039	-0.031	0.069	0.243	-0.257
	MAM	-0.208	-0.039	-0.196	0.080	0.241	-0.183	0.313	-0.138
	JJA	-0.441*	0.291	-0.367	-0.215	0.399	-0.268	0.184	0.076
SO	DJF	-0.133	0.147	-0.282	-0.201	-0.164	-0.318	-0.051	0.087
	MAM	-0.190	-0.106	0.111	-0.124	-0.214	0.289	-0.252	0.142
	JJA	-0.379	-0.285	0.298	-0.157	-0.382	0.624**	-0.471*	-0.173
Eurasian snow:	DJF	-0.017	0.222	-0.347	0.486*	0.139	0.194	0.156	0.265
	MAM	0.165	0.116	-0.492*	0.160	0.004	-0.005	0.038	-0.085
Stratospheric winds									
at 50 mb:	MAM	-0.290	-0.108	-0.122	0.166	-0.221 -	0.418*	-0.179	0.021
	JJA	0.133	-0.181	-0.307	0.177	-0.188	0.293	-0.412*	-0.058
at 30 mb:	MAM	0.140	0.033	0.099	0.286	-0.032	0.183	-0.293	0.012
	JJA	0.188	0.112	0.334	0.009	0.104	-0.081	-0.182	0.046

Correlation coefficient values for various levels of significance with 22 degrees of freedom (Data: 1967-1990)
* 5% level of significance; 0.404.
** 1% level of significance; 0.515.

6. CONCLUDING REMARKS

The El Niño and La Niña events in conjunction with droughts and floods in the Indian monsoon are shown to have significant impact on world grain yields. In his classical paper, Bjerknes (1969) brought out the importance of monitoring east equatorial regions of the Pacific for long-range forecasting over North and South America. The present study emphasize the importance of monitoring ENSO events in conjunction with Indian monsoon droughts and floods, for foreshadowing of world grain yields. Suitable indices of large-scale flow pattern that govern the broad features of the Indian monsoon can be developed and monitored on a year-round basis.

The ENSO events in association with the Indian monsoon provide the strongest climatic signal of the coupled atmosphere-ocean system. Suitable monitoring of this signal can provide a valuable guidance in foreshadowing of world grain yields.

ACKNOWLEDGEMENTS

This paper was presented as a keynote paper at the fifth biennial symposium of the International Natural Hazards Society, held in Qingdao, China, in August 1993. I wish to express my sincere gratitude to the scientific committee of the Natural Hazards Society for inviting me to the Symposium and for providing partial cost of the travel. I also wish to thank Ray Garnett of the Canadian Wheat Board, Winnipeg, for providing world grain yield data. Comments and suggestions from two anonymous reviewers are gratefully acknowledged. Finally appreciation is expressed to Ms. Ruth Tung of the Atmospheric Environment Service for expertly typing the manuscript.

REFERENCES

Angell, J.K., (1981). Comparison of variations in atmospheric quantities with sea surface temperature variations in the equatorial eastern Pacific. *Monthly Weather Review*, 109, 230-243.

Banerjee, A.K.,. Sen P.N and Raman C.R.V., (1978). On foreshadowing southwest monsoon rainfall over India with mid-tropospheric circulation anomaly of April. *Indian J. Met. Hydrology and Geophy.*, 29, 425-431.

Barnett, T.P., Dümenil L., Schlese U., Roeckner E. and Latif M., (1989). The effect of Eurasian snow cover on regional and global climate variation. *J. Atmospheric Sciences*, 46, 661-685.

Berlagè, H.P., (1966). The Southern Oscillation and the World Weather. Mededeelingen en Verhandelingen, No. 88, Koninklijk Nederlands Meteorologisch Instituut, 152pp, 1966.

Bhalme, H.N. and Mooley D.A., (1980). Large-scale drought/floods and monsoon circulation. *Monthly Weather Review*, 108, 1197-1211.

Bhalme, H.N., Rahalkar S.S. and Sikdar A.B., (1987). Tropical quasi-biennial oscillation of the 10mb wind and Indian monsoon rainfall -Implications for forecasting. *J. Climatology*, 7, 345-358.

Bhanu Kumar, O.S.R.U., (1988). Interaction between Eurasian winter snow cover and location of the ridge at the 500 hPa level along 75°E. *J. Meteorol. Soc. of Japan*,66, 509-514.

Bjerknes, J., (1966). A possible response of the atmospheric Hadley circulation to equatorial anomalies of ocean temperatures. *Tellus*, 18, 820-829.

Bjerknes, J., (1969). Atmospheric Teleconnections from the equatorial Pacific. *Monthly Weather Review*, 97, 163-172.

Blanford, H.F., 1884). On the connection of Himalayan snowfall and seasons of drought in India. *Proc. Royal Soc. London*, 37, 3-22.

Dey, B., and Bhanu Kumar O.S.R.U., (1983). Himalayan winter snow area and summer monsoon rainfall over India. *J. Geophy. Res.*,88, 5471-5474.

Dickson, R.R., (1984). Eurasian snow cover versus Indian monsoon rainfall - An extension of Hahn-Shukla results. *J. Clim. and Appl. Meteorology*, 23, 171-173.

Elliott, W.P., and Angell J.K., (1987). The relation between Indian monsoon rainfall, the southern oscillation and hemispheric air and sea temperature: 1884-1984. *J. Clim. and Appl. Meteorology*, 26, 943-948.

Garnett, E.R. and Khandekar M.L., (1992). The impact of large-scale atmospheric circulations and anomalies on Indian monsoon droughts and floods and on world grain yields - A statistical analysis. *Agricultural and Forest meteorology*, 61, 112-128.

Gowarikar, V., Thapliyal V., Sarker R.P., Mandal G.S. and Sikka D.R., (1989). Parametric and power regression models: New approach to long range forecasting of monsoon rainfall in India. *Mausam*, 40, 115-122.

Gray, W.M., Sheaffer J.D. and Knaff J.A., (1992). Influence of the stratospheric QBO on ENSO variability. *J. Meteorological Soc. of Japan*, 70, 975-995.

Hahn, D. and Shukla J., (1976). An apparent relationship between Eurasian snow cover and Indian monsoon rainfall. *J. Atmospheric Sciences*, 33, 2461-2462.

Handler, P. (1984). Corn yields in the United States and sea surface temperature anomalies in the equatorial Pacific Ocean during the period 1868-1982. *Agricultural and Forest Meteorology*, 31, 25-32.

Khandekar, M.L., (1979). Climatic teleconnections from the equatorial Pacific to the Indian monsoon - Analysis and implications. *Arch. für Meteor. Geoph. Bioklimatology*, A28, 159-168.

Khandekar, M.L., (1991). Eurasian snow cover, Indian monsoon and El Niño/Southern Oscillation - A synthesis. *Atmosphere-Ocean*, 29, 636-647.

Khandekar, M.L., and Neralla V.R., (1984). On the relationship between the sea surface temperatures in the equatorial Pacific and the Indian monsoon rainfall. *Geophy. Res. Letters*, 11, 1137-1140.

Krishna Kumar, K., Rupakumar K. and Pant G.B., (1992). Pre-monsoon ridge location over India and its relation to monsoon rainfall. *J. Climate*, 5, 979-986.

Meehl, G.A., (1987). The annual cycle and interannual variability in the tropical pacific and Indian ocean regions. *Monthly Weather Review*, 115, 27-50.

Mooley, D.A., and Paolino D.A., (1989). The response of the Indian monsoon associated with the change in sea surface temperature over the eastern south equatorial Pacific. *Mausam*, 40, 369-380.

Mooley, D.A., Parthasarathy B. and Pant G.B., (1986). Relationship between Indian summer monsoon rainfall and location of the ridge at 500mb level along 75°E. *J. Clim. and Appl. Meteorology*, 25, 633-640.

Mooley, D.A., and Parthasarathy B., (1984). Indian summer monsoon and the east equatorial Pacific sea surface temperature. *Atmosphere-Ocean*, 22, 23-35.

Mukherjee, B.K., Reddy R.S. and Ramana Murty Bh.V., (1979). High-level warmings, winds and Indian summer monsoon. *Monthly Weather Review*, 107, 1581-1588.

Nicholls, N. (1985). Impact of the Southern Oscillation on Australian crops. *J. Climatology*, 5, 553-560.

Pant, G.B., and Parathasarathy B., (1981). Some aspects of an association between the southern oscillation and the Indian summer monsoon. *Arch. für Met. Geoph. und Bioklim.*, B29, 245-252.

Philander, S.G.H., (1985). El Niño and La Niña. *J. Atmospheric Sciences*, 42, 2652-2662.

Rasmusson, E.M., and Carpenter T.H., (1983). The relationship between eastern equatorial Pacific sea surface temperatures and rainfall over India and Sri Lanka. *Monthly Weather Review*, 111, 517-528.

Reid, G.C. and Gage K.S., (1985). Interannual variations in the height of the tropical tropopause. *J. Geophysical Research*, 90, 5629-5635.

Ropelewski, C.F. and Halpert M.S., (1987). Global and regional scale precipitation patterns associated with El Niño/Southern Oscillation. *Monthly Weather review*, 115, 1606-1626.

Sikder, A.B., Patwardhan S.K. and Bhalme H.N., (1993). Tropical Stratospheric circulation and monsoon rainfall. *Advances in Atmospheric Sciences*, 10, 379-385.

Shukla, J., and Paolino D.A., (1983). The Southern Oscillation and long-range forecasting of monsoon rainfall over India. *Monthly Weather Review*, 111, 1830-1837.

Thapliyal, V. and Kulshrestha S.M., (1992). Recent models for long-range forecasting of southwest monsoon rainfall over India. *Mausam*, 43, 239-248.

Troup, A.J., (1965). The Southern Oscillation. *Q.J. Royal Met. Soc.*, 91, 490-506.

Walker, G.T., (1923). Correlation in seasonal variation of weather, VIII, A preliminary study of world weather. *Memoirs, India Met. Dept.*, 24, 75-131.

Walker, G.T., (1924). Correlation in seasonal variation of weather, IX, A further study of world weather. *Memoirs, India Met. Dept.*, 24, 275-332.

Walker, G.T., and Bliss E.W., (1932). World Weather V, *Memoirs, Royal Met. Soc.*, 4 53-84.

Wright, P.B., (1977). The souther oscillation - Patterns and mechanisms of the teleconnections and the persistence. Hawaii Inst. of Geophysics, Univer. of Hawaii (HIG-77-13), 107pp.

Yasunari, T., Kitoh A. and Tokioka T., (1991). Local and regional responses to excessive snow mass over Eurasia appearing in the northern spring and summer climate - A study with MRI GCM. *J. Meteor. Society of Japan*, 4, 473-487.

EL-NIÑO AND STORM SURGES IN THE BAY OF BENGAL

T.S. MURTY
National Tidal Facility, Flinders University of South Australia, GPO Box 2100, Adelaide SA 5001, Australia

and

V.R. NERALLA
Atmospheric Environment Service, 4905 Dufferin Street, Downsview, Ontario M3H 5T4, Canada

Abstract: The teleconnections between the occurrence of El-Niño events and the generation of tropical cyclones in various ocean basins have been considered by several authors. Here we present data linking storm surges in the Bay of Bengal with El-Niño. In fact this preliminary study shows that there is no correlation between El-Niño occurrences and Storm Surge Events in the Bay of Bengal.

1. INTRODUCTION

A possible linkage between the occurrence of El-Niño events and the genesis of tropical cyclones has been studied by several authors starting in the late 1980s. For example, Dong (1988) found that El-Niño events have a definite suppression effect on tropical storm or typhoon activity in the western Pacific area (about 120°-160°E in the north and about 145°-165°E in the south). He attributed this suppression effect to the variations in the Walker Circulation.

Wu and Lau (1992) used a General Circulation Model (G.C.M.) to simulate the relationship between tropical storm formation and ENSO (El-Niño and Southern Oscillation). They showed that in the western part of the North Pacific, South Pacific and North Atlantic, below normal frequency of tropical cyclone formation was simulated during El-Niño years. They found no significant correlation between ENSO and TS (Tropical Storm) formation over the South Atlantic and the eastern part of the North Atlantic. However, over the Indian Ocean, the relationship between storm formation and ENSO exhibited a seasonal dependence.

Evans and Allan (1992) investigated the effect of the phase of ENSO on tropical cyclone activity in the Australasian region. They found that during ENSO phases tropical cyclone

M. I. El-Sabh et al. (eds.), Land-based and Marine Hazards, 97–111.

activity is reduced to the northeast of Australia, but increases to the north and northwest of the continent. Coastal crossings of tropical cyclones are enhanced in Western Australia and the Northern Territory, but suppressed in Queensland.

Gray and Sheaffer (1991) studied hurricane tracks in the western part of the North Atlantic for the period 1900 to 1976 and conclusively showed that fare fewer hurricanes occurred during El-Niño years.

Next turning to the Bay of Bengal (Figure 1), Gupta and Muthuchami (1991) studied the tracks of tropical cyclones over the Bay of Bengal for the post-monsoon season (October to December). They showed that during El-Niño years, most will landfall south of 17°N (i.e. on the coasts of Andhra and Tamilnadu), whereas in (El-Niño-1) years, most will recurve and landfall north of 17°N (i.e. Orissa, West Bengal and Bangladesh).

In the present study we relate storm surges in the Bay of Bengal directly to the El-Niño events. To our knowledge, there is no study in the open literature directly linking storm surges and El-Niño occurrences either in the Bay of Bengal or any other water body.

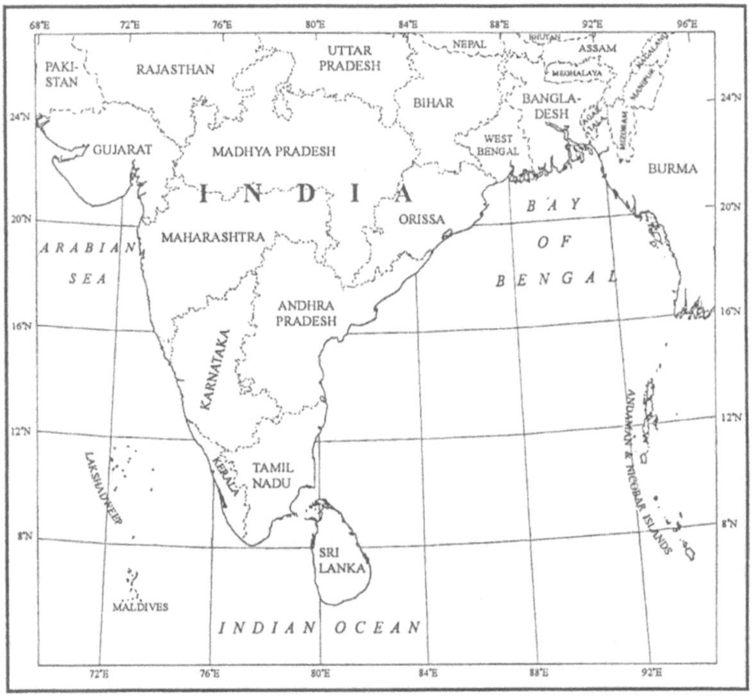

Figure 1: Geographical map of the Indian subcontinent showing the Bay of Bengal. The states mentioned in Table 2 are also shown.

2. DATA ON EL-NIÑO EVENTS

Table 1 lists the El-Niño occurrences for the period 1525 to 1993, mostly based on Quinn et al (1987). This table is also consistent with Gray and Sheaffer (1991) and Gupta and Muthuchami (1991). Even though, in principle one could have used more number of classifications than just three, considering the uncertainties both in storm surge data as well as El-Niño occurrences, it was felt that it is not realistic to have more than just three categories to designate the strength of the El-Niño events; namely Strong, Moderate and Weak.

3. STORM SURGE DATA FOR THE BAY OF BENGAL

Table 2 lists storm surges in the Bay of Bengal for the period 1584 to 1991. As can be seen there are a total of 144 events. Out of these, 98 surges occurred in years when there was no El-Niño, there were 15 surges during weak El-Niño years, and during moderate and strong El-Niño years, there were respectively 17 and 14 surge events. The number of surges associated with non El-Niño is 98 out of 144 (slightly greater than 68%).

An examination and analysis of the data from Tables 1 and 2 clearly indicates that there is no correlation at all between the occurrences of El-Niño events and storm surges episodes in the Bay of Bengal. Analysis of the data separately for the 19th and 20th Centuries reveals the same inescapable result, that there is no statistically significant evidence of a correlation between these two events.

Even though this is a negative result, nevertheless it is a very useful result to have, especially in the context of pronouncements in the media of attributing every storm surge disaster to an El-Niño event.

REFERENCES

Dong K. (1988). El-Niño and Tropical Cyclone Frequency in the Australian Region and Northwest Pacific. *Australian Meteorological Magazine*. Vol. 36, No. 4, 219-225.

Evans J.L. and Allan R.J. (1992). El-Niño/Southern Oscillation Modification to the Structure of the Monsoon and tropical Cyclone Activity in the Australasian Region. *International Journal of Climatology*. Vol. 12, 611-623.

Gray. M. and Sheaffer J.D. (1991). El-Niño and QBO Influences on Tropical Cyclone Activity, in "*Teleconnections Linking Worldwide Climate Anomalies: Scientific Basis and Social Impact*". edited by M.H. Glantz, R.W. Katz and N. Nicholls. Cambridge University Press. New York. 258-284.

Gupta A. and Muthuchami A. (1991). El-Niño and Tropical Storm Tracks Over Bay of Bengal During Post-Monsoon Season. *Mausam*. Vol. 42, No. 3, 257-260.

Jayanthi N. and Sarma A.K.S. (1986). A catalogue of Storm Surges in India. Rept 86-1, Regional Meteorological Centre, Madras, India, 25 p.

Katsura J., Hayashi T., Nishimura H., Isobe M., Yamashita T., Kawata Y., Yasuda T. and Nakagawa H. (1992). Storm Surges and Severe Wind Disasters Caused by the 1991 Cyclone in Bangladesh. March, 1992, Rept. No. B-2, Research Rept. on Natural Disasters, Supported by the Japanese Ministry of Education, Science and Culture, Grant No. 03306020, 101 p.

Khalil, G.M. (1992). Cyclones and Storm Surges in Bangladesh: Some Mitigative Measures. *Natural Hazards*, Vol. 6, No. 1, 11-24.

Murty T.S., Flather R.A. and Henry R.F. (1986). The Storm Surge problem in the Bay of Bengal. *Progress in Oceanography*, Vol. 16, 195-233.

Murty T.S. and Flather R.A. (1994). Impact of Storm Surges in the Bay of Bengal. *Journal of Coastal Research*, in press.

Quinn W.H., Neal V.T. and. Antunez de Mayolo S.E (1987). El-Niño Occurrences Over the Past Four and a Half Centuries. *Journal of Geophysical Research*. Vol. 92, NO. 13, 14,449 - 14,461.

Talukder J., Roy G.D. and Ahmad M. (1992). Living with Cyclones: Study on Storm Surge Prediction and Disaster Preparedness. Community Development Library, Dhaka, Bangladesh, 124 p.

Wu, G. and. Lau N.C (1992). A G.C.M. Simulation of the Relationship Between Tropical Storm Formation and ENSO. *Monthly Weather Review*. Vol. 13, 958-977.

Table 1: The El-Niño Years

Year	Strength	Year	Strength	Year	Strength
1525-26	Strong	1803-04	Strong	1911	Strong
1531-32	Strong	1806-07	Moderate	1914	Moderate
1539-41	Strong	1812	Moderate	1917	Weak
1552	Strong	1814	Strong	1918	Strong
1567-68	Strong	1817	Moderate	1823	Weak
1574	Strong	1819	Moderate	1925	Strong
1578	Strong	1821	Moderate	1929	Moderate
1591-92	Strong	1824	Moderate	1932	Weak
1607	Strong	1828	Strong	1939	Moderate
1614	Strong	1832	Moderate	1941	Strong
1618-19	Strong	1837	Moderate	1943	Weak
1624	Strong	1844-45	Strong	1946	Weak
1634	Strong	1850	Moderate	1948	Weak
1652	Strong	1857-58	Moderate	1951	Weak
1660	Strong	1860	Moderate	1953	Moderate
1671	Strong	1864	Strong	1957	Strong
1681	Strong	1866	Moderate	1963	Weak
1687-88	Strong	1867-68	Moderate	1965	Moderate
1696	Strong	1871	Strong	1969	Weak
1701	Strong	1874	Moderate	1972	Strong
1707-08	Strong	1877-78	Strong	1975	Weak
1714-15	Strong	1880	Moderate	1976	Moderate
1720		1884	Strong	1982	Strong
1728	Strong	1887-89	Moderate	1983	Strong
1747	Strong	1891	Strong	1986	Moderate
1761	Strong	1896-97	Moderate	1987	Moderate
1775	Strong	1899-1900	Strong	1992	Strong
1785-86	Strong	1902	Moderate		
1791	Strong	1905	Moderate		

Table 2: List of Storm Surges in the Bay of Bengal for the period 1584 to 1991 based on Jayanthi and Sarma (1986), Murty et al. (1986), Katsura et al. (1992), Khalil (1992), Talukder et al. (1992) and Murty and Flather (1994). For more details see the original papers. It is quite likely that several events till the 1900's have not been reported.

No.	Date	Location	Damage
1	1584	Near Backergunge, Bangladesh	200,000 deaths.
2	10 November, 1681	Nagapattinam, Tamilnadu, India	14,000 deaths
3	18 April, 1700	Visakhapatnam, Andhra Pradesh, India	Several ships lost at sea. No estimate of number of deaths available.
4	December, 1706	Coringa, Andhra Pradesh, India	No exact estimate of the number of deaths, except a statement that "people in great number were lost".
5	18 December, 1709	Coastal Ramnad, Tamilnadu, India	Extensive inundation of coastal Ramnad. Major surge. Deaths in thousands.
6	7-11 October, 1737	Mouth of Hoogly River, West Bengal, India	12m surge (probably includes tide and wind waves). Surge penetrated 100 Km inland through the Hoogly River. 300,000 deaths in India and Bangladesh (could be a slight exaggeration).
7	30 December, 1760	Pondicherry, Tamilnadu, India	Several ships destroyed. 1,100 Europeans died in the surge.
8	13 October, 1779	Masulipatnam, Andhra Pradesh, India	4m surge. 20,000 deaths.
9	29 May, 1787	Coringa, Andhra Pradesh, India	20,000 deaths.
10	December, 1789	Near Coringa, Andhra Pradesh, India	3 major waves in the surge. Several ships lost. 20,000 deaths. Inundation of the whole city of Coringa of 1 to 2m.

Table 2: (Continued)

No.	Date	Location	Damage
11	May - June, 1797	Chittagong, Bangladesh	2 ships sank. No estimate of deaths.
12	10 December, 1807	Madras, Tamilnadu, India	Extensive destruction in Madras. No estimate of deaths.
13	2 May, 1811	Near Madras, Tamilnadu, India	No estimate of deaths.
14	May, 1820	Nellore District, Andhra Pradesh, India	inundation. No estimate of deaths.
15	6 June, 1822	Barisal/ Backergunge, Bangladesh	100,000 deaths. 100,000 cattle lost.
16	27 May, 1823	Balasore, Orissa, India	Inundation up to 10 Km inland. Several ships and whole villages disappeared.
17	31 October, 1831	Balasore, Orissa, India	2 to 5m surges. Extensive inundation. 22,000 deaths. 50,000 cattle lost. This might have been the same storm that also generated a surge at Barisal, Bangladesh.
18	10 May, 1832	Coringa, Andhra Pradesh, India	Extensive damage at Coringa.
19	October, 1832	Balasore, Orissa, India	More violent storm than in 1823 (No. 16) but the surge was less destructive
20	21 May, 1833	Sagar Island/ 24 Paraganas, West Bengal, India	3m surge. 50,000 deaths. 100,000 cattle lost.
21	16 November, 1839	Between Visakhapatnam and Narsapur/ Coringa, Injaram, Andhar Pradesh, India	Over 6m surge. 20,000 deaths.

Table 2: (Continued)

No.	Date	Location	Damage
22	22 May, 1843	Masulipatnam, Andhar Pradesh, India	Extensive damage.
23	October, 1854	Mouth of Hoogly River, West Bengal, India	Up to 12m water level increase near Calcutta and environs. 50,000 deaths.
24	5 October, 1864	Near Contai, West Bengal, India	Up to 12m surges. Flooded up to 13 Km on either side of the Hoogly River. Number of deaths by drowning: 50,000. Number of deaths from disease as a result of inundation: 30,000.
25	2 November, 1864	Masalipatnam, Andhar Pradesh, India	About 130 Km long stretch was subjected to surges up to 4m. Surge penetrated up to 27 Km inland. 40,000 deaths.
26	1 November, 1867	East of Calcutta, West Bengal, India	2m surges. Damage at Port Canning. 13m surge at Hatia, Bhola Islands.
27	16 May, 1869	Khulna	Surge entered through the Pussur and Haringhata Rivers and inundated a vast area causing damage to life and property. 250 people died in Morelganj.
28	4 November, 1870	Viskhapatnam, Andhar Pradesh, India	Lower parts of the town flooded. Also whole of Anakapalle town flooded. 7 deaths.
30	31 October, 1876	Backergunge, Bangladesh	This was the infamous Backergunge cyclone. Peak water levels in Bangladesh varied from 3 to 15m, including tide and wind waves. 100,000 deaths in Bangladesh (then part of India). The India Meteorological Department was expanded following this surge. GILL (1975) and KHALIL (1992) put the death toll at 400,000, but this could be an over-estimate.

Table 2: (continued)

No.	Date	Location	Damage
31	3 December, 1878	Kakinada, Andhra Pradesh, India	Local storm surge and inundation.
32	22 September, 1885	Cuttack (False Point), Orissa, India	7m surge. 5,000 deaths by drowning. 300 deaths by falling trees. 50,000 houses destroyed. 10,000 cattle lost.
33	26 May, 1887	False Point, Orissa, India	Major surges.
34	18 June, 1980	North of Gopalpur and Cuttack, Orissa, India	Extensive damage by the surge.
35	27 October, 1893	Noakhali	Lot of destruction. Inundation subsided in 2 hours.
36	September, 1895	False Point, Orissa, India	UP to 7m water levels (tide and surge) on the north Orissa coast. More than 5,000 deaths.
37	October, 1895	Sunderban/ Bagerhat, Subdivision, Bangladesh	Damage estimates not available.
38	24 October, 1897	Chittagong, Kutubdia Island, Bangladesh	175,000 deaths. Thousands died in the epidemic which followed the calamity.
39	May, 1898	Teknaf, Bangladesh	Damage estimate not available.
40	November, 1901	Wester Sunderbans, Bangladesh	Damage estimate not available.
41	October, 1909	Chittagong, Bangladesh	Damage estimate not available.
42	December, 1909	Cox's Bazar, Bangladesh	Damage estimate not available.
43	April, 1911	Teknaf, Bangladesh	Damage estimate not available.
44	1912		40,000 deaths in Bangladesh.
45	May, 1917	Sunderbans, Bangladesh	Damage estimate not available.
46	20-25 September, 1919	Barisal, Bangladesh	40,000 deaths in Bangladesh.
47	20 May, 1925	Masulipatnam, Andhra Pradesh, India	The surge penetrated up to 6 Km inland.
48	May, 1926		More than 5,000 deaths in India.

Table 2: (Continued)

No.	Date	Location	Damage
49	May, 1926	Cox's Bazar, Bangladesh	Damage estimate not available.
50	1 November, 1927	Nellore, Andhra Pradesh, India	629 deaths.
51	17 November, 1933	Ongole, Andhra Pradesh, India	Moderate surge inundated low-lying areas.
52	20 April, 1936	Burma	Surges up to 5 m.
53	28 October, 1936	Masulipatnam, Andhra Pradesh, India	Surge penetrated up to 6 Km inland. Destruction of agriculture due to saline water intrusion.
54	25 November, 1938	Coastal districts, Andhra Pradesh, India	2m surges. Railway bridge washed away. Extensive damage to waterworks due to saline water intrusion
55	26 May, 1941	Between Barisal and Noakhali, Bangladesh	4m surge. 7,500 deaths.
56	16 October, 1942	Contai, West Bengal, India	5m surge at Midnapore (64 Km upstream in Hoogly River). 15,000 deaths.
57	31 October, 1943	Kalingapatnam, Andhra Pradesh, India	Surge combined with floods in River Vamsadhara caused great damage.
58	18 October, 1945	Masulipatnam, Andhra Pradesh, India	Up to 5m surges penetrated 24 Km inland.
59	17-19 May, 1948	Between Chittagong and Noakhali, Bangladesh	Damage estimate not available.
60	29 October, 1949	Between Masulipatnam and Harsapur, Andhra Pradesh, India	5m surges penetrated 16 Km inland. 800 deaths.
61	15-20 November, 1950	Patuakhali, Bangladesh	Damage estimate not available.
62	30 November, 1952	South of Nagapattinam, Tamilnadu, India	1.3m surge penetrated 8 Km inland. 400 deaths.

Table 2: (Conitnued)

No.	Date	Location	Damage
63	11 October, 1955	Kalingapatnam, Andhra Pradesh, India	Surge up to 1.7m.
64	1 December, 1955	Tanjavore District, Tamilnadu, India	Surges up to 5m penetrated 8 Km inland. 500 deaths.
65	1 June, 1956	Near Calcutta, West Bengal, India	Inundation in Midnapore District. Damage to agriculture due to saline water intrusion.
66	16-19 May, 1958	East Meghna Estuary, Bangladesh	Damage estimate not available.
67	21-24 October, 1958	Noakhali and West Meghna Estuary, Bangladesh	UP to 2m surges in Bangladesh and West Bengal State of India.
68	29 May, 1960	Sunderbans, Bangladesh	Peak surge of 3.2m.
69	9 October, 1960	Noakhali, Meghna Estuary, Bangladesh	1.5m tide and 5.1m surge. 3,000 deaths. 200 Kph winds.
70	30 October, 1960	Noakhali, Chittagong, Bangladesh	6.6m surge. 8,149 deaths.
71	9 May, 1961	Noakhali, West Meghna Estuary, Bangladesh	3m surges. 11,468 deaths.
72	30 May, 1961	Chittagong, Cox's Bazar, Noakhali, offshore islands, Bangladesh	6 to 8m surge. Peak water levels up to 9m. 10,466 deaths.
73	26 - 30 October, 1962	Feni, Bangladesh	Damage estimate not available.
74	28 May, 1963	Noakhali, Chittagong, Bangladesh	Peak water levels up to 8.1 m. 11,520 deaths.
75	5 - 8 June, 1963	Jessore, Bangladesh	Peak surge 3.1m.
76	21 October, 1963	Cuddalore, Tamilnadu, India	Almost 7m surge.
77	25 - 29 October, 1963	Teknaf, Bangladesh	Damage estimate not available.
78	11 April, 1964	Bangladesh	196 deaths.

Table 2: (Continued)

No.	Date	Location	Damage
79	23 December, 1964	Tondi (Rameswaram Island), Tamilnadu, India	6m surges. 1,000 deaths.
80	11 May, 1965	Barisal and Noakhali, Bangladesh	6m water levels. 19,270 deaths.
81	31 May, 1965	Chittagong, Bangladesh	Up to 7.6m surges. 12,000 deaths.
82	5 November, 1965	Bangladesh	Up to 3.7m surges.
83	11-15 December, 1965	Cox's Bazar, Bangladesh	Up to 8.8m water levels. 873 deaths.
84	1 October, 1966	Chittagong, Bangladesh	Peak water levels up to 10m. 850 deaths.
85	1 November, 1966	Bangladesh	Peak water levels up to 10 m
86	12 December, 1966	Cox's Bazar, Bangladesh	Damage estimate not available.
87	10 October, 1967	Noakhali, Bangladesh	Peak water levels up to 9.3m.
88	9-11 October, 1967	Puri, Orissa, India	Water Levels up to 9m.
89	23-24October, 1967	Cox's Bazar, Bangladesh	Up to 7.6m surges.
90	8 December, 1967	Nagapattinam, Tamilnadu, India	Moderate surges. 7 deaths.
91	10 May,. 1968	Akyab, Burma	1,037 deaths in Burma. 4.7m surges in Bangladesh.
92	17 April, 1969	Bangladesh	75 deaths
93	10 October, 1969	Khulna, Bangladesh	Peak water levels up to 8m.
94	7 November, 1969	Kakinada, Andhra Pradesh, India	3.1m surge. 900 deaths.
95	7 May, 1970	Cox's Bazar, Bangladesh	Up to 5m surges.
96	22 - 23 October, 1970	Chandpur, Bangladesh	4.7m surge. 300 deaths.
97	12 November, 1970	Khulna to Chittagong and offshore islands, Bangladesh	Up to 9m surges. 275,000 deaths. The often quoted 500,000 deaths could be an over-estimate.
98	8 May, 1971	Meghna Estuary, Bangladesh	Up to 5m water levels.
99	30 September, 1971	Chandpur, Bangladesh	Surge plus tide of 5m.

Table 2: (Conitnued)

No.	Date	Location	Damage
100	30 October, 1971	Paradip, Orissa, India	Water levels up to 6m. The surge penetrated 25 Km inland. 10,000 deaths.
101	5 - 6 November, 1971	Chittagong, Bangladesh	Water levels up to 5.5m.
102	28 - 30 November, 1971	Sunderban coast, Khulna, Bangladesh	1m surge.
103	10 September, 1972	Barua, Orissa, India	3.4m surge. 0.8m tide.
104	22 September, 1972	Gopalpur, Orissa, India	Inundation in Puri District.
105	15 - 23 November, 1972	South of Nellore, Andhra Pradesh, India	Minor Surge.
106	11 October, 1973	Chandbali, Orissa, India	Mild surge in river estuaries caused saline water intrusion in the coastal areas of north Orissa and West Bengal.
107	16 - 18 November, 1973	Barisal, Bangladesh	1.0m surge. Peak water level including wind waves 3.8m.
108	9 December, 1973	Patuakhali coast and offshore islands, Bangladesh	Peak water level 6.2m. Peak surge 4.5m. 183 deaths.
109	13 - 15 August, 1974	Khulna, Bangladesh	6.7m surge.
110	20 August, 1974	Contai, West Bengal, India	Inundation of low-lying areas of Digha and Juneput. 7 deaths
111	28 November, 1974	Chittagong, Bangladesh	4.9m surge. 20 deaths, 50 injured, 280 persons missing, 1,000 cattle killed, 2,300 house destroyed.
112	5 - 7 May, 1975	Gwa (between Sandoway and Bassein), Burma	303 deaths. Damage US$78 million.
113	9 - 12 May, 1975	Bhola, Cox's Bazar, Khulna, Bangladesh	5 deaths. 4 injured. 36 fishermen missing.
114	5 - 7 June, 1975	Chittagong, Bangladesh	Peak surge 4.0m.
115	24 - 28 June, 1975	Bangladesh	Tide and surge 4.8m.

Table 2: (Conitnued)

No.	Date	Location	Damage
116	8 - 12 November, 1975	Barisal, Noakhali, Bangladesh	Maximum surge 3.1m.
117	29 April - 2 May, 1976	Sandoway, Burma	Moderate surge.
118	11 September, 1976	Contai, West Bengal, India	2.5m surge. 1.4m tide. 40 deaths.
119	20 October, 1976	Meghna Estuary, Bangladesh	Tide plus surge 5.0m at Comapaniganj.
120	20 November, 1976	Chittagong, Bangladesh	1.0m surge. 2.1m tide.
121	12 - 13 May, 1977	Sunderban, Bangladesh	0.6m surge. 0.7m tide.
122	19 November, 1977	Chirala, Andhra Pradesh, India	Peak surge 5.0m, Tide 0.3m. Divi and surroundings totally inundated. 10,000 deaths.
123	12 - 17 May, 1978	Myebon (between Akyab and Kyaukpyu), Burma	90 to 95% of the city of Kyaukpyu was damaged.
124	30 September - 3 October, 1978	Khulna and Sunderban coast, Bangladesh	Damage estimate not available
125	24 November, 1978	Between Kilakkarai and Rochemary Island, Tamilnadu, India	4m surges on the coasts of Tamilnadu and Sri Lanka. Extensive damage on the northeast coast of Sri Lanka. 373 deaths in Sri Lanka. 10 deaths in India.
126	12 May, 1979	Kavali, Andhra Pradesh, India	3m surge. 0.6m tide. 700 deaths.
127	10 December, 1981	Bangladesh	2m surge 15 deaths.
128	2 - 5 May, 1982	South of Gwa, Burma	4m surge along the south Arakan coast. 31 deaths. US$3.8 million damage.
129	1 - 4 June, 1982	Between Paradip and Chandbali, Orissa, India	2m surges along the Orissa and West Bengal coasts. Peak surge of 4.8m 35 Km north of Dhamra harbour. 245 deaths.

Table 2: (Conitnued)

No.	Date	Location	Damage
130	15 October, 1983	Chittagong coast near the Feni River, Bangladesh	43 deaths. Substantial damage.
131	9 November, 1983	Chittagong, Cox's Bazar, Bangladesh	300 fishermen missing. Substantial damage.
132	24 May, 1985	Cox's Bazar, Chittagong, Sandwip, Bangladesh	4.3m surge. 11,069 deaths. 94,000 houses destroyed.
133	29 June, 1985	Sri Lanka	8 deaths
134	20 September, 1985	Close to Puri, Orissa, India	2m surges. Inundation lasted for 3 days. Substantial damage.
135	16 October, 1985	Near Balasore, Orissa, India	Up to 4m surge. Damage due to saline water inundation.
136	9 November, 1986	Chittagong, Bangladesh	14 deaths. Substantial damage.
137	14 - 19 October, 1987	North of Ongole, Andhra Pradesh, India	Moderate surge. 17 deaths. Substantial damage.
138	31 October - 3 November, 1987	Nellore, Andhra Pradesh, India	50 deaths. 26,000 cattle killed. Substantial damage.
139	29 November, 1988	Khulna coast near Raimangal, Bangladesh	4.4m surge. 5,708 deaths. 6,000 missing. 65,000 cattle killed. 15,000 deer killed. 9 royal Bengal tigers killed. Substantial crop damage.
140	4 - 7 November, 1989	Kavali, Andhra Pradesh, India	Mild surge caused destruction in Nellore and Kavali.
141	4 - 10 May, 1990	Mouth of Krishna River, Andhra Pradesh, India	600,000 house destroyed. 21,600 cattle killed. 3,500,000 poultry killed. 42,700 goats and sheep killed. Substantial damage to agriculture.
142	7 - 8 October, 1990	Barisal, Bangladesh	150 fishermen missing. Substantial damage.
143	29 April, 1991	Chittagong, Cox's Bazar, Bangladesh	4 to 8m surges. 150,000 deaths. 70,000 cattle killed. Great damage.
144	2 June, 1991	Bangladesh	1.2m surge.

STORM SURGES AND RIVER FLOODING IN BANGLADESH : A RISING CHALLENGE IN A CHANGING GLOBAL CLIMATE

SELINA BEGUM
Water Resources Engineering Department,Bangladesh University of Engineering and Techonology, Dhaka-1000, Bangladesh.

Abstract: Bangladesh is a delta of the rivers Ganges, Brahmaputra and Meghna at the head of Bay of Bengal. The three main rivers the Ganges, the Brahmaputra and the Meghna rise in the Himalayan hills and drains a catchment area of about 1.5 million Km^2 of which only 7.5% lies within Bangladesh. Numerous tributaries and distributaries of the river system covers entire Bangladesh. There is heavy rainfall within the catchment area of the G-B-M rivers concentrated within the four month monsoon period (June-Sept.).

Bangladesh is prone to meteorological and geological natural disasters due to its geographical location, variable climate, variable topography, dynamic river system and exposure to the sea. Floods and cyclones are the natural disasters that Bangladesh faces recurrently. The steady increase in population continuously increases the potential for natural disaster in Bangladesh.

The present scenarios of natural disasters in Bangladesh may aggravate due to possible changes in global climate as a consequence of global warming. Increase in precipitation and rise in sea level will increase the intensity and frequency of extreme events. The long-term implications of global warming with the addition of population explosion could make Bangladesh more vulnerable to natural disasters in years to come. This paper mainly focusses on the causes and occurrence of natural disasters in Bangladesh (especially 1988 flood and 1991 cyclone) and the weakness and strength of the existing measures in disaster prevention, mitigation and management. The counter measures and strategies which should be undertaken in disaster prevention/mitigation in Bangladesh have been recommended.

1. INTRODUCTION

Bangladesh is one of the largest and wettest river deltas in the world at the head of Bay of Bengal. The total land area of Bangladesh is 144,000 Km2 and it has a population of about 120 million. The average land elevation ranges from mean sea level to 30 meters above, while more than half of the country has an elevation of less than 8 meters above mean sea level. There are only a few hills in the South-east and North-east part of the country. The average slope is approximately 3-5 cm/km.

Bangladesh is a delta of the rivers Ganges, Brahmaputra and Meghna. Numerous tributaries and distributaries of the river system interlace and cover the entire country. This river system drains a catchment area of about 1.5 million Km2 of which only

M. I. El-Sabh et al. (eds.), Land-based and Marine Hazards, 113–138.

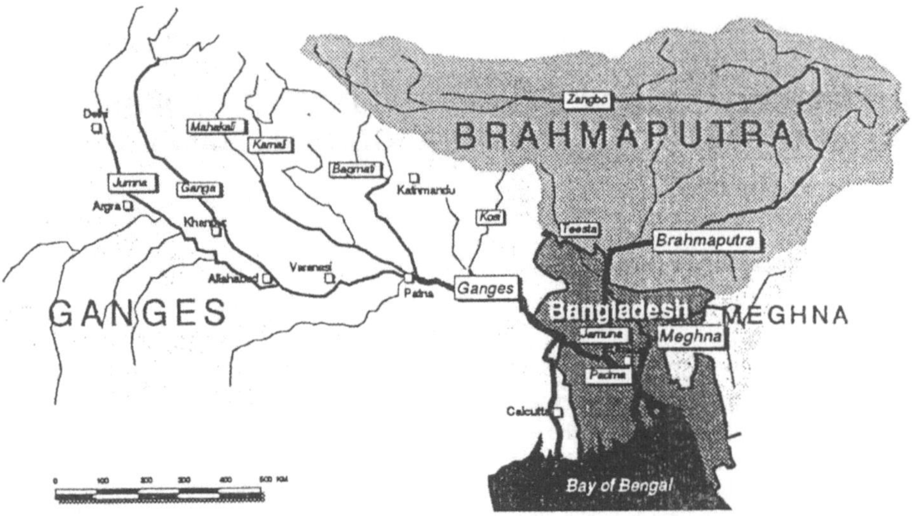

Figure 1: The Ganges, The Brahmaputra and the Meghna basins

7.5% lies within Bangladesh. These rivers rise in the Himalayan hills and spread over China, Nepal, Bhutan, India, Bangladesh and finally flow into the Bay of Bengal (Figure 1).

The rivers of Bangladesh possess very active morphodynamic systems characterized by low gradient of the rivers, pronounced meandering, channel shifting and severe bank erosion. The Ganges - Brahmaputra - Meghna river system carries a large volume of water and sediment load. The distribution of these loads varies with variation between very wet monsoon season (June-Sept.) and the relatively drier winter months.

The climate of Bangladesh is of tropical monsoon variety, characterized by high temperature, heavy rainfall and excessive humidity with marked seasonal variation. There is heavy rainfall within the catchment area of the G-M-B river system. About 80% of total annual rainfall is concentrated in the four month monsoon period (June-Sept.). The mean annual rainfall of Bangladesh ranges from 1400 mm in the west to over 6000 mm in the extreme east. However streamflow within Bangladesh depends heavily on inflows which originate outside the country. At Cherrapunji, on the Meghalayan plateau of northeast India, which is near the north-eastern boundary of Bangladesh, more than 11 meters of rain falls in an average year. In a wet year rainfall exceeds 25 meters. Most of this water finds its way into the rivers and flows into Bangladesh. The total flow drains through the single outlet to the Bay of Bengal. Total inflow to Bangladesh from outside the boundary is approximately 90% of total available streamflow in the country.

Bangladesh is prone to natural disasters due to its geographical location, variable climate, flat topography, active river system and exposure to the sea. This physical system is highly sensitive to the input of flow which comes from outside the boundary and high tides and surges from the sea. Floods and cyclones are the disasters that

Bangladesh faces recurrently. Occurrence of other natural disasters resulting from droughts, river erosion, tornadoes and earthquakes are not unusual.

The present scenarios of natural disasters in Bangladesh may aggravate due to possible changes in global climate resulting from the greenhouse effect. If the current trend of greenhouse gas emissions continues, there may be an increase of global mean surface temperature between 1.5 and 4.5¡C by the middle of the next century. A global warming could led to a sea level rise of 20-140 cm by the middle of the next century. For each degree of global warming, the worldwide precipitation may be expected to increase by 2-3% and substantial changes may occur in the timing and distribution of regional rainfall. The long-term implications of global warming with the addition of population explosion could make Bangladesh more vulnerable to flooding in years to come. So occurrence of more extreme events will cause an exponential increase in loss of life.

2. NATURAL DISASTERS IN THE WORLD

Natural disasters are claiming the lives of approximately 150,000 people every year. Record shows that more than 3 million persons have been lost due to natural disasters, with more than 800 million persons being injured and at least 23 billion dollars in damages directly lost within the past 20 years in the world (United Nation, 1987).

In recent years disasters injuring several thousands or several tens of thousands of people at a time have frequently occurred, such as the Spitak Earthquake in the Republic of Armenia, USSR in 1988; earthquakes in Iran and Philippines in 1990; cyclone and storm surge disasters in Bangladesh in 1991; the eruption of Pinatubo Volcano and a typhoon in Layte Island in the Philippines in 1991.

Looking at the situation concerning disaster occurrence by region, it can be seen that more than 80% of the total deaths are in Asia and South and West Pacific countries. These countries are prone to meteorological and geological natural disasters. Moreover most of them are developing countries and thus their social and economic development is considerably damaged by natural disasters. The frequency of disasters in the South -East Asia region has increased at an alarming rate during the last thirty years (Table I). Bangladesh is ranking on top over other countries of the S-E Asia region. Continuous increase in population, concentration of population in Urban areas and rapid population migration to different parts of the country, including high-risk areas have increased the extent of damage. Future risks of geological and meteorological natural disasters in the world may increase due to global warming and its consequences.

Table I Countries with Highest Prevalence of Disasters from 1960 to 1989 by Ten Year Period (the figures are the number of disasters)

Country	1960 - 1969	1970 - 1979	1980 - 1989
Bangladesh	18	37	77
India	34	102	172
Indonesia	20	46	88
Myanmar	10	10	24
Nepal	7	8	19
Sri Lanka	5	8	25
Thailand	4	5	25

Source: WHO, 1991

3. OCCURRENCE OF NATURAL DISASTERS IN BANGLADESH

Bangladesh is prone to meteorological and geological natural disasters. Floods and cyclones are the disasters that Bangladesh faces recurrently. Bangladesh also experience disasters due to drought, river erosion, tornadoes and earthquakes.

3.1 Floods

3.1.1 General Causes of Flood

Bangladesh is a land of rivers. About 18-20% of the land area of Bangladesh is vulnerable to normal flooding and the area increases to about 40% in major floods. From June to September each year the warm moist air of the monsoon sweeps up the Bay of Bengal from the Indian Ocean producing some of the highest recorded rainfalls in the world over Bangladesh and particularly in the upstream catchment of the G-B-M rivers. The large rainfall - runoff leads to a concentration of high discharge and sediment load in the rivers and causes the typical flood. Since more than 90 percent of the catchment area of these rivers lies outside Bangladesh, it is mainly the rainfall in the upper catchment that contributes to the floods in Bangladesh. Rainfall within Bangladesh contributes only 10-15 percent to the total streamflow. Locally concentrated rainfall of more than 300 mm in 2-6 hours leads to local floods of severe intensity for short durations. High monsoon rainfall coupled with snow-melt runoff over the Himalayas causes flooding in the delta.

In addition to this a) the low terrain, b) the back water effect of tides in the Bay of Bengal, c) synchronization of peak floods in the Ganges and the Brahmaputra river, e) siltation of river bed, e) tectonic activity and g) human intervention in the natural system (e.g. deforestation, flood control activities in the upper reaches) plays a major role in aggravating the flood situation within Bangladesh.

3.1. 2 History of Floods and Recent Major Floods

Available records of floods in the lower reaches of the Ganges - Brahmaputra - Meghna rivers show the occurrence of major floods. As a result of floods in 1787 (Morgan, 1959) the river Teesta changed its course. Recurrent floods between 1787 and 1830 also changed the old course of the Brahmaputra (now called old Brahmaputra) to its present course. Statistical analysis of available records, revealed that severe floods can occur every 7 years, and catastrophic floods every 33 to 50 years. Available records of flood affected areas show the occurrence of major floods in 1954, 1955 and exceptional major floods occurred in 1987 and 1988. About 18-20% of the total land area of Bangladesh is vulnerable to normal flooding and this increases to about 40% in major floods. In 1987, about 50% of the land was inundated. The flood of 1988 is the worst flood of Bangladesh in living memory. About 80% of the land was inundated and 50 million people were affected.

The flood of 1987 exceeded all the previous records. There had been an abnormally high rainfall throughout the country, particularly in the northern part of Bangladesh. However in general rainfall within Bangladesh contributes 10-15% to its streamflow. Hence it is very important to observe the rainfall in the entire basin of the Ganges, Brahmaputra and Meghna especially in the upper catchment. In 1987 rainfall exceeded all previous records in many places within Bangladesh. Intensity of rainfall is another important factor in addition to total amount of rainfall. Records showed periods of intense rainfall for some stations in Bangladesh for the monsoon in 1987. The record high 1987 flood peak on the Ganges at Hardinge Bridge was 76,000 cumec.

In 1988 the problem started outside Bangladesh. Concentrated heavy rainfall on the Meghalayan plateau caused an exceptionally high flow in the Meghna river in early June. Further rainfall over north and northeast Bangladesh, India, Bhutan and Nepal brought the Brahmaputra and its tributaries Dharla and Teesta to flow full. The Brahmaputra remained in flood stage from early July and had two other flood peaks on July 10th and July 30th, before the devastating flood of August - September.

After the 3rd week of August the Brahmaputra started rising very fast. On the 28 August there was a full moon at perigee and there was high tide in the ocean. On the 30th of August the Brahmaputra flow at Bahadurabad reached 99,500 cumec breaking all previous records. Three days later, flow in the Ganges at Hardinge Bridge reached 72,300 cumec, which is not much less than the highest record at Ganges in 1987. In early September of 1988, water spread rapidly across the flood plain and Dhaka, the capital of Bangladesh, was seriously flooded for the first time.

The peak flood of 19,800 cumec at Bhairab Bazar on the Meghna, occurred on 10 September. Table II shows the recorded highest flow and that of 1987 and 1988 at Ganges, Brahmaputra and Meghna in Bangladesh. In 1988 Flood level at 10 out of the 34 flood monitoring stations exceeded the recorded highest level while at 22 other stations it exceeded the level of 1987 flood (Abbas, 1989). The spring tide due to the new moon, in conjunction with the solar eclipse on 11 September, resulted in a considerable delay in the recession of flood water from the southern part of

Bangladesh. Construction of unplanned roads without culverts also caused serious drainage congestion. The main parts of Dhaka city, including its commercial area, was under 2-3 meters of water during peak flow. All communication with the outer world was disrupted for more than two weeks.

Table II Recorded Highest Flow and that of 1987 and 1988 at Ganges, Brahmaputra and Meghna

River	Station	Peak Flow (cumec)	Peak Flow (cumec) 1987	Peak Flow (cumec) 1988
Ganges	Hardinge Bridge	73, 150 on 1-9-61	76, 000 on 19 Sept.	72, 300 on 2 Sept.
Brahmaputra	Bahadurabad	91, 100 on 6-8-74	67, 200 on 12 Aug.	99, 500 on 30 Aug.
Meghna	Bhairab Bazar	19, 500 on 8-8-74	16, 000 on 16 Aug.	19, 800 on 10 Sept.

In 1988 heavy rainfall occurred in the upper catchment of the Ganges, Brahmaputra and Meghna rivers particularly of the Brahmaputra. Record shows that the rainfall within Bangladesh was not a major factor for the 1988 flood, and rainfall was less than the 1987 in most of the places. The rainfall in the upper catchment was the heaviest in 50 years. It takes a few days for the flood water in upper Assam to travel to lower Assam, and thereafter to Bangladesh. The peak flow at Tseladzong the most downstream station in Tibet of Tsangpo (Brahmaputra in Tibet) was 11, 909 cumec on 27 August (Matin et al., 1989). This flow is only 11% of the peak flow of Brahmaputra at Bahadurabad. However, the peak flow at Tseladzong in 1988 was about 70% higher than that of 1987 peak flow (Matin et al., 1989). In addition to heavy rainfall in the upper catchment, synchronization of peak flows in the major rivers of Bangladesh also caused heavy onrush of water near the confluence of Ganges and Meghna, resulting in the catastrophic flood throughout the country. The time difference of the peak flow between Brahmaputra at Serajgang and Ganges at Hardinge Bridge during the major floods in 1954, 1974 and 1987 varied between 27 and 32 days. In the 1988 flood this time difference was only 3 days. Large scale deforestation in the head water regions have increased sedimentation in recent years. This could be an additional major cause of aggravation to flood situation.

It was a catastrophic flood of unprecedented dimension, resulting in heavy loss of life and property. Crops in 4.26 million hectares of land were totally damaged, while those in 3.28 million hectares were partly damaged. Innumerable dwelling houses, schools, roads, bridges and industries were damaged. Within two weeks the economic programme was pushed back by several years. People suffered from scarcity of drinking water and sanitation problems. Very soon water borne disease had spread into many places.

3.2. River Erosion

The delta formed by the quarternary sediments deposited by the Ganges, Brahmaputra and Meghna river is young and undergoes continuous changes. The major rivers of Bangladesh are moderate to very unstable. The Ganges and Brahmaputra show high rates of lateral migration. Periodic disastrous earthquakes caused rapid subsidence and uplift of the land of this region. As a result major changes in the courses of the rivers Ganges and Brahmaputra occurred within the past 200 years. The oldest reliable picture of the rivers of the Bengal delta is shown in Renell's map (Figure 2).

Comparison of the Rennell's map with the maps of the 19th century show major morphological changes in the river system of the delta. The major diversion of Brahmaputra into its present course occurred due to a sudden increase in the volume of water as a result of the abnormal flood of 1787 (Morgan et al., 1959) which is

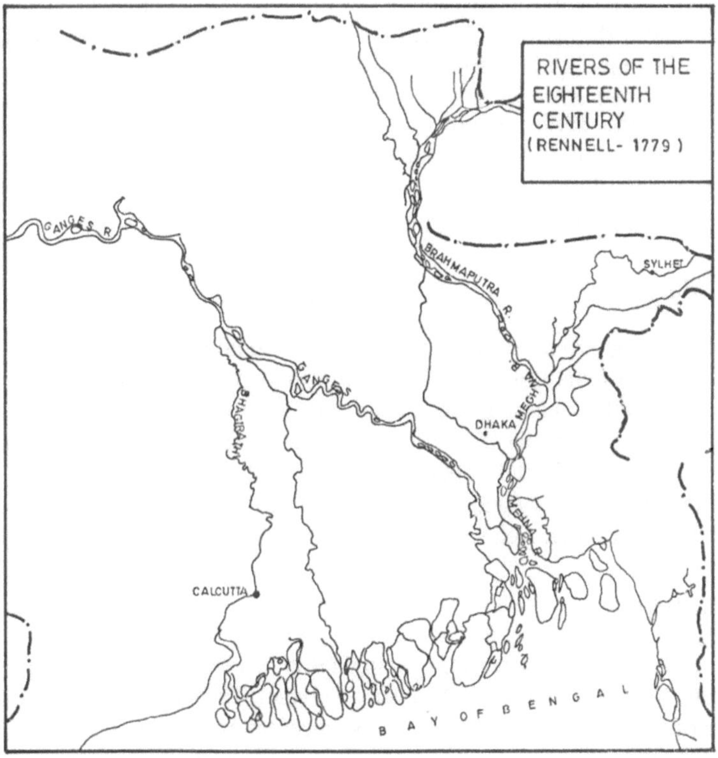

Source : Chatterjee, 1949

Figure 2: Rivers of the nineteenth century.

shown in Rennell's map (Figure 2).

Rennell's map also shows that the Ganges and the Meghna had two separate courses all along upto the Bay of Bengal which indicates that their present meeting near Chandpur took place at a later stage.

Numerous major earthquakes affected the region in 1762, 1775, 1812, 1888 and 1897 (Morgan et al., 1959). These earthquakes caused tilting of the block resulting in uplift and subsidence of the land.

In Bangladesh, every monsoon period brings significant changes in the courses of the Ganges and Brahmaputra. As a result of severe erosion and river migration, approximately 1 million people get affected every year. Erosion and river migration is a severe problem especially in the Brahmaputra.

3.3. Drought

Although there is abundance of water in the monsoon season (June - Sept), Bangladesh faces seasonal aridity for four months which may increase up to a maximum of six months. A considerably large area of land in the north west and some part of south-west region is susceptible to drought. Drought induced famines were major disasters in the past. By increasing irrigation facility with the aid of improved technology, the impacts of hazards due to drought have been reduced.

In recent years the north-western part of Bangladesh has faced severe aridity and drought conditions due to the withdrawal of Ganges water in the low flow season by Farakka Barrage in India. The most severe drought occurred in 1993 and the Ganges in Bangladesh experienced the lowest dry season flow.

3.4. Tornadoes

Disasters resulting from tornadoes are also common in Bangladesh. Tornadoes normally hit in the month of April. The central part of Bangladesh is mostly affected. Two severe tornadoes caused widespread damage over Dhaka. A tornado with an estimated wind speed greater than 500 km/hr caused severe damage in an area east of Dhaka in 1969. Approximately 1000 people were killed with an economic loss of 40-50 million taka (Chowdhury and Rahman, 1992). In 1989 another tornado with a wind speed of 400 km/hr hit an area west of Dhaka and caused widespread damage.

3.4. Earthquakes

The north and north-eastern region of Bangladesh has experienced several earthquakes of moderate to high intensity and severe damage. The earthquake of 1987 with a magnitude of 8.7 in the Richter scale had its epicenter at Shillong plateau of India, near the northeastern border of Bangladesh. It caused widespread damage throughout Bangladesh. Two other major earthquakes, the Bengal earthquake of 1885 and

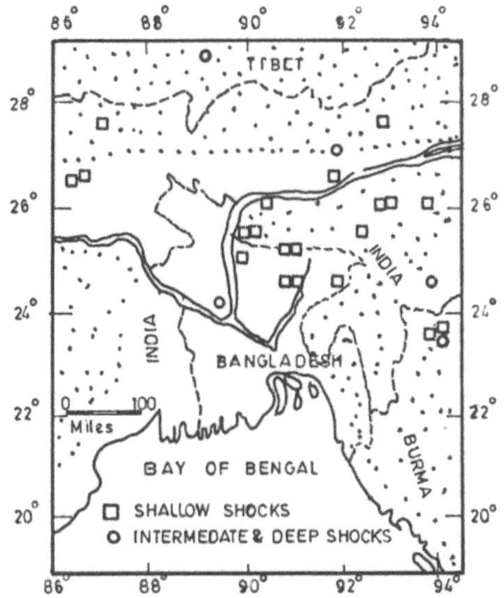

Figure 3: Epicentres of earthquakes in the Bengal Basin and adjacent areas (Morgan et al., 1959)

Srimangal Earthquake of 1918 caused severe damages on limited areas surrounding their epicenters. In 1950, another earthquake of 8.5 Richter magnitude caused widespread damage in Assam, India and minor damage in Bangladesh. Figure 3 shows epicenters of recorded earthquakes in the Bengal basin and adjacent regions.

Bangladesh has been divided into four seismic zones (Figure 4). The north and north eastern part of the country including the towns of Sylhet, Mymensing and Rangpur are in the most active seismic zone. Dhaka and Chittagong, the two largest cities are in Zone II, where the shocks could be moderate. Minor shocks are possible in Zone III which covers Rajshahi, Pabna, Faridpur, Noakhali, Chittagong and Cox's Bazar. The southwestern part of Bangladesh including Jessore, Khulna, Barisal and Patuakhali fall in Zone IV where negligible shocks are observed.

3.5 Cyclones

3.5.1. Causes and Occurrence

The coastal belt of Bangladesh is occasionally struck by severe cyclones, the most violent natural disaster. Being in the path of tropical cyclones, Bangladesh is hit by cyclonic storms and associated storm surges causing loss of human lives and livestock and severe devastation of crops and property. Tropical cyclones generally form in the region between 5¡ and 30¡ latitudes on both sides of the equator. The Bay of Bengal and the adjoining Andaman sea is one of the most favourable areas for the generation of tropical cyclones. Bangladesh is on the Tropic of Cancer and geographically belongs to the tropical zone.

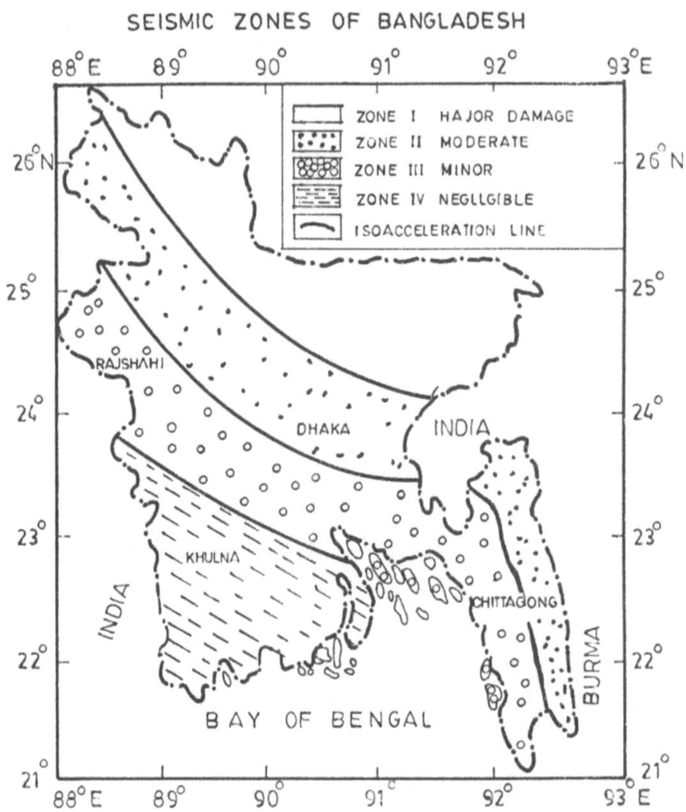

Figure 4: Seismic zones of Bangladesh (Bangladesh country report, 1989)

About 10 percent of the global total of tropical cyclones occur in the Bay of Bengal (Gray, 1968). About 60 percent of all death due to storm surge have occurred in the low lying arable coastal areas of the countries bordering the Bay and the Andaman Sea, with Bangladesh alone counting for about 40 percent (Murty et al., 1986). Between 1877 and 1980, 63 out of 392 tropical cyclones formed in the Bay of Bengal had landfall on the Bangladesh coast (Matsuda, 1991). The ten years distribution of the number of cyclones that struck Bangladesh is shown in Figure 5.

The reasons for the nonuniform distribution and large impact of cyclones and storm surges in Bangladesh are the following :

a) The phenomenon of recurvature of tropical cyclone in the Bay of Bengal.
b) Shallow continental shelf, especially in the eastern part of Bangladesh.
c) Converging bay i.e., triangular shape at the head of Bay of Bengal.
d) High tidal range.
e) Almost sea level orography at the Bangladesh coast.
f) High density of population.

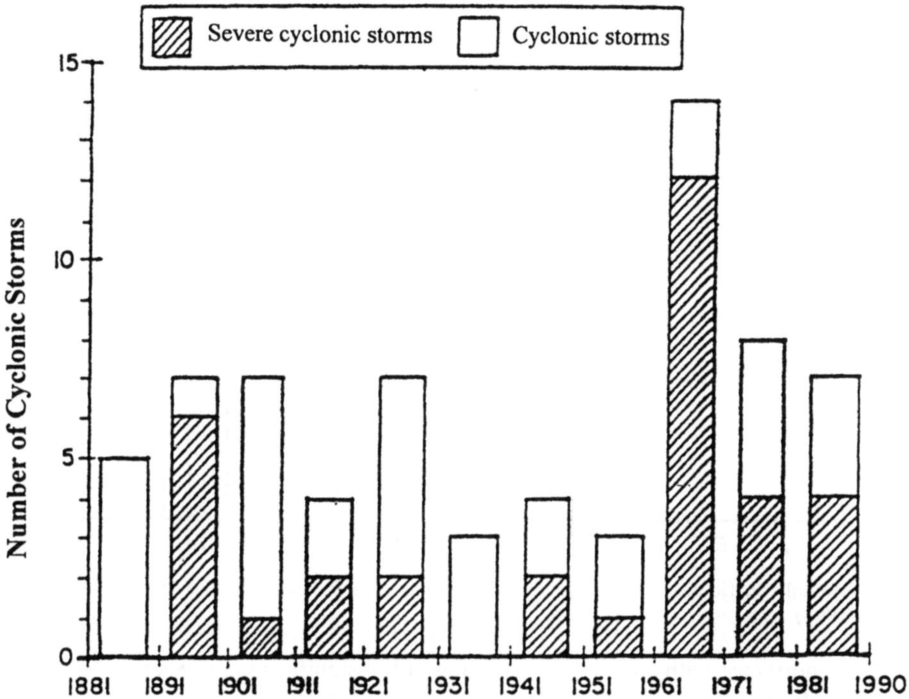

Figure 5: Ten years distribution of the number of cyclones that struck Bangladesh (Matsuda, 1991)

Genesis of cyclones is most favoured by an ocean temperature of 26¡C or more, weak vertical wind shear (between 90 hpa and 200 hpa levels), and other thermodynamical and dynamical factors. The formation of cyclonic storms in the Bay of Bengal has two distinct periods the pre-monsoon (April-June) and post-monsoon (Sept-Dec). It is also evident from the monthly distribution of cyclonic storms of various intensity which hit the Bangladesh coast during the period 1877 to 1990 as shown in Figure 6.

The thermodynamical conditions necessary for the development of cyclonic storms exists in the pre-monsoon and post-monsoon months. The required minimum energy for the development of cyclone is available at the air sea interface during this time. Cyclones and depressions form in the Bay of Bengal during pre-monsoon, monsoon and post-monsoon. However, cyclones which form during pre-monsoon periods, are most destructive due to the great instability of the atmosphere and weak vertical wind shear. The cyclonic storms, which form during monsoon period known as monsoon depression are not so much destructive in nature due to large vertical wind shear during south-west monsoon. Monsoon depressions travel west / north-westwards along the trough of low and mainly cross Orissa and West Bengal coasts.

Usually the recurvature of tropical cyclones occurs between 10 and 18¡N (Murty et al., 1986). About 12-18h before recurvature, the active front turns from west or northwest

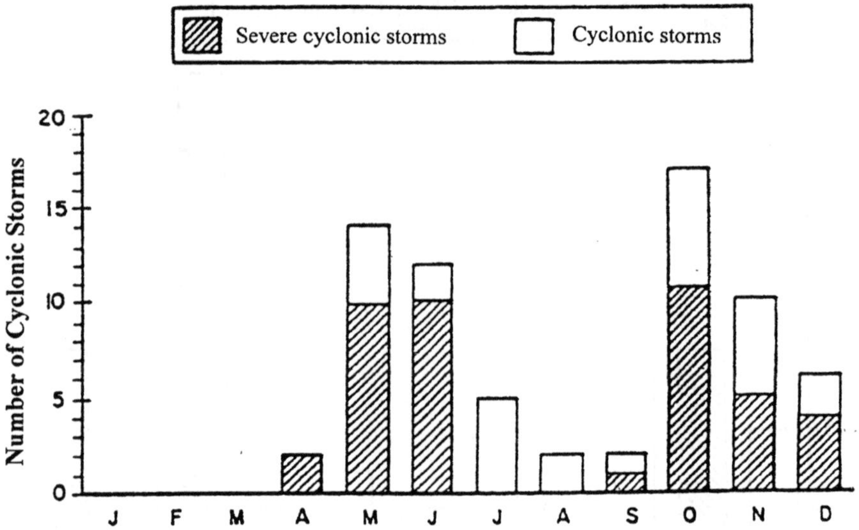

Figure 6: Monthly distribution of cyclonic storms that landed on Bangaldesh coast 1877-1990 (Matsuda 1991)

to north or northeast with an associated turning of the significant precipitation belt and cloud field from west or north west to north or northeast. About 80 percent of recurvature showed this feature (Siefert et al., 1991). Another feature is a reduction in the movement before recurvature. Also before recurvature the upper atmospheric structure changes in such a manner that the average steering force will be directed progressively to the east (i.e. westerlies at about 3000m will strengthen before recurvature).

The storm surge is not appreciable in deep sea until it reaches the continental shelf which is 150km to 300km long near the coast line of Bangladesh. Before the continental shelf, there is relatively steeper part of the sea bed which rises rapidly from a depth of 2000m to a depth of about 200m at the edge of the shelf (Figure 7). Then the depth decreases gradually towards the coast line. This significantly increases the wind drift effect that is also important in the generation of storm surge. The amplification of a cyclonic storm surge in the Bay of Bengal occurs due to long and shallow continental shelf and the converging Bay. Bangladesh is at the top of the inverted V-shaped Bay of Bengal; therefore, storm surges are easily amplified due to the concentration of energy.

When a cyclone moves north - easterly to the north of Chittagong, strong wind blows from off shore to the coast and together with the suction effect of low pressure, the wind drift effect (amplified by the shallow water depth) brings about a large storm surge along the coast. The displacement of water surface during a cyclonic storm surge also depends on the tide. Tides in the Bay of Bengal originates in the Indian Ocean and gets amplified due to non-linear shallow water effect. The average tidal

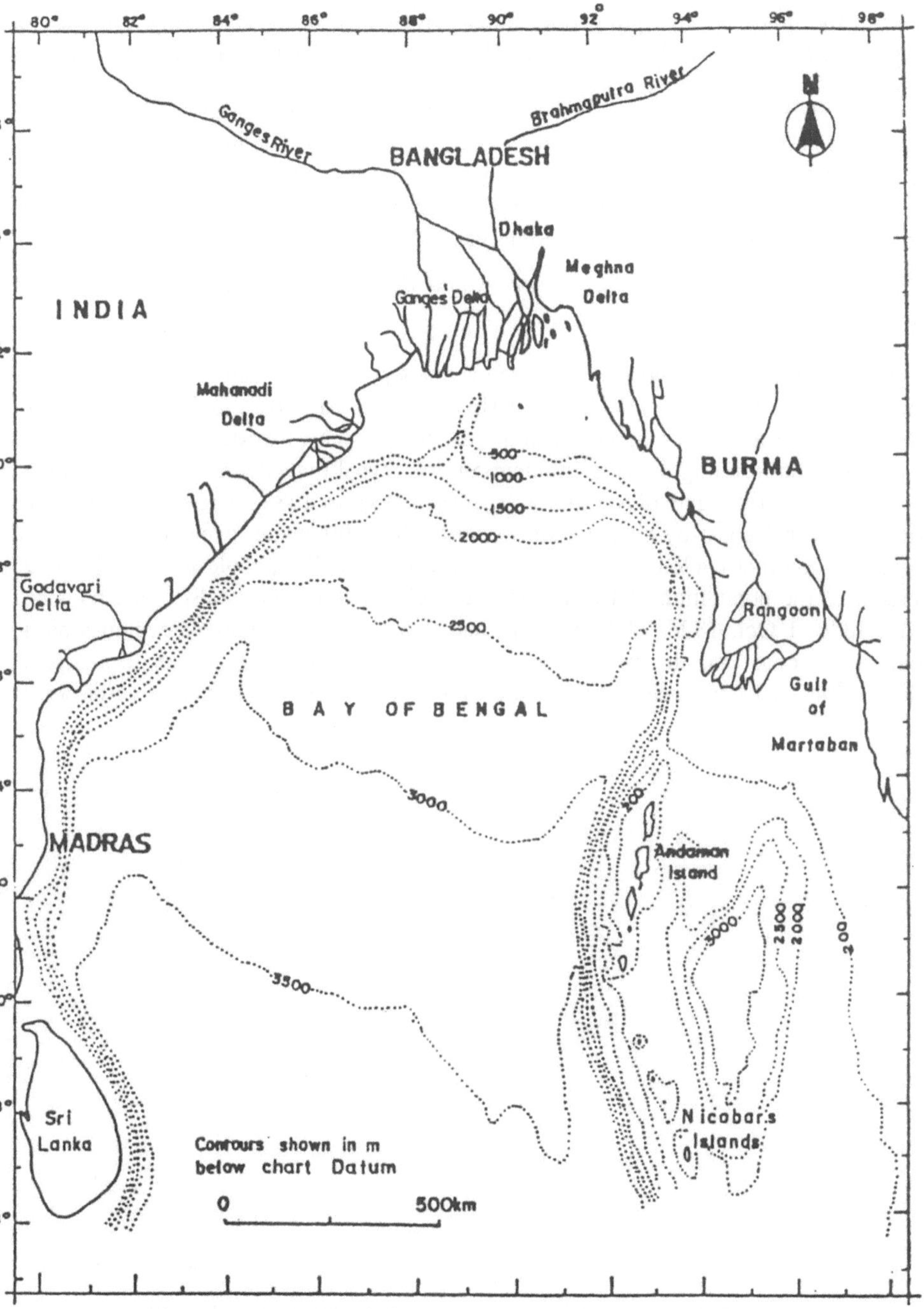

Figure 7; Sea chart of the Bay of Bengal (Katsura et al., 1992)

range along the Bangladesh coast varies between 3 to 6m. Several studies have estimated the effect of astronomical tide on storm surges (Das et al., 1974, Johns et al., 1985). The displacement of water surface is largest when the cyclonic storm surge reaches the coast during spring tide. Such coincidence occurred during cyclones of Nov. 12, 1970, Dec. 10, 1981 and April 29, 1991 (BUET - BIDS, 1993).

The Bay of Bengal coast has been classified into three zones based on the combined amplitude of the storm surge and wind waves (Rao, 1968). The most dangerous zone, occurs along the coastal belt around the head of the Bay of Bengal, approximately to the north of 20¡ N. Several studies have been undertaken to predict the storm surges in the Bay of Bengal (Das et al, 1994, Johns et al., 1980).

3.5.2. History of Cyclones / Storm Surges and Recent Catastrophic Storm Surges in Bangladesh

In the coastal belt of Bangladesh, severe cyclones periodically caused serious casualty and property damage. As Bangladesh is prone to storm surges because of meteorological and topographical conditions, record shows evidence of many storm surges and the resulting damage. Ancient reports reveal that severe cyclones occurred in 1584, 1790, 1822 and 1876 (BUET - BIDS, 1993). In 1584, about 200,000 people were killed in Barisal by cyclonic storm surge. In 1790 severe cyclones occurred in Noakhali. Another cyclone in 1822 killed more than 70,000 people in Barisal and 95 percent of the population of Hatiya. Severe cyclonic storm surge also occurred on 1 Nov. 1876, 1 Oct, 1896, 21 Sept, 1919 and 3 May 1947. Tracks of cyclones along Bangladesh coast for the period 1905 to 1991 is shown in Figure 8. Bangladesh Meteorological Department classified the cyclones of various intensity as described in

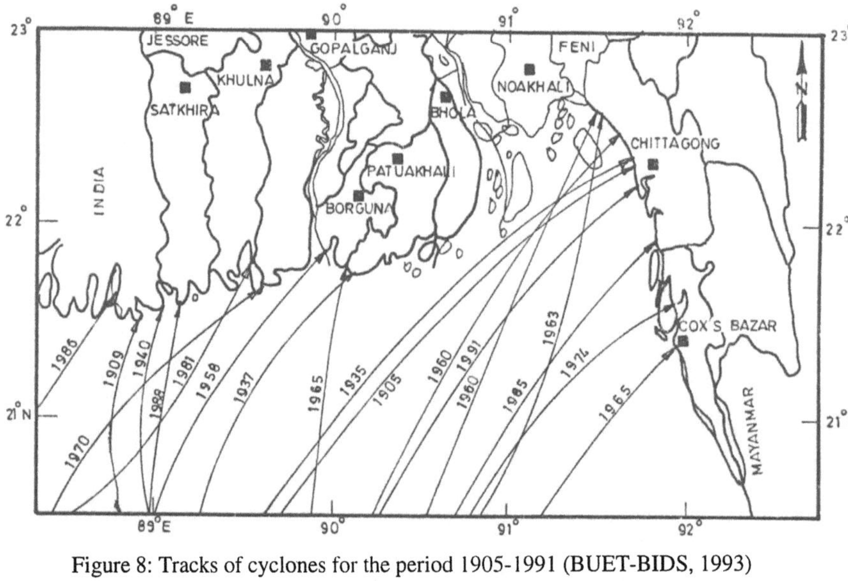

Figure 8: Tracks of cyclones for the period 1905-1991 (BUET-BIDS, 1993)

Table III. The distribution of the occurrence of various classes of cyclones which hit the three coastal region of Bangladesh since 1960 is shown in Table IV. Available data on cyclonic storm surge height and wind speed is very scanty. A list of severe cyclones during 1960-1991 along with their maximum sustained wind speed and reported surge height is shown in Table V. Storm surge heights are given as maximum levels above mean sea level (M. S. L.). The surge levels can vary from 1.5 to about 9 m as evident from Table V. Central pressures for Bangladesh, which are available only for the last ten years, do not show very low values (usually 980 to 990mb) except the 945 mb recorded on April 29, 1991.

Table III Classification of Cyclones

Cyclone Intensity		Range of Pressure Drop (mb)	Range of Maximum Sustained Wind Speed (km/hr)
A	Depression	3.0 to < 3.5	44 to < 52
B	Deep depression	3.5 to <6.0	52 to < 63
C	Cyclonic storm	6.0 to < 9.0	63 to < 89
D	Severe cyclonic storm	9.0 to < 16.0	89 to < 119
E	Severe cyclonic storm of hurricane intensity	16.0 or more	119 and above

Source : BUET- BIDS, 1993

Table IV Distribution of the Number of cyclones which Hit the Three Coastal Regions of Bangladesh Since 1960

Intensity of Cyclone	Chittagong-Cox's Bazar Coast	Meghna Delta Coast	Barguna - Khulna Coast
Cyclonic storm	4	1	5
Severe cyclonic storm	3	5	4
Severe cyclone with hurricane intensity	7	6	2

Source : BUET- BIDS, 1993

In a comprehensive review (Murty, 1984) valuable information on storm surges in the Bay of Bengal and elsewhere have been described. Cyclones of Nov 12, 1970, May 25, 1985 and April 29, 1991 were the most catastrophic cyclones of the century. Of the cyclones listed in Table V, those of Nov 12, 1970 and April 29, 1991 rank the highest for maximum wind speed, storm surge height and the number of deaths. According to official statistics, the cyclone of April 29, 1991 devastated about 2000sq. km of coastal area having a coast line of over 400 km, about 140,000 people were killed and more than 200 families became homeless.

Table V Maximum Sustained Wind Speed and Reported Surge Height of Severe Cyclones During 1960 - 1991

Day (d/m/y)	Maximum wind speed (km/hr)	Storm surge height (m)	Date (d/m/y)	Maximum wind speed (km/hr)	Storm surge height (m)
9-10-60	162	-	3-5-71	-	2.5 - 4.0
30-10-60	210	4.5 - 6.0	30-9-71	-	2.5 - 4.0
9-5-61	146	2.5 - 3.0	6-11-71	-	2.5 - 5.5
30-5-61	146	6.0 - 9.0	18-11-73	-	2.5 - 4.0
28-5-63	203	4.0 - 5.0	9-12-73	122	1.5 - 7.5
11-4-64	-	-	15-8-74	97	1.5 - 6.5
11-5-65	162	3.5	28-11-74	162	2.0 - 5.0
31-5-65	-	6.0 - 7.5	21-10-76	105	2.0 - 5.0
14-12-65	210	4.5 - 6.0	13-5-77	122	-
1-10-66	146	4.5 - 9.0	10-12-81	97	2.0
11-10--67	-	2.0 - 8.5	15-10-83	97	-
24-10-67	-	1.5 - 7.5	9-11-83	122	-
10-5-68	-	2.5 - 4.5	3-6-84	89	-
17-4-69	-	-	25-5-85	154	3.0 - 4.5
10-10-69	-	2.5 - 7.0	29-11-88	162	1.5 - 3.0
7-5-70	-	3.0 - 5.0	29-4-91	225	6.0 - 7.5
23-10-70	-	-	2-6-91	100	2.0
12-11-70	223	6.0 - 9.0			

Source : BUET - BIDS, 1993

Recorded information on the meteorological and hydrodynamic characteristics of the 1991 cyclone are available in detail. A pressure depression was recognized in the middle of the Bay of Bengal by a meteorological satellite on 23 April. It turned into a cyclone on 25 April. Initially, the cyclone moved slightly northwest; but its speed gradually decreased and remained at about 12¡N and 87¡E on 26 April. It again began to move northward on 27 April. The direction of movement changed from north to northeast and it struck the coast of Chittagong during the night of 29 April.

Three tracking paths of the 1991 cyclone are shown in Figure 9. These tracks were charted by different organizations (Katsura et al, 1992):

1. U. S. Army and Air Force Joint Typhoon Warning Center (JTWC).
2. Bangladesh Red Crescent.
3. Space Research and Remote Sensing Organization (SPARSO), Bangladesh.

The pressure data for the center of the cyclone were recorded by the Indian Meteorological Department. Figure 10 shows the pressure data recorded on the Bangladesh Navy ship. The pressure had dropped below 930 hpa at 01:45 (local time) but increased to 935 hpa at 02:00. Heavy rainfall was recorded at some station along the coastal area ranging from 80mm to 200mm from 06Z on 29 to 06Z on 30 April.

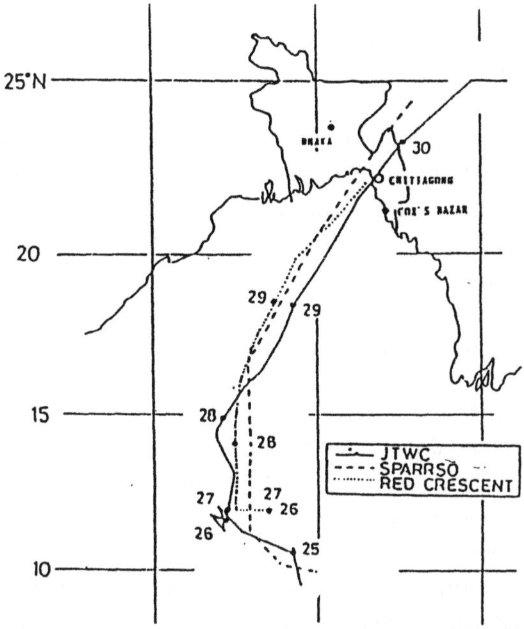

Figure 9: tracking paths of 1991 cyclone (Katsura et al., 1992)

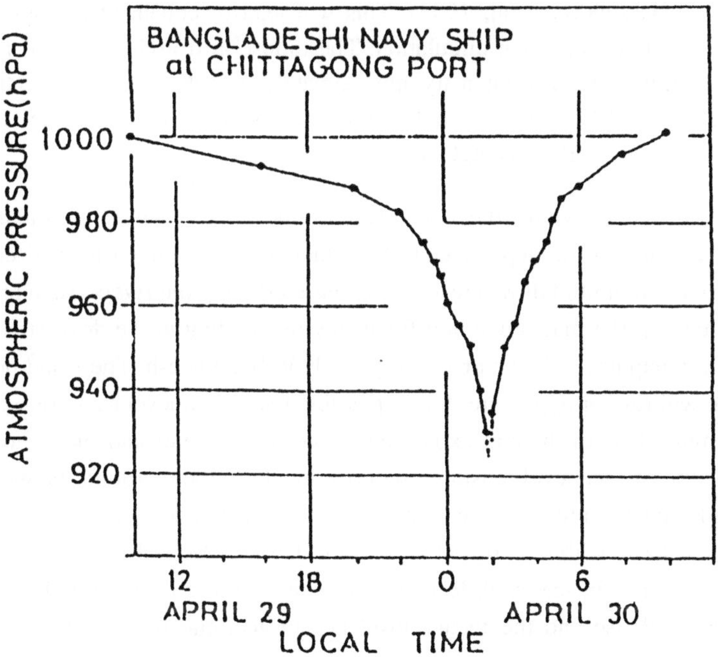

Figure 10: Pressure data during 1991 cyclone (Katsura et al., 1992)

4. CYCLONES, RIVER FLOODING AND RAINFALL PATTERN IN A WARMER ATMOSPHERE

The global mean surface temperature may increase between 1.5 and 4.5°C by the middle of next century, if the current trend of greenhouse gas emission continues. A rise in annual mean surface temperature of between 2.0 and 3.5°C over the Indian subcontinent is expected during the next hundred years (Chakraborty et al, 1992). The most significant consequence of higher surface temperatures could be changes in the rainfall distribution of the subcontinent and increase in the intensity and frequency of flooding and tropical cyclones in the Bay of Bengal.

The atmospheric low pressure that develops over the subcontinent during summer is likely to get intensified and drive the monsoon winds strongly. Changes in the thermal gradient as a result of higher warming in the high latitudes than at the equator also contributes to the increase in the intensity of S. W. monsoon. The low pressure area, known as 'monsoon trough' during the summer season over the Indian subcontinent is roughly oriented from northwest to southeast. In a warmer atmosphere, the monsoon trough is likely to advance further northward as a result of an enhanced ocean-continent pressure gradient. There will be an increase in mean sea level pressure to the south of 20¡N and a decrease in pressure to the north of 20¡N due to rise in temperature (Chakraborty et al, 1992). This will lead to enhanced pressure gradient and hence a stronger monsoon circulation. The monsoon trough may shift northward by 5¡ to 7¡ latitude and also intensify approximately by 10% in a warmer atmosphere (Chakraborty et al, 1992). There would be an increase in rainfall over the central and northeast India and adjoining Bangladesh.

Heavy rainfall occurs within the catchment of the rivers Ganges - Brahmaputra - Meghna (especially in the upper catchment). Increase in the intensity and changes in spatial distribution of rainfall will result in increased flow in the rivers. As a result there will be increase in the magnitude and frequency of flooding in the downstream of the Ganges - Brahmaputra - Meghna rivers especially in Bangladesh. The flood problem of Bangladesh will further aggravate due to possible rise in sea level as a consequence of global warming. Due to thermal expansion of the ocean water and the melting of the Antarctic Ice Sheet, sea level (eustatic) will rise by 20 - 140cm around the globe by the middle of the next century. More important for local sea level rise generally is the effect of subsidence within a coastal area. Subsidence in Bengal Delta can accentuate greatly the local rise in sea level. Relative sea level rise at a location is the sum of the eustatic sea level rise and the local subsidence or tectonic behaviour at the site. A subsidence rate of 1 cm/year across the Bengal Delta has been estimated (Milliman et al, 1989).

A rise in sea level will lead to high tides in the Bay of Bengal causing more inundation inland. As a result of coincidence of high flow in the river and high tide in the sea, a vast area of Bangladesh goes under water. The combined effect of enhanced precipitation and rise in sea level as a consequence of global warming will increase the depth and area of flooding due to changes in the hydrodynamics of flow (Begum, 1992). Hence Bangladesh will face serious flood problems both ways.

A one meter rise in sea level could turn a moderate storm into a catastrophic one. The tropical cyclones in the Bay of Bengal form only when Sea-surface Temperatures (SSTs) are 26¡C or higher. Tropical cyclones in the Bay of Bengal generally form in the pre-monsoon (April - June) and post-monsoon (Sept - Dec.) season as evident from Figure 6.

An increase in the surface air temperature might lead to a more widespread occurrence of cyclones in the Bay of Bengal. However, a number of environmental factors also influence the formation of tropical cyclones. Thus the criteria that the lower boundary of SST of 26¡C is necessary for the development of cyclones in the Bay of Bengal may not remain constant with changes in climate. Moreover, enhanced evaporation over the Bay of Bengal during the pre-monsoon season (due to increase in surface temperature), leading to increased moisture convergence and latent heat release (which is converted to locally transient kinetic energy, resulting in stronger winds) may also increase the number and duration of cyclones in a warmer atmosphere. However, the increase in local energy generation and conversion depends on the cloud feedback process which is a very complex phenomena (Chakraborty et al., 1992). Hence an increase in the intensity, frequency and seasonal distribution of tropical cyclones in the Bay of Bengal may occur due to global warming.

The homes and lives of millions of people living in the flood plains of the Ganges, the Brahmaputra and the Meghna rivers are threatened due to increase in the intensity and frequency of extreme events. The people of Bangladesh already live at the margin of survival. The long term implication of global warming with the addition of population explosion could make Bangladesh more vulnerable to natural disasters in years to come. So occurrence of more extreme events will cause an exponential increase in loss of life. By the end of next century, Bangladesh as it is known today may virtually have ceased to exist.

5. PROBLEMS AND MEASURES IN DISASTER PREVENTION, MITIGATION AND MANAGEMENT IN BANGLADESH

Floods and cyclone are the most crucial disasters that Bangladesh has to face recurrently. There are several structural and non-structural measures which can be undertaken for the prevention and mitigation of floods and cyclones as indicated in Table VI and Table VII.

Table VI : Structural and Non-Structural Measures in Flood Mitigation

Reservoirs
Embankments
Dikes, Levees and Flood walls
Channel Alterations
Flow Diversions
Flood Proofing
Flood Insurance
Bank Protection
Stormwater Management
Floodplain Zoning
Flood Forecasting, Warning System and Emergency Plans
Flood Shelters
Land Treatment Measures

Table VI : Structural and Non-Structural Measures in Flood Mitigation

Reservoirs
Embankments
Dikes, Levees and Flood walls
Channel Alterations
Flow Diversions
Flood Proofing
Flood Insurance
Bank Protection
Stormwater Management
Floodplain Zoning
Flood Forecasting, Warning System and Emergency Plans
Flood Shelters
Land Treatment Measures

At present several structural and non-structural measures e.g. embankments, bank protection and river training (including some local methods), channel alterations, flood forecasting are in practice for flood and erosion control. The existing measures can provide some protection during normal floods only. In case of severe floods, e. g. the floods in 1987 and 1988, these measures could not provide any protection and were severely damaged with the onrush of water. Embankments, coastal afforestation, cyclone shelters and warning systems are the existing cyclone protection and mitigation measures. Coastal plantation has proven as an effective measure in some part of the coastal belt of Bangladesh. During the April 1991 cyclone, coastal communities of Mirsarai and Sitakunda suffered the least casualties and property damage, primarily due to protection provided by a one to two kilometer wide belt of plantations along the coastal shore (BUET - BIDS, 1993). There are at least 706 km of sea-facing and similarly exposed embankments in the coastal districts where there are no trees or very little trees on the embankment (BUET - BIDS, 1993). Short term plans should be taken to raise plantations on embankments to reduce damage. Cyclone shelters are another most effective measure to reduce casualties. Existing public and private buildings which can be used as shelters during cyclone are extremely insufficient to provide protection of the population in the cyclone prone area.

Disaster preparedness encompasses plans for mitigation, warning and emergency operations, training and public information activities. Improvement of the existing system in disaster mitigation is very important before adopting new measures. Warning systems and accompanying emergency response have long been recognized as effective ways to save lives and reduce damages in both river flooding and cyclones. Weather forecasting and hence flood forecasting can be improved using remote sensing and real time data. Improved and reliable cyclone warning system can reduce the loss of life and property damage. Emergency activities during and immediately after a flood may include removing people and property from areas about to be flooded, sandbagging around individual structures and constructing emergency dikes to direct water away from vulnerable areas, search and rescue, and steps to protect the health and safety of residents. Similar emergency activities such as removing people to safer place and search is especially important before and after cyclone.

One of the major problem in cyclonic disaster management is an inadequate warning system. Currently there are fifteen signal numbers for cyclone warning, eleven for the sea ports and four for the river ports. The definitions of these signals are given in Table VIII.

These signal numbers are meant for port authorities only; there is no warning system for the vast majority of the general public who reside outside the port area. Moreover, the highest signal number ten is raised even for a moderate storm. As the intensity of the storm increases from severe to very catastrophic, the signal number does not change correspondingly. Thus the present warning system fails to indicate the severity of the calamity. The public in general fail to understand the magnitude of the calamity in their area and hence are reluctant to go out of their home to safer places. If the people could be made aware of the magnitude of the calamity well ahead, they could take adequate measures to meet the situation. The inherent meaning of the warning system could be explained to the people in simple language. If possible, people should be trained like war time emergency preparation, for effective management to cope with the situation. Cyclone warning forecasts need to be improved so that the credibility of the forecasts are reduced and their reliability improved.

Table VIII Storm Warning Alerts Issued at Maritime Ports in Bangladesh

Distance	Level	Storm Warning Signals	Description
Distant signals	Cautionary	1	A region of squally weather exists in which a storm may be formed .
	Warning	2	A storm has formed .
Local signals	Cautionary	3	Port threatened by squally weather.
	Warning	4	Port threatened by a storm, but the danger is not yet sufficiently great to justify extreme precautionary measures.
	Danger	5-7	Port will experience severe weather from a storm of slight to moderate intensity.
		5	Storm expected to cross the coast to the south of the port.
		6	Storm expected to cross the coast to the north of the port.
		7	Storm expected over or near a port.
	Great Danger	8-10	Port will experience severe weather from a storm of great intensity.
		8	Storm expected to cross the coast to the south of the port.
		9	Storm expected to cross the coast to the north of the port.
		10	Storm expected over or near a port.
Failure of communication			Communication with the Meteorological Warning Center has broken down, and the local officer considers that there is danger of bad weather.

Source : Katsura, J. et al., 1992

After each severe cyclone and flood, intensive national and international efforts are mobilized to meet the immediate requirements but the subsequent natural processes take a longer time to repair the damages sustained by the coastal natural resources and establish the equilibrium. There is lack of adequate public and private buildings in the coastal area where people can take shelter during severe cyclones. Following the disastrous cyclone and surge in April 1991, a Task Force was formed in collaboration with UNDP and World Bank to find out the adequacy of existing shelters in the cyclone prone areas.

On the basis of data on all types of public and private buildings, the Task Force observed that only 11.48% of the population in the cyclone prone area can take shelter during a severe cyclone (BUET - BIDS, 1993). The Multipurpose Cyclone Shelter Programme (MCSP) is an intervention with the primary intention to protect human life and livestock in the coastal area during severe cyclones. Shelters recommended under the MCSP will at normal time, function as primary and secondary schools during early hours of the day and as centres for community activities at other time of the day. Provisions of shelters are made for estimated population in the year 2002.

After the catastrophic flood in 1988, World Bank coordinated a Flood Action Plan (FAP) for Bangladesh based on the five major studies of the UNDP, USAID, France, Japan and China. The USAID preferred non-structural solution to the flood problem while the others suggested structural solution. In December 1989, the five year (1990-95) Flood Action Plan involving 26 components was launched. The main objective of the first five-year phase is to establish the principles and criteria for sustainable flood mitigation, to undertake comprehensive planning studies and to begin the implementation of high priority projects (Dempster et al., 1992). A concept of controlled flooding through compartmentalization forms the main component of FAP. There are important hydrologic, morphologic, agricultural, fisheries, ecological, social, economic and institutional issues related to the concept of controlled flooding and compartmentalization. FAP Comprises of construction of large scale embankments. FAP has generated an unprecedented public debate among a wide spectrum of the society. In fact, total embankment is not desired by the majority of flood-plain inhabitants at present. The social dimension of FAP depends on the response of the local population and in particular on the participation of the farming community. Thus this multimillion dollar Flood Action Plan should be studied carefully both from environmental and socio-economic point of view for the benefit of the people and effective utilization of the resources.

Efforts towards control/mitigation of flood in Bangladesh can not be effective without regional cooperation with upper riparian countries. Solution to the problems within Bangladesh with abundance or scarcity of water cannot be a successful measure when more than 90% of the catchment area of the rivers Ganges-Brahmaputra-Meghna lies outside the country. It is necessary to explore regional water management prospects before going for domestic solution. Bangladesh is the territory through which the water of the entire basin of all the three great rivers draining to the sea. Thus Bangladesh deserves regional co-operation in flood control/mitigation. Flood forecasting and warning are the only area of co-operation with upper riparian countries. There are several other options which can be adopted through regional co-operation for water management in Bangladesh. Solution of another most critical natural disaster which Bangladesh is facing due to scarcity of water in the dry season as a result of unilateral withdrawal of Ganges water by constructing Farakka Barrage in the upstream of Ganges river also requires urgent regional co-operation in sharing the Ganges flow on rational and equitable basis.

Pre-disaster planning is important in disaster management. Hazard assessment, vulnerability and risk assessment, are the vital steps in pre-disaster planning, which are lacking in Bangladesh. Resource constraint and lack of co-ordination is a major handicap in effective disaster management in Bangladesh.

The population of Bangladesh is growing at more than 2% per annum: it is expected to double by the year 2030. High population density along with dependance on agricultural activities is forcing people to live in more vulnerable areas. It is necessary to slow down the rate of population growth in order to be successful with any new plan or policy and more job opportunities need to be created in less vulnerable areas.

6. CONCLUSIONS

The following counter measures should be undertaken for effective prevention, mitigation and management of natural disasters in Bangladesh :

1. In view of the anticipated climate change and sea level rise, the present hydrometric and coastal measuring stations have to be strengthened.

2. Adequate and reliable cyclone warning system should be introduced since the existing warning system fails to indicate the severity of the calamity.

3. Coastal afforestation should be increased to reduce the damage from cyclone and storm surge. Tree planting programme should be taken along coastal embankments and various categories of settlements.

4. Public awareness, participation and self reliance should be increased. Peoples could be made aware of the magnitude of the calamity and the emergency measures to meet the situation.

5. In order to quickly and adequately respond to disasters, training programmes can be arranged for staff and officials of local communities in information collection and transmission, advice and instruction for evacuation, rescue and relief activities. The emergency measures can be taught in school at all levels and can also be reached to a wide audience through television and radio.

6. Resource constraint and lack of co-ordination is a major handicap in disaster management in Bangladesh.

7. Efforts towards prevention/mitigation of flooding in Bangladesh cannot be effective/successful without regional co-operation with upper riparian countries. International collaboration is also required in disaster management in Bangladesh.

8. Scientific and technological research and development activities related to disaster prevention and mitigation should be conducted for changing scenarios of global climate with regional prospects.

9. It is very important to slow down the population growth in order to be successful with any new plan or policy.

REFERENCES :

Abbas, B. M. (1988). International Cooperation for Flood Control. Bangladesh Today, Quarterly from London.

Ahmed, M. (1989). Deluge in the Delta. In: Flood in Bangladesh. Edited by Mohiuddin Ahmed.

Begum, S. (1992). The Flood Problem of Bangladesh : Numerical Simulation of the Effects of Possible Global Climate and Sea Level Changes. Unpublished PhD Thesis, University of Strathclyde, U. K.

Bingham, A. (1989). Floods of Aid for Bangladesh. *New Scientist*, 2 December.

Chakraborty, B., Bhaskaran, B. and Lal, M. (1992). Bay of Bengal Cyclones in a Warmer Atmosphere. Paper presented in the International Conference on Coastal Zone Management, Dhaka, December.

Chatterjee, S. P. (1949). Bengal in Maps : A Geographical Analysis of Resource Distribution in West Bengal and East Pakistan.

Das, P. K., Sinha, M. C., and Balasubramanyam, V. (1974). Storm Surges in the Bay of Bengal. *Quart. Journal of Royal Meteorological Society*, Vol. 100, pp. 437-449.

Das P. K. (1994). On the Prediction of Storm Surges. Sadhana-Academy Proceedings in Engineering Sciences, New Delhi, Vol. 19, Part 4, pp. 583-595.

Dempster, J. I. M. and Brammer, H. (1992). Flood Action Plan - Bangladesh. *Outlook on Agriculture*, Vol.21, No.4, pp.301-305.

Gray, W. M. (1968). Global View of the Origin of Tropical Disturbances and Storms; *Monthly Weather Review*, Vol. 96.

Johns, B., Ali, A. (1980). The Numerical Modelling of Storm Surges in the Bay of Bengal. *Quart. Journal of Royal Meteorological Society*, Vol. 106, pp. 1-18.

Johns, B., Rao, A. D., Dube, S. K., Sinha, P. C. (1985). Numerical Modelling of Tide-Surge Interaction in the Bay of Bengal. *Philos. Trans. Royal Meteorological Society*, A313, pp. 507-535.

Katsura, J. et al. (1992). Storm Surge and Severe Wind Disasters Caused by the 1991 Cyclone in Bangladesh. Research Report on Natural Disasters by Japanese Group for the Study of Natural Disaster Science, No. B - 2, March.

Matin, M. A. and Husain, M. A. (1989). Hydrological Aspects of 1988 Flood. In: Flood in Bangladesh, Edited by Mohiuddin Ahmed, April.

Matsuda, I. (1991). Standing Problems in Countermeasures for Cyclone Disaster : How it is Possible for Bangladesh to Help Itself?; Report on Field Study and Investigations on the Damage Caused by the Cyclone in Bangladesh in 29-30 April 1991; United Nations Centre for Regional Development (UNCRD), Japan.

Milliman, J. D., Broadus, J. M. and Gable, F. (1989). AMBIO, *Royal Sweedish Academy of Sciences*, Vol. 18, No. 6.

Morgan, J. P. and McIntire, W. G. (1959). Quaternary Geology of the Bengal Basin, East Pakistan and India. *Bulletin of the Geological Society of America*, Vol. 70, pp. 319 - 342, March.

Murty, T. S. (1984). Storm Surges, Meteorological Ocean Tides. Canadian Bulletin of Fisheries and Aquatic Sciences 212. Dept. of Fisheries and Oceans, Ottawa, Canada.

Murty, T. S., Flather, R. A. and Henry, R. F. (1986). The Storm Surge Problem in the Bay of Bengal. *Prog. Oceanog.* Vol. 16, pp. 195 - 233.

Siefert, W. and Murty, T. S. (1991). Strom Surges, River Flow and Combined Effects. State of the Art Report Prepared for the UNESCO Workshop "Storm 91" in Hamburg, Federal Republic of Germany.

Rao, N. S. B (1968). On Some Aspects of Local and Tropical Stroms in the Indian Area PhD Thesis University of Jadavpur, India. In : The Storm Surge Problem in the Bay of Bengal by Murty et al. (1968).

United Nations (1966). A Compendium of Major International Rivers in the ECAFE Region, Water Res. Series No. 29, N. Y.

BUET - BIDS (1993). Multipurpose Cyclone Shelter Programme. Final Report, Part 1, Volume 1, Main Report. UNDP/World Bank/GOB Project BGD/91/025, July.

World Bank (1991). Flood Control in Bangladesh. A plan for Action. World Bank Technical Paper No. 119.

NUMERICAL STUDY OF NONLINEAR TIDE-SURGE INTERACTION IN THE COASTAL WATERS OF SHANGHAI

ZENGHAO QIN AND YIHONG DUAN
Shanghai Typhoon Institute,166 Puxi Road, Shanghai 200030, P.R. China

Abstract. Using conventional two-dimensional nonlinear shallow water equations and the associated semi-momentum finite difference schemes, the nonlinear interaction between storm surges and tides, caused by eight tropical cyclones of different tracks over the coastal waters of Shanghai is numerically studied in this paper. The contributions of various nonlinear terms in the dynamic equations to the water level heights is diagnostically analyzed in detail. It is shown that the nonlinear bottom friction term is of prime importance in controlling the water level and nonlinear tide-surge interaction. The correction to the water level and storm surge due to tide-surge interaction is significant.

1. INTRODUCTION

In shallow waters and regions of great tidal range, such as the coastal waters of Shanghai, located on the East China Sea, the nonlinear coupling of storm surges and tides is quite pronounced. In this region the residual curve obtained by subtracting the tide from the observed water level usually exhibits oscillation with main period of tides. They exert a great influence on the storm surge and water level simulations and predictions. Since 1950's several authors (e.g. Proudman 1955a, 1955b; Bank, 1974; Wolf, 1978, 1981; Prandtle and Wolf, 1978, 1986; Das et al, 1980; Murty and El-Sabh, 1981; Zhang and Wang, 1989) have investigated the mechanism related to the tide-surge interaction in different water bodies, using both analytical and numerical approaches. However, we know nothing about such phenomenon in continental shelves of the East China Sea, except in the Bohai Sea region (Zhang and Wang, 1989). Due to its practical aspects in terms of better understanding the dynamics of storm surges, this present paper aims at the study of this problem in the waterbody around Shanghai.

The numerical models used to incorporate the tide-surge interaction in water level simulations are described by (Qin et al, 1994) and are briefly described in section 2. The results of tide simulations in Shanghai during the eight tropical cyclone periods are described and compared with those obtained by the conventional harmonic analysis in section 3. Comparisons of the storm surge simulations with observations, the characteristics of water level caused by the nonlinear tide-surge (T-S) interaction ζ_{TS} as

M. I. El-Sabh et al. (eds.), Land-based and Marine Hazards, 139–156.

well as the contributions of various nonlinear terms in the dynamic equations to the water levels and ζ_{TS} are analyzed in detail in sections 4 and 5, respectively. Our primary finding is summarized in the last section.

2. NUMERICAL MODELS

The following two-dimensional surge, tide and water level models with the semi-momentum finite difference scheme (Qin et al, 1994) have been computed for the domain covering the region 27°39'N-34°15'N, 119°48'E-127°40'E, i.e., the main part of the continental shelves of the East China Sea (Figure 1),

$$\left(\frac{\partial}{\partial t} + f\vec{K}x\right)\vec{V} + \Delta\left(g\zeta + \frac{P_a}{\rho}\right) - \underset{(D)}{\frac{\tau_a}{\rho(h+\zeta)}} = \underset{(C)}{-\vec{V}\bullet\Delta\vec{V}} - \underset{(A)}{\frac{C_b|\vec{V}|\vec{V}}{h+\zeta}} \qquad (1)$$

$$\frac{\partial\zeta}{\partial t} + \Delta\bullet(hV) = \underset{(B)}{-\Delta\bullet(\zeta\vec{V})} \qquad (2)$$

where V(u,v) denotes the horizontal current vector with its (u,v) components along (x,y) axis, ζ is the elevation of sea surface relative to the undisturbed surface, P_a and τ are the atmospheric pressure and wind stress on the sea surface, respectively, h water depth, f

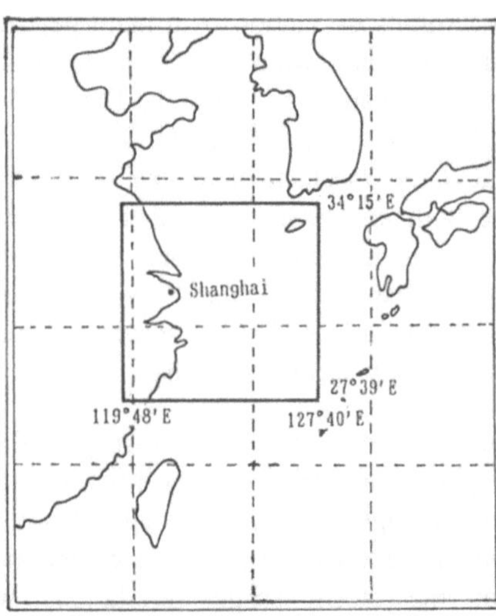

Figure 1: Computational domain

Coriolis parameter, Cb the drag coefficient of the bottom stress, g acceleration of gravity, t time and ∇ horizontal nabla.

The mesh size of the scheme is chosen to be 33 km while the time step 200 Sec. Computations of storm surges caused by the eight tropical cyclones(TCs), i.e. Marge(1951), Grace(1954), Carmen(1960), Mary(1974), Judy(1979), Anges(1981), Forrest(1983), and Vera(1986), impinging Shanghai seriously, are carried out.

The water depths in the most part of the computational domain are within 100m. The deepest region is located at the southeast of the domain to be about 1500m, while the shallowest is in Hangahou Bay, 8m (Figure 2)

The axisymmetric surface air pressure model suggested by Fujita(1952) and asymmetric surface wind model proposed by Ueno(1964) are applied to compute the surface air pressure and wind around the TC. The TC model wind is composed of two parts, i.e., an xi-symmetric wind which is proportional to the gradient wind with a factor C, and anasymmetric wind caused by the TC translation whose speed is proportional to the moving speed of TC. Experiments show that the storm surges are quite sensitive to the factor C. The magnitude of C differs from one storm to another. Surface wind stress can be easily calculated by using quadratic-velocity parameterization in usual way. We start to run surge model computations at calm sea state. The vanishing normal components of the currents at the coast and the inverse barometric effect at the open boundary for the surge are required. The open boundary is far enough away from the study sea area around

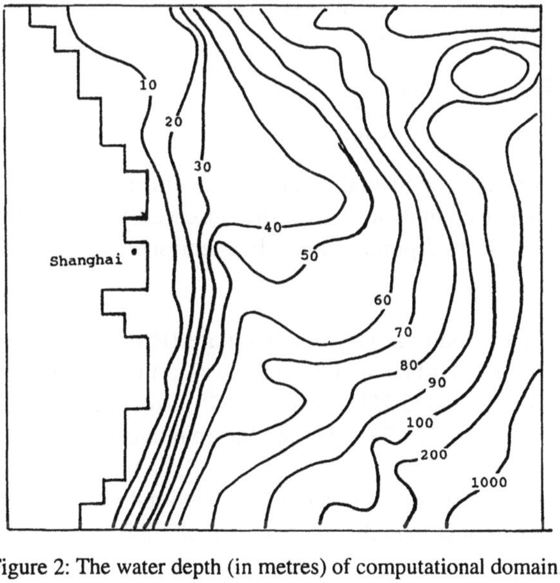

Figure 2: The water depth (in metres) of computational domain.

Shanghai and radiation boundary conditions for the current and water level height is used. To filter out the erroneous short waves due to computation, the appropriate field smoothing is necessary.

3. TIDE SIMULATIONS

In addition to the storm surge simulations, the accurate tide simulations are necessary in order to get the accurate water level resulted from the nonlinear T-S interaction. In simulating tides, the atmospheric forcing is set to be zero with the tide-generating force excluded in the model equations. However, the prescribed eight predominate tidal constituents(i.e., M2, S2, N2, K2, Q1, O1, P1, and K1) at the open sea boundary grids of the computational domain should be added to the inverse barometric effect of water level in consideration of the irregular shallow water semi-diurnal tidal pattern of Shanghai harbor.

Table 1 Comparison of simulated tides with observed ones(cm)

TC Name	HTL(o)			HTL(s)			LTL(o)		LTL(s)	
	I	II	III	I	II	III	I	II	I	II
Marge(1951)	386	415		393(-7)	402(13)		112		44(68)	
Grace(1954)	384	441	391	382(2)	412(29)	396(-5)	99	82	45(54)	33(49
Carmen(1960)	330	397	350	325(5)	360(37)	343(7)	106	90	110(-4)	80(10
Mary(1974)	388	434	397	398(-10)	421(13)	408(-9)	79	68	34(45)	18(50
Judy(1979)	414	369	408	377(37)	360(9)	378(30)	101	119	64(37)	76(43
Anges(1981)	376	413	381	382(-6)	403(10)	385(-4)	104	103	50(54)	42(61
Forrest(1983)	373	358	370	382(-9)	372(-14)	379(-9)	112	101	58(54)	71(30
Vera(1986)	304	317		304(0)	321(-4)		130	165	128(2)	144(21

HTL(o)...observed high tide level
HTL(s)...simulated high tide level
LTL(o)...observed low tide level
LTL(s)...simulated low tide level

The eight simulations of tides for Shanghai during the storm period are plotted in Figures 3-10 and Tables 1-4. Their comparisons with the observed values and the corresponding error statistics are given. Several points should be mentioned here based on these figures and tables.

(1) Minimum and maximum root-mean-square (RMS) errors (absolute errors) of the 22 high tide level (HTL) simulations are 4 cm (0 cm) and 36.8 cm(37.0 cm), respectively. The simulated HTL whose RMS errors (absolute errors) are within 30 cm amounts to 88%(91%), whereas the minimum and maximum RMS errors (absolute errors) of the fifteen low tide level (LTL) simulations are 10.7 cm (4.0 cm) and 81.4 cm (68.0 cm), respectively. The simulated LTL whose RMS errors (absolute errors) are less than 30 cm is only 25%(33%).

Table 2. Errors of simulated tide

	HTL	LTL	HHW	LLW
Height RMS error(cm)	4.0-36.8	10.7-81.4		
Average RMS error(cm)	18.0	57.6		
Height absolute error(cm)	0-37.0	4.0-68.0	4.0-37.0	2.0-68.0
Average absolute error(cm)	14.3			
Average RMS error less than 30cm	88%	25%		
Average absolute error less than 30cm	91%	33%	75%	38%
Time absolute error(hrs)	0-3	0-4	0-3	0-4
Time absolute error within 1hr	82%	73%	75%	3%

HHW...higher high water
LLW...lower low water

(2) The minimum and maximum absolute errors of the high water (HHW) simulations are 4 cm and 37 cm . 75% of them are within 30 cm, whereas the minimum and maximum absolute errors of lower low water (LLW) are 2 cm and 68 cm, respectively. Only 38% of them are within 30cm.

(3) The LTL and LLW give smaller simulated values than the observed ones because of the positive absolute errors of LTL and LLW.

(4) The absolute errors of the minimum and maximum time lags of HTL simulations are 0 hour and 3 hours, respectively. 82% of them are within one hour.

(5) The absolute errors of the minimum and maximum time lags of HHW simulations are 0 hour and 4 hours, respectively. 75% of them are within one hour.

(6) Most of the simulated temporal variation in LTL and LLW occurs later than the observed ones and most of simulated HTL and HHW appear earlier relative to the observed ones.

On the whole, the simulations of the high tide levels(HTL) and the times at which the HTL occur give better results than those of the low tide levels. These results are consistent in some degree with those obtained by the tide harmonic analysis with 63, 119 and 170 constituents, respectively.

Table 3 Comparison of tidal hour (hr.)

TC Name	HTT(o)			HTT(s)			LTT(o)		LTT(s)	
	I	II	III	I	II	III	I	II	I	II
Marge(1951)	6	18		6(0)	20(-2)		13		12(1)	
Grace(1954)	2	15	27	5(-3)	18(-3)	30(-3)	9	23	10(-1)	3(4)
Carmen(1960)	3	15	28	4(-1)	16(-1)	29(-1)	10	23	9(1)	22(1)
Mary(1974)	4	16	29	5(-1)	17(-1)	29(0)	12	25	11(1)	23(2)
Judy(1979)	4	17	29	5(-1)	17(0)	30(-1)	12	23	12(0)	23(0)
Anges(1981)	5	17	29	5(0)	17(0)	29(0)	12	25	12(0)	24(1)
Forrest(1983)	6	18	30	6(0)	18(0)	30(0)	13	27	16(-3)	23(4)
Vera(1986)	1	14		2(-1)	15(-1)		9	21	8(1)	21(0)

HTT(o)...observed high tide hour
HTT(s)...simulated high tide hour
LTT(o)...observed low tide hour
LTT(s)...simulated low tide hour

The initial time used for calculation in Table 3 is as follows:

TC Name	Initial time
Marge(1951)	08:00 Aug. 20, 1951
Grace(1954)	08:00 Aug. 15, 1954
Carmen(1960)	08:00 Aug. 20, 1960
Mary(1974)	08:00 Aug. 19, 1974
Judy(1979)	02:00 Aug. 23, 1979
Anges(1981)	08:00 Aug. 31, 1981
Forrest(1983)	08:00 Sept. 26, 1983
Vera(1986)	02:00 Aug. 27, 1986

Table 4. RMS errors of water levels and tide levels

TC Name	HWL	LWL	HTL	LTL
Marge(1951)	16.4	57.3	14.7	68.0
Grace(1954)	32.3	75.0	20.9	80.6
Carmen(1960)	45.0	37.5	26.8	10.7
Mary(1974)	23.4	55.3	13.9	67.3
Judy(1979)	54.0	21.8	36.8	63.7
Anges(1981)	29.0	27.6	7.6	81.4
Forrest(1983)	19.7	49.8	19.2	67.8
Vera(1986)	23.0	36.0	4.0	21.1
Aveage RMS	30.4	45.24	18.0	57.6

HWL...high water level
LWL...low water level
HTH...high tide level
LTH...low tide level
Here, RMS errors for simulated water level refer to observed water level whereas RMS errors for tide level refer to predicted tide level with 63 constituents.

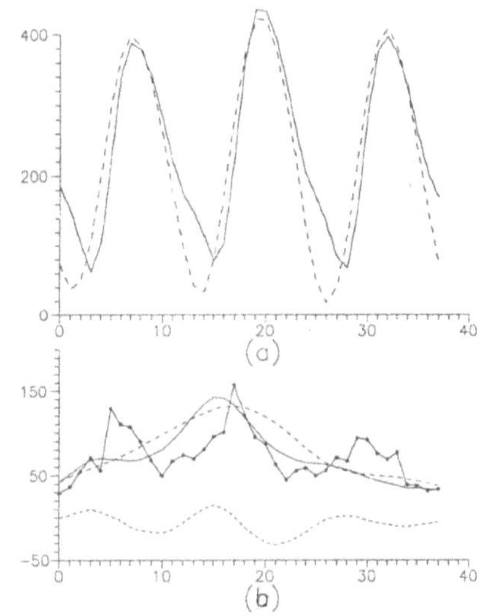

Figure 3: Simulations of tide, storm surge and water level caused by nonlinear T-S coupling during TC Mary (1974). (a) Tide level : Solid line - tidal level prediction with 63 constituents (observed), dash line - tidal level prediction with 8 constituents (simulated). (b) Storm surge and ζ_{TS} : solid line with symbols - observed surge ζ_R, solid line without symbols - simulated surge comprising T-S coupling effecs ζ_{T+S}, long dash line: simulated pure surge ζ_S, short dash line: simulated ζ_{TS}

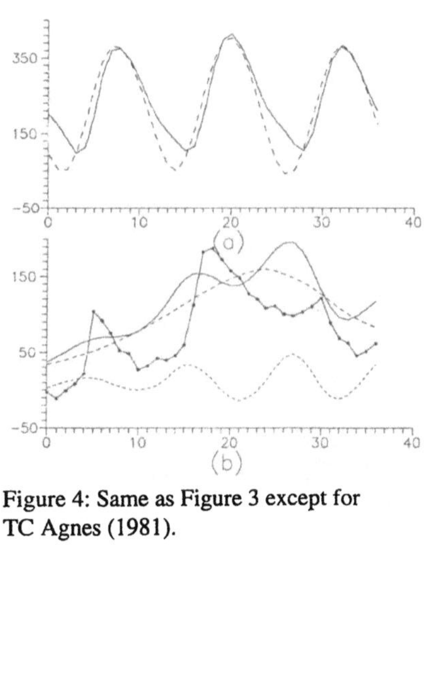

Figure 4: Same as Figure 3 except for TC Agnes (1981).

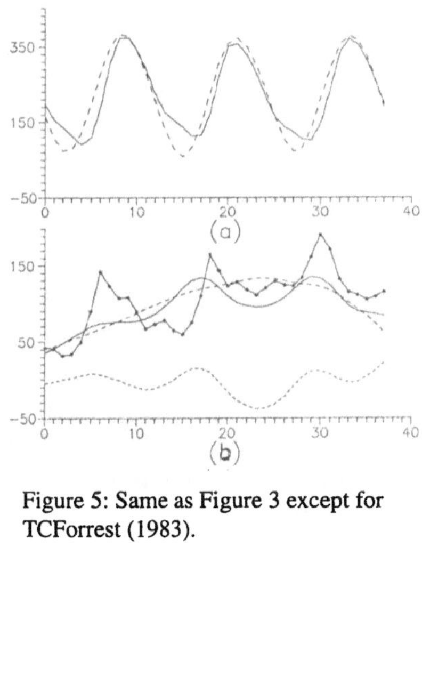

Figure 5: Same as Figure 3 except for TCForrest (1983).

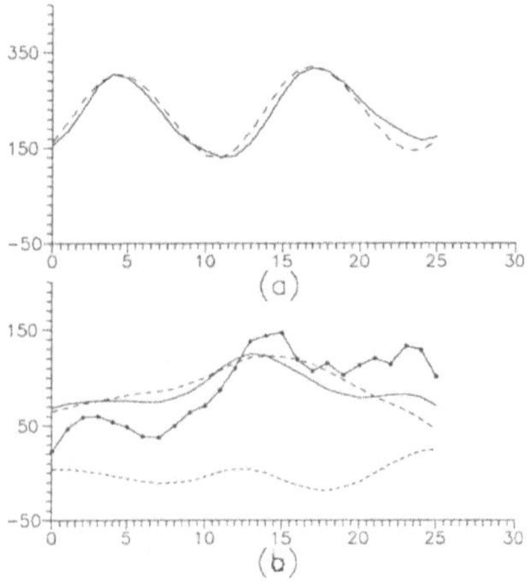

Figure 6: Same as Figure 3 except for TC Vera (1986).

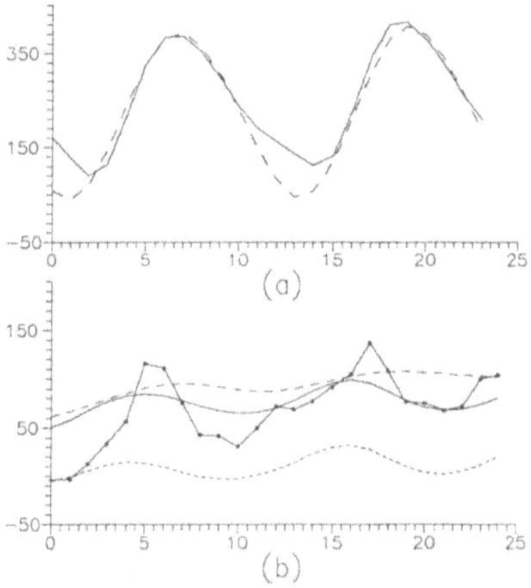

Figure 7: Same as Figure 3 except for TC Marge (1951).

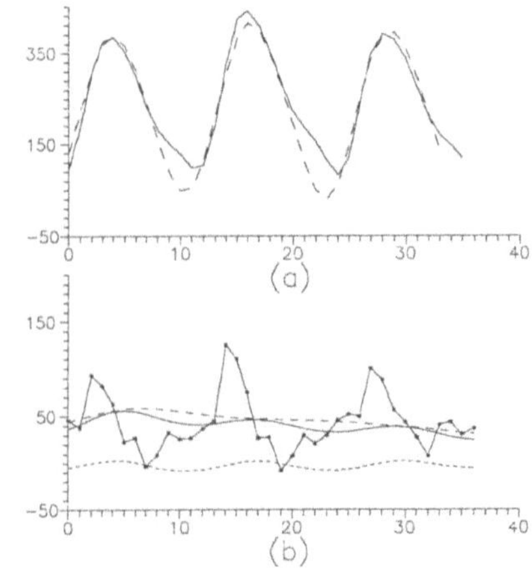

Figure 8: Same as Figure 3 except for TC Grace (1954).

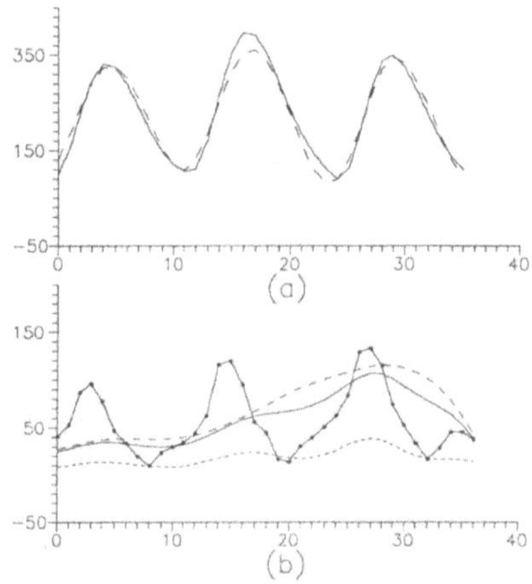

Figure 9: Same as Figure 3 except for TC Carmen (1960).

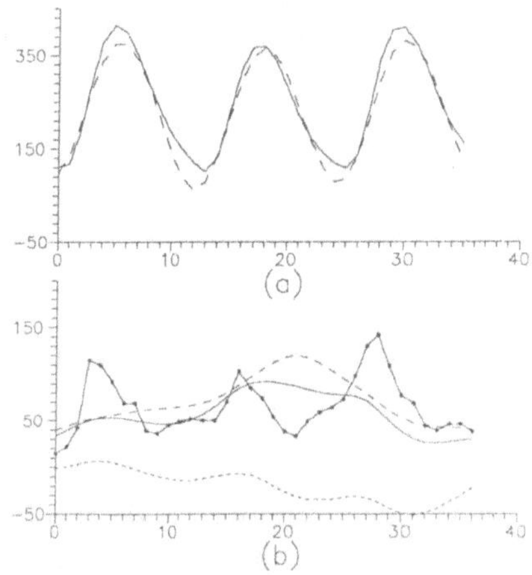

Figure 10: Same as Figure 3 except for TC Judy (1979).

4. STORM SURGE SIMULATIONS

Comparisons of the simulations of the pure storm surge ζ_S and storm surge simulation comprising nonlinear T-S interaction $\zeta_S+\zeta_{TS}$ with the observed ones ζ_R will be given in this section. ζ_S can be reproduced by integrating the surge dynamic equations numerically without tide waves propagating across the open boundary of the computational domain. $\zeta_S+\zeta_{TS}$ can be obtained by subtracting the simulated tides ζ_T from the simulated total water levels. z_R the residuals, result from the subtraction of the predicted tides, in terms of the tide harmonic analysis from the observed water levels. $\zeta_S+\zeta_{TS}$ includes the wind set-up caused by both the storm and nonlinear T-S effect.

(1) It is shown from Figures 3-10 that the simulated storm surges comprising the nonlinear T-S effect $\zeta_S+\zeta_{TS}$ give better results than the pure storm surge simulations both in their peak values and the temporal variations. Specially the former can reproduce the main tidal periods of 11-13 hours except for TC Anges(1981). This implies that the results will be improved when the effect of the nonlinear T-S interaction is taken into account.

(2) In general, the peak surges incorporating nonlinear T-S coupling are stronger than those of the pure storm surges during the rising tides, whereas they become weaker during the falling tides because of the nonlinear T-S coupling. The simulated peak values of $\zeta_S+\zeta_{TS}$ have a lag of 1-3 hours behind the simulated pure peak surges.

(3) Most of the observations show that the residual ζ_R has a lag of 3-5 hours in phase in relation to the tides, i.e., the peak residual is located at the rising tides. The simulations of

the tides and $\zeta_S+\zeta_{TS}$ agree well with the above observational facts except for TC Anges 1981. However, the peak values of $\zeta_S+\zeta_{TS}$ occur in 2-3 hours behind the time at which the tides rise.

5. ANALYSIS OF NONLINEAR TIDE-SURGE INTERACTION

The total water level ζ_{T+S} , the tide ζ_T, the pure storm surge ζ_S and the water level caused by the nonlinear T-S interaction (ζ_{TS}) can be by the following equation:

$$\zeta_{TS} = \zeta_{T+S} - \zeta_T - \zeta_S \qquad (3)$$

As shown from Figures 3-10 the temporal variations in the nonlinear T-S effect, ζ_{TS} can also be characterized by their main tidal periods of 11-13 hours, which result from the similar characteristics of $\zeta_S+\zeta_{TS}$. The magnitude of the nonlinear T-S effect varies quite different in each case. It depends on the tidal state, the intensity of storm surges and wind set-up stage at which it locates. In general, the nonlinear T-S coupling will intensify with increasing intensity of storm surges. For examples, the intense nonlinear T-S coupling occur in TCs Mary(1974), Anges(1981), and Forrest(1983) because of strong storm surges caused by both high wind and intense air pressure drop during the storm periods. On the contrary, the weak nonlinear T-S coupling effects associated with weak storm surges in TC Vera(1986). In the eight cases we discussed, the maximum amplitudes of the nonlinear T-S coupling effect change rang from 35 cm to 50 cm.

The storm surges and tides interact nonlinearly in the sea. There are two categories of the nonlinear terms. One belongs to the interior of the sea and another is on the air-sea interface as indicated by the parenthesis in dynamic equations (1) and (2). There are three nonlinear terms in the interior of the sea, i.e., the momentum convective term (C), the shallow-water term (B) the quadratic bottom friction term (A), and wind stress-depth term (D). The contributions of these nonlinear terms to both total water level and storm surge in the area of interest are discussed. in the following sections (see Figures 11-18). Nevertheless, we will confine ourselves to the subject of the nonlinear interaction in the interior of the sea, so that the wind stress-depth term is excluded in our discussion.

5.1 Nonlinear bottom friction effect

(1) The amplitude of the water level is diminished by the quadratic bottom friction. Consequently, high water level rises and low water level decrease. The amplitude of the water level attenuates with increasing strength of friction. Basically, its phase remains unaltered. However, there are slight changes in the phase in some special cases, for example TC Anges(1981).

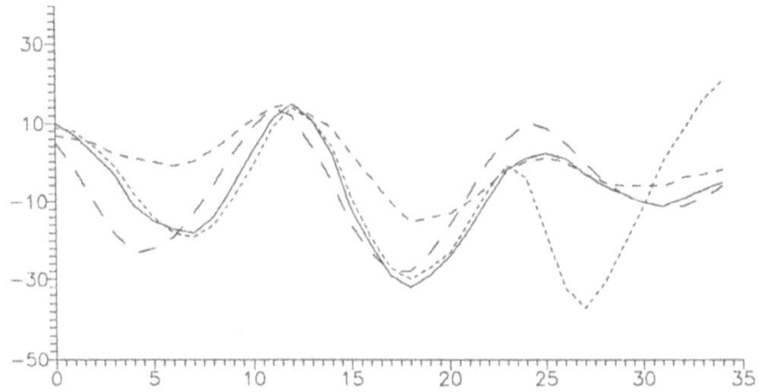

Figure 11: Numerical contributions of the bottom friction term and shallow-water term to ζ_{TS} for TC Mary(1974). Solid line - Normal simulation ($Cb=2x10^{-4}$), long-dash line - intense bottom friction ($Cb=2x10^{-3}$), short-dash line - weak bottom friction ($Cb=8x10^{-5}$), shorter long-dash line - no shallow-water term case.

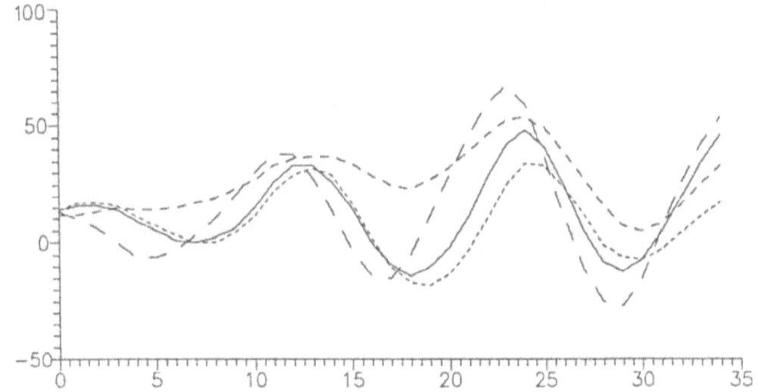

Figure 12: Same as Figure 11 but for TC Agnes (1981)

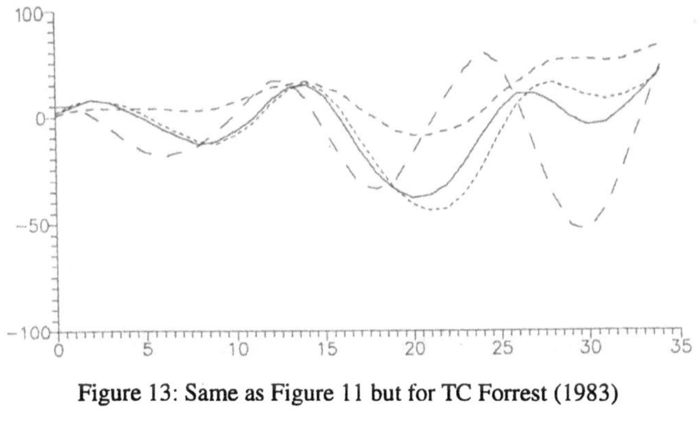

Figure 13: Same as Figure 11 but for TC Forrest (1983)

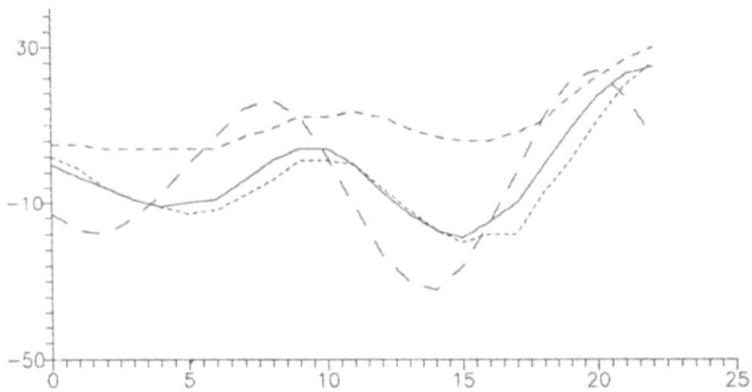

Figure 14: Same as Figure 11 but for TC Vera (1986)

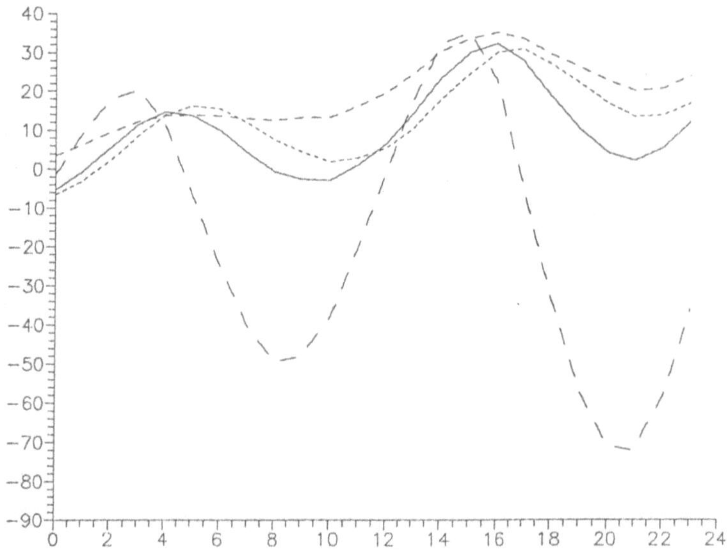

Figure 15: Same as Figure 11 but for TC Marge (1951)

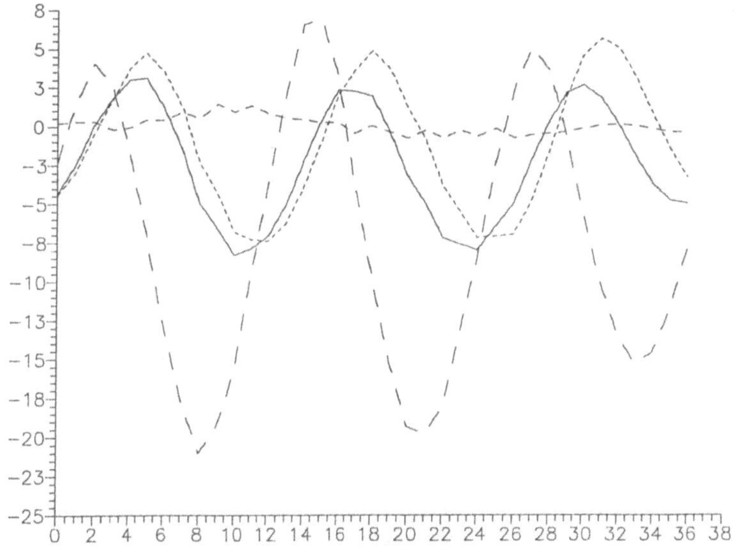

Figure 16: Same as Figure 11 but for TC Grace (1954)

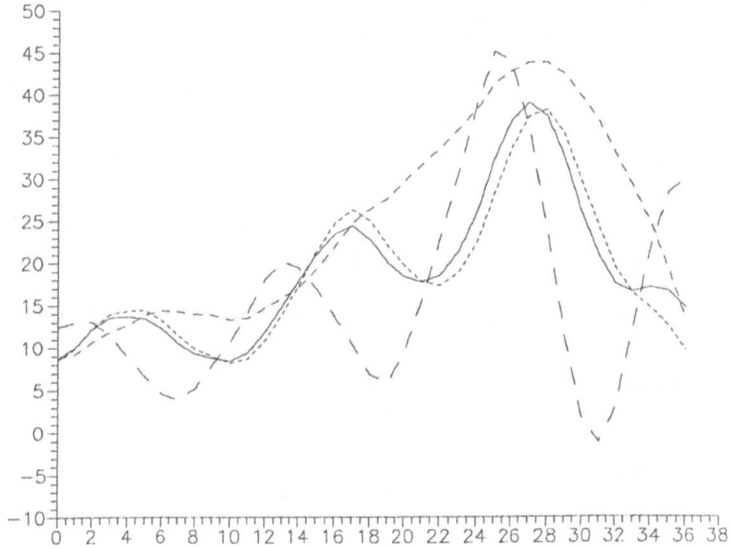

Figure 17: Same as Figure 11 but for TC Carmen (1960)

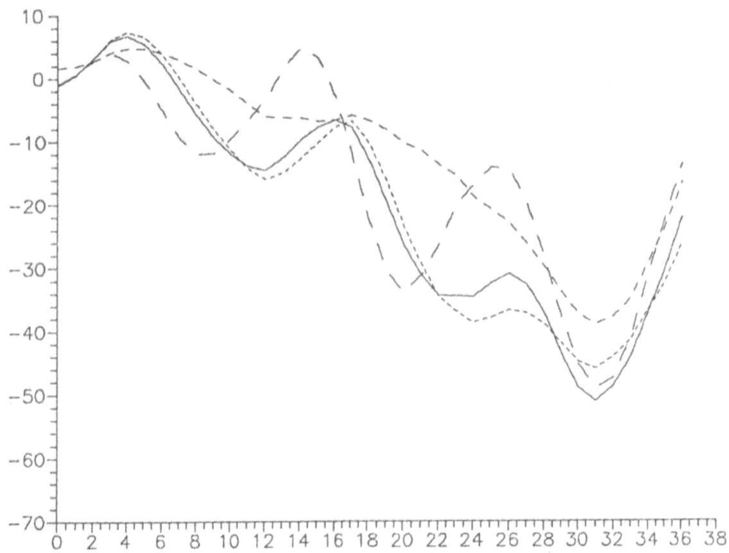

Figure 18: Same as Figure 11 but for TC Judy (1979)

As expected when the bottom friction coefficient Cb increases to ten times of its initially prescribed value 2 x 10^{-4}, the high water level will decrease by 7.6-15.9%, depending on the feature of the time variations of the water levels.

(2) The amplitudes of the ζ_{TS} increase with increasing bottom friction. The phase of z_{TS} versus time curves undergoes considerable change. The relatively intense friction (Cb=2*10^{-3}) makes both high and low ζ_{TS} levels occur earlier than otherwise, whereas the very weak friction (Cb=8*10^{-5}) makes both high and low ζ_{TS} levels occur later than otherwise. When the bottom friction coefficient increases to ten times its initially prescribed value of 2*10^{-4}, the maximum amplitude of ζ_{TS} will increase by 10.6-48%.

To ascertain the contribution of the quadratic bottom friction term to the total water level and ζ_{TS}, it is also noteworthy that all variables in this term are not only changeable from point to point on the computational domain but also from case to case. In fact, the bottom friction term in Eq. (1) can be expanded as follows:

$$\frac{C_b\,|\vec{V}|\,\vec{V}}{h+\zeta} = \frac{C_b}{h}\,|\vec{V}|\,\vec{V}[1-\frac{\zeta}{h}+(\frac{\zeta}{h})^2+\ldots],\ as\ \frac{\zeta}{h}\gg 1 \qquad (4)$$

Where the factor ζ/h is a well-known small nondimensional parameter, which characterizes nonlinear effects in the dynamic problems relevant to the gravity waves, such as tides. This means that the results obtained not only illustrate the current-water level height but also current-current interactions. We have examined the current-current interaction and the bottom friction term under different values of ζ/h and Cb. That is the sensitivity test of ζ_{TS} to ζ/h and bottom friction strength(Cb).

In consideration of dependence of ζ/h on grid point and tropical cyclone, a changeable multiplication factor a, ranging from zero to unity should be conveniently added in front of term ζ/h in equation (4) in order to test the contribution of ζ/h in the bottom friction term to ζ_{TS}. ζ_{TS} tends to increase slightly with decreasing ζ/h whatever the bottom friction strength is. In eight experimental tropical cyclones, the contribution of ζ/h to ζ_{TS} varies from 0 to 45.5%, depending on bottom friction strength . In general, ζ/h is responsible for an increase in ζ_{TS} under the influence of intensifying bottom friction, whereas ζ/h causes a decrease in ζ_{TS} under the effect of bottom friction. Occasionally, there are a few exceptions. When the bottom frictional coefficient Cb increases to ten times of its initially prescribed value($2*10^{-4}$) and a is set to be unity, the peak value of ζ_{TS} increases from 4% to 400%. On the contrary, when Cb reduces to 25% of its initially prescribed value and a is still assumed to be unity, the peak value of ζ_{TS} decrease from 0 to 110% . Consequently, it can be concluded that the bottom friction strength contributes greatly to the peak value of ζ_{TS}.

5.2 Nonlinear shallow-water effect

(1) The calculated high water level without taking the shallow-water term into account overestimates the real one. This implies that the shallow-water term makes the high water level lower than otherwise. The contribution of this term to the high water level is ranging between 1.1 and 6.0%. The role of the shallow-water term is to make high water level occur earlier than otherwise.

(2) When the shallow-water term is absent from the dynamic equations, the amplitudes of tides become smaller and both high and low tidal levels become larger whichever time variation in tides will be. At the same time, no phase change occurs, with the exception of certain low tidal level in the case of TC Mary(1974). Roughly speaking, this implies that the shallow-water term makes both high and low tidal levels lower than otherwise without significant phase change. It is found that the contribution of the shallow water term to the maximum amplitude of ζ_{TS} is ranging between 22 and 44%. In a few cases, the shallow-water term makes the high ζ_{TS} level occurs earlier and low ζ_{TS} latter than otherwise.

5.3 Effect of nonlinear convective term

The contribution of momentum convective term to the high water level is about 0-1.8% and therefore is negligible. However, it contributes to the maximum amplitude of ζ_{TS} by 8-12%, and therefore 14.5% on average. In summary the largest contribution of the the effects of above three nonlinear terms on ζ_{TS} comes from the quadratic bottom friction term. The contribution of the shallow-water term accounts for 33% on average, the contribution of the momentum convective term is 14.5%, and the remaining contribution comes from the quadratic bottom friction term. This result, agrees quite well both qualitatively and quantitatively with that obtained by Wolf(1978) in his study of the interaction between tide and surge on the east coast of the United Kingdom.

6. STORM SURGE PHASE EXPERIMENT IN RELATION TO TIDES

Extreme water levels occur when the peak storm surges coincide with the high tides.. When this occurs the peak surge frequently coincides with the rising tide. It is of practical value to examine different surge phases associated with the tides. Here, the impacts of six phase lags, with one hour interval apart, of the peak surge behind the associated high tides on the total water level are numerically studied.

As shown from the numerical experiments described earlier whichever lag(0-6hr) is adopted, the majority(97%) of water levels, resulted from the algebraic sum of the pure peak surge ζ_S and high tides, are higher than those from linear superposition of the peak surges incorporating T-S coupling and tides. This result agrees well with the simulated one mentioned above. The correction of the nonlinear T-S interaction to the water level varies from 1 cm to 93 cm.

The maximum difference between them occurs in lags of 5 hours and 1 hour, respectively. This implies that the effect of the nonlinear T-S coupling does not always play a minor role in the water level numerical simulation and prediction.

7. SUMMARY

Using two-dimensional shallow water dynamic models, the mechanism of nonlinear interaction between storm surge and tide as well as the effect of nonlinear T-S interaction on surge and water level in the coastal waters of Shanghai is numerically investigated under the combined action of the atmospheric forcing and external progressive wave of tides during the eight selected storm period. Errors of the tide simulations are generally within the permissive limit. This enables us to study nonlinear T-S coupling and make the result convincible.

In order to match the model results to the observation, smaller bottom friction coefficients are taken in this article than those in the oceanographic literature.

To sum up, the main conclusions from this study are as follows:

(1) The oscillation with the main tidal period occurring in temporal variation of the water level residuals during tropical cyclone passage is caused by the nonlinear T-S interaction. The magnitude of the nonlinear T-S interaction tends to increase when the storm surge amplitude enhances. The time-varying nonlinear T-S interaction also oscillates with the main tidal period. The nonlinear T-S interaction exerts a considerable effect on the water level. Taking this effect into account, both storm surge simulation and water level modelling are remarkably improved. The maximum correction to the water level resulting from this effect is about 100 cm.

(2) In general, the peak surge occurs on the rising tide. The simulated peak surge occurs nearer the time of tidal low water than that of tidal high water. This is because of the effect of nonlinear T-S interaction. They occur in 2-3 hours after the time of tidal low water. The nonlinear T-S interaction makes the pure storm surge enhanced during the rising tide. That is to say, the traditional pure peak surge simulation overestimates the real one because of nonlinear T-S interaction. In a quite few cases, the peak surges can occur on the falling tide where the pure surge will be lower under the influence of the nonlinear T-S interaction.

(3) Roughly, the maximum differences in the simulated water levels between those calculated with and without taking the nonlinear T-S effect into consideration emerge in where the variations of storm surges have lags of 5 hours and 1 hour, respectively, relative to those of tides.

(4) Among three primary internal nonlinear term, the largest contribution to z_{TS} comes from the quadratic bottom friction term. The second and third contributions are from the shallow-water term and convective term, respectively. The quadratic bottom friction not only makes the amplitude of time variation of z_{TS} enhanced but also considerably changes its phase. The maximum amplitude of z_{TS} increases with increasing strength of the bottom friction.

(5) The role of the shallow-water term is generally to make both high and low levels for z_{TS} lower than otherwise without significant change in the phase. The effect of the convective term on z_{TS} water level is of minor importance.

(6) The amplitude of the water level is diminished by the quadratic bottom friction without change in phase. This is probably attributed to the removal of energy from storm surges by friction. The high water level reduces its value and occurs earlier than otherwise under the effect of the nonlinear shallow-water term.

(7)The mechanism of the nonlinear T-S interaction in Shanghai water has been preliminary clarified throughout this study. However, many problems remain to be solved because of the complexity of the nonlinear T-S effect. First, only eight tropical cyclone examples for tidal and nonlinear T-S interaction simulations have been chosen in this paper. It is necessary to carry out more numerical simulations to verify the results obtained. Second, the tidal simulation needs to be improved because it is closely related to the accurate description of the nonlinear T-S interaction. Third, the complex and fine topography of the sea-bed and coast configuration, particularly, in the Yangtze River estuary should further be carefully treated. Finally, the bottom friction coefficient Cb is assumed to be constant in all eight case calculations. This seems to be not suitable for highly accurate simulation.

REFERENCES

Bank, J. E: 1974, A mathematical model of river-shallow sea system used to investigate tide, surge and their interaction in the Thames, *Geophy. J. Roy. Astro. Soc.*, **55**, 203-216

Das, P.K: 1980, Storm surge in the Bay of Bengal. *Proceeding WMO Symposium on Typhoon, Shanghai* , 171-183.

Chai Fei and Wang Jingyong: 1990, The nonlinear interaction of surge and tide in the East China Sea, *J.Ocean University of Qingdao*, **20(3)**, 56-62 (in Chinese with English abstract)

Fujita, T.: 1952, Pressure distribution in typhoon. *Geophys. Mag.,* **23**, 437-452.

Murty, T.S. and M.I. El-Sabh: 1981, Interaction between storm surge and tides in shallow waters, *Mar. Geodesy.* **5**, 19-33.

Prandle, D. and Wolf, J.: 1978, The interaction of surge and tide in the North Sea and River Thames, *Geophys. J. R. Astro. Soc.,* **55**, 203-216.

Prandle, D. and Wolf, J.: 1986, Surge-tide interaction in the South North Sea, *Proc. 9th Int. Liege Colloq. Ocean Hydrodyn,* 161-185, edited by J.C.J. Nihoul, Elsevier, Amsterdom, The Netherlands.

Proudman J.: 1955a, The propagation of tide and surge in an estuary, *Proc. Roy. Soc.* ,London Ser. **A231**, 8-24.

Proudman, J.: 1955b, The effect of friction on a progressive wave of tides and surges in an estuary, *Proc. Roy. Soc.*,London Ser. **A233**, 407-418.

Qin, Z., Y. Duan, Y. Wang, Z. Shen and K. Xu: 1994, Numerical simulation and prediction of storm surge and water level in Shanghai Harbour and its vicinity, *Natural hazards,*, **9**, 167-188.

Ueno, T.: 1964, Nonlinear numerical studies on tides and surges in the central part of Sato Inland Sea, *Oceanogr Mag.*, **16**, 53-124.

Wolf, J.: 1978, Interaction of tide and surge in a semi-infinite uniform channel, with application to surge propagation down the East Coast of Britain, *Math. Modeling*, **2**, 245-253.

Wolf, J.: 1981, Surge-tide interaction in the Nirth Sea and River Thames, *Floods due to high wind and tides*, 75-94, edited by D.H. Pergrine, Academic Press, London U.K.

Zhang Yanting and Wang Yijiao: 1989, Numerical modelling of the sea level under the action of the tide and strong wind in the Bohai Sea. *Acta Oceanologia Sinica*, **8**, 511-520.

A NUMERICAL MODEL TO SIMULATE THE MOVEMENT OF SUSPENDED SEDIMENT OFF THE HUANGHE RIVER (YELLOW RIVER) DELTA

LE KENTANG, SHI JIUXIN AND LIU XINGQUAN
Institute of Oceanology, Chinese Academy of Sciences, 7 Nan-hai Road, Qingdao,P.R.China

Abstract: The Huanghe River is known as a river with an extraordinary enormous sediment (about 14×10^8 ton/a) being carried out into its estuary. The enormous sediment has frequently caused the deposition, extension and avulsion of the tail of the Huanghe River since 1855. From 1855 to 1984, the area of the Huanghe River Delta had increased by 2530.4 km^2 due to the deposition, extension and avulsion, which have not only made the lower reaches suffered from disasters, but also had significant influence on the ecosystem in the coastal water region. To bring the development of the channel of its estuary under control, it is necessary to predict the distribution and evolution of the sediment transport off the river mouth quantitatively. This paper presents a 2D numerical model to simulate the movement of suspended sediments off the mouth. The hindcast results of the model reproduce the main observed features of the sediment distribution.

1. INTRODUCTION

The Huanghe River (Yellow River) is the second longest of China's rivers, which flows from the eastern highlands of the Tibetan Autonomous Region 5464 km to the Bohai Sea. More than 150,000,000 people live in its basin which drains an area of 745,000 km^2.

The Huanghe River is known as a river with an extraordinary enormous sediment (about 14´108 ton/a) being carried out into its estuary. The name Huang (Yellow) comes from the fine, loess soil that it brings down with it. Such enormous sediment is bound to deposit at its mouth and then to change the local base level of the lower reach. This will inevitably result in a change of longitudinal profile and transport capacity of flow to conform with the incoming sediment load in order to maintain the relative equilibrium. Such a change will in turn lead to aggregation and degradation, which have frequently caused the deposition, extension and avulsion of the tail of the Huanghe River.

From 1855 to 1984, the Huanghe River Delta had increased its area by 2530km^2 due to the deposition, extension and avulsion, which have not only made the lower reaches

M. I. El-Sabh et al. (eds.), Land-based and Marine Hazards, 157–177.

suffered from disasters, but also had significant influence on the ecosystem in the coastal water region. To bring the development of the channel of its estuary under control, much surveys and investigations have been carried out in the Huanghe River Delta and its adjacent sea since 1949 (Le, 1993; Le and Huang, 1993; Le and Zhou, 1993; Pang and Shi, 1979, 1980,1982). Many of these papers dealt with the sedimentation off the Huanghe River Delta in a qualitative manner.

To tackle the environment of the current Huanghe River estuary in a comprehensive way, first of all, it is necessary to predict the distribution and development of the sediment transport off the current Huanghe River Mouth quantitatively. As a first step, a numerical model to simulate the movement of suspended sediments off the Huanghe River Delta is presented here.

2. MODEL EQUATIONS

A two dimensional tidal model with a run-off input is applied to simulate the fluid dynamics of the estuary. Since the actual transport of water and suspended sediment is a Lagrangian process, a numerical approach is also incorporated to the 2D-model to simulate the Lagrangian motion. The model is expressed in a vertically-averaged two dimensional system. The Cartesian axes x and y are taken counterclockwise in a horizontal plane of the undisturbed water surface with the x, y, and z axes being eastward, northward and vertically upward, respectively. The components of the velocity are u and v, which are parallel to the coordinate axis at the point (x,y).

2.1 Equations of motion

The large scale horizontal motion in an estuary can be described by the following shallow water equations:

$$\frac{\partial u}{\partial t}+u\frac{\partial u}{\partial x}+v\frac{\partial u}{\partial y}-fv=-g\frac{\partial \zeta}{\partial x}+A_l\left(\frac{\partial^2 u}{\partial x^2}+\frac{\partial^2 v}{\partial y^2}\right)-\frac{u\sqrt{u^2+v^2}}{C_0^2(H+\zeta)} \quad (1)$$

$$\frac{\partial v}{\partial t}+u\frac{\partial v}{\partial x}+v\frac{\partial v}{\partial y}+fu=-g\frac{\partial \zeta}{\partial y}+A_l\left(\frac{\partial^2 v}{\partial x^2}+\frac{\partial^2 v}{\partial y^2}\right)-\frac{v\sqrt{u^2+v^2}}{C_0^2(H+\zeta)} \quad (2)$$

$$\frac{\partial \zeta}{\partial t}+\frac{\partial[u(H+\zeta)]}{\partial x}+\frac{\partial[v(H+\zeta)]}{\partial y}=0 \quad (3)$$

where z is the surface elevation above the undisturbed water surface; H, the depth of water; t, time; f, the Coriolis parameter being a function of the latitude; g, the acceleration due to gravity; C_0, Chezy coefficient; A_l,horizonal eddy viscosity coefficient. The bottom stress is proportional to the velocity squared.

2.2 Lagrangian motion

A Lagrangian viewpoint is usually followed in dealing with advection-dominated transport processes because the displacement of dissolved and suspended substances is in principle a Lagrangian phenomenon (Cheng and Casulli, 1982). In order to analyze the Lagrangian motion of a labeled water parcel, the Lagrangian velocity components $\vec{u}_l = (u_l , v_l)$ in (x, y) directions are defined as the velocity of a water parcel at time t, which is released from $\vec{X} = (x_0, y_0)$ at time t_0. Let $\vec{Y} = (\vec{X}_0 , t)$ be the trajectory of the labeled fluid parcel can be determined by

$$\vec{Y}(\vec{X}_0 , t) = \vec{X}_0 + \int_{t_0}^{t} \vec{u}[\vec{Y}(\vec{X}_0 , t'), t']dt' \tag{4}$$

A Lagrangian residual velocity can be defined by the net Lagrangian drift displacement over a tidal period T and divided by T, thus

$$\vec{u}_{LY}(\vec{X}_0 , t_0) = \frac{\vec{Y}(\vec{X}_0 , t_0 + T) - \vec{Y}(\vec{X}_0 , t_0),}{T} = \frac{1}{T}\int_{t_0}^{t} \vec{u}[\vec{Y}(\vec{X}_0 , t'), t']dt' \tag{5}$$

2.3 Mathematical model of total sediment transport

Suppose that non-uniform suspended loads can be classified into N groups in size, then

$$C_j = P_j C$$

$$C = \sum_{j=1}^{N} C_j$$

where C is the total sediment concentration; P_j is the ratio of the sediment concentration in the jth group to the total sediment concentration. Following Dou et al. (1987), the fundamental equation of non-equilibrium transportation of non-uniform suspended load is as follows

$$\frac{\partial[(H+\zeta)C_j]}{\partial t}+\frac{\partial[(H+\zeta)uC_j]}{\partial x}+\frac{\partial[(H+\zeta)vC_j]}{\partial y}+\alpha_j\omega_j\beta_j(C_j-C_j^*)=0 \quad (6)$$

$$(j=1,2,\ldots,N)$$

where C_j is the sediment concentration of the jth group; C^*_j, sediment transport capacity of the flow; w, settling velocity of the sediment; a, settling probability of the sediment particle; N, number of the size group; subscript "j" denotes the sequence number of the size groups. And,

$$\beta_j=\begin{cases}1 & C_j\geq C_j^* \\ P_{bj} & C_j<C_j^*\end{cases} \quad (7)$$

where P_{bj} is the size distribution of the bed material. If the deposition occurred, P_{bj} is taken to be 1, otherwise 0.5.

Equation of the sea bed deformation can be written by

$$\frac{\partial\eta_j}{\partial t}=\alpha_j\omega_j\beta_j(C_j-C_j^*)/\gamma_j \quad (8)$$

$$(j=1,2,\ldots,N)$$

where η_j is the elevation of sea bed silted up by the suspended load of the jth group (a negative η_j denotes the scoured depth); γ_j is dry unit weight of the jth group.

2.4 Boundary Conditions

At the coast, flow and suspended load can not penetrate the solid wall, i.e.,

$$\vec{v}_n = 0 \quad (9)$$

where $\vec{v}_n$ is the flow normal to the wall.

Because the Bohai Sea is a semi-closed sea, the open sea boundary is located at the Bohai Strait. The tidal elevation at the boundary can be taken as

$$\zeta=\Sigma H_i\cos(\sigma_i\ t-g_i) \quad (10)$$

where H, σ and g are amplitude, angular frequency and phase lag, respectively. Subscript i consists of main diurnal and semi-diurnal constituents. Sediment concentrations are taken to be zero at the Bohai Strait.

At the mouth of the Huanghe River the velocity and concentration are

$$u = \frac{Q}{b(H + \zeta)} \tag{11}$$

$$C = C_m(t)$$

respectively, in which b and Q are the breadth and the runoff discharge, and $C_m(t)$ is the sediment concentration to be a known function of time at the mouth.

3. NUMERICAL APPROACH

Historical data show that the sediment loads from the Huanghe River spread mainly around the current Huanghe River Delta, i.e. basically within the region (118°12'-119°42'E,37°12'-38°42'N) since the 1950's. To better simulate the sediment movement off the Huanghe River mouth, the open boundary is set at a section across the Bohai Strait, which suggest that the computation area for the present study should cover the entire Bohai Sea. Thus, a nested grid net system is used for our study as shown in Fig. 1. In the system the large grid spacing is taken to be 14.84km and the fine grid spacing is one fifth of the big one. The bathymetry of the fine grid net region is given in Fig. 2.

To solve Equations (1)-(3), a space-staggered scheme, where velocities, water levels and depths are described at different grid points, and the ADI method developed by Leendertse (1967) are used. Equation (6) is discretized by full implicit scheme, in which the convective term is approximated by an upwind difference scheme (Patankar, 1980). According to the C.F.L. condition, time step is taken to be

$$\Delta t = 207.01002 \text{ s.}$$

Furthermore, the following parameters and formulae are employed in solving Equations (1)-(3) and (6):

Chezy coefficient is

$$C_0 = \frac{1}{n} H^{1/6} \tag{12}$$

where 1/n is the Manning roughness factor, which is taken to be 17. Horizontal eddy viscosity coefficient Al is taken to be 103 m^2/s.

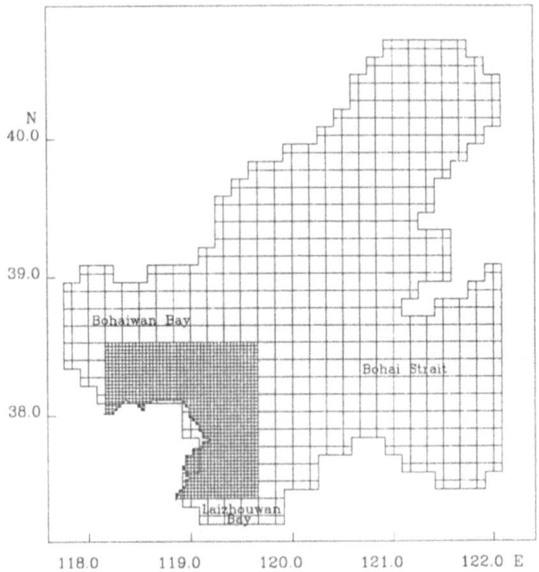

Figure 1: Nested grid system over the Bohai Sea.

Following Dou et al. (1987), sediment transport capacity can be expressed as

$$C_j^* = P_j^* \frac{\gamma \gamma_s (u^2 + v^2)^{3/2} K_1}{(\gamma_s - \gamma)\, gH \omega_m \; C_0^2} \tag{13}$$

where

$$P^* = \frac{(\frac{P_j}{\omega_j})^{\alpha}}{\sum_{j=1}^{N} (\frac{P_j}{\omega_j})^{\alpha}} \tag{14}$$

$$\omega_m = \sum_{j=1}^{N} P_j \omega_j \tag{15}$$

in which K_1=0.034, $(\gamma_s-\gamma)/\gamma$=1.65, and specific weight of a grain γ_s=0.265(kg/m^3).

According to the Prandtl's mixing length theory, shear velocity can be formulated (Fang and Ichiye, 1983) by

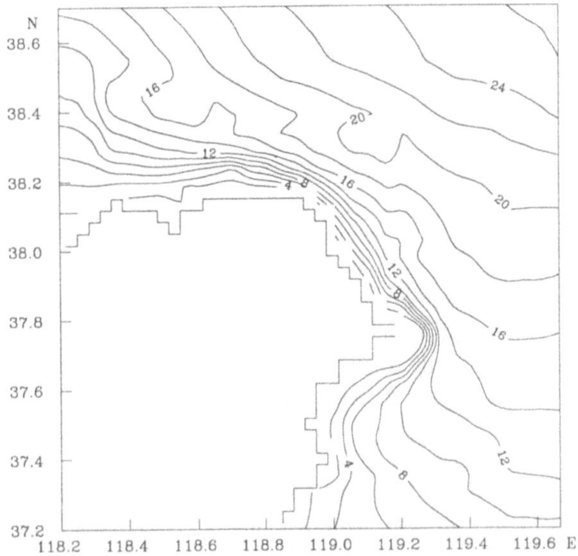

Figure 2: Bathymetry (in metres) around the Huanghe river delta.

$$(u_*, v_*) = \frac{\kappa_0 (u, v)}{\ln \frac{4h}{z_0} - \frac{8}{3}} \tag{16}$$

where $h = H + \zeta$, κ_0, the Karman constant, approximately is equal to 0.4 and z_0 is the roughness length factor taken as 0.12 (cm). Following Dou (1963), settling probability reads

$$\alpha_j = 0.49 + 0.48\left(\frac{\omega_j}{\sigma}\right) - 0.126\left(\frac{\omega_j}{\sigma}\right)^2 \tag{17}$$

where $\sigma = 1.25\,(u_*^2 + v_*^2)^{1/2}$.

Burt (1986)suggested that the following median settling velocity can be expressed as

$$\omega_{50} = 1.341 x\, 10^{-7}\, C^{1.37} \tag{18}$$

where ω_{50} (m/s) is the settling velocity below which 50% by weight of the sample settled out.

In the modelling described below, all computations were started with the water levels, velocities and concentrations at all grid points taken to be zero and ended when the difference between computation values at two time steps in succession was<0.001.

4. RESULTS

4.1 Tidal Current Field

Le et al. (1993) developped a numerical model to study the influence of changes of the Huanghe River estuary on the tidal wave movement and their changes in the past century. The current Huanghe river mouth formed in 1976 has undergone three stages. In its second stage (1980-1986) the tail of the Huanghe River is in a relative stable state. Le and Zhou (1993) studied the basic features of the current field during this stage on the basis of field data. It has been found that the tidal current field exhibits to regular semi-diurnal pattern in the adjacent area off the Huanghe River mouth and there is a high speed region to the southeast of the River mouth as shown in Fig. 3. In this section, we give our exclusive attention to the tide-induced residual current field and its effects on suspended sediment transport. However, to reveal clear, unambiguous definition for various residual variables, only a single tidal constituent M2 in the simulation was used. The realistic tides in the delta system are much more complicated. The simulated tidal circulation begins with all water mass at rest. A dynamic steady state is reached after approximately 25 complete tidal cycles when the solution becomes periodic. Once the tidal circulation has reached a dynamic steady state, a massless tracer particle is released from each computational grid point. The tidal circulation is then computed for another complete tidal cycle while tracer particles travel along the trajectory prescribed by Eq. (4). At the end of the tidal cycle both the Eulerian residual velocity and the Lagrangian residual velocity are obtained.

To understand the major observed features of the tidal current and turbid water distributions, the distribution of the maximum possible tidal current and the distribution of the turbid waters are respectively given in Figs 3 and 4 (Le and Huang, 1993). As shown in Fig. 3, the maximum ebb current exceeds 180 cm/s. Fig. 4 obtained from satellite pictures shows that the variations of tidal current velocity significantly affect the transport rate, track of suspended sediment and sedimentation process. As shown in Fig.4, the turbid waters run northeastward in the ebb period and southeastward in the flood period. Considering that the turbid waters are of heavy suspended load and current velocity decreases very rapidly, the sediment transport capacity decreases rapidly, which may in turn cause suspended sediment to accumulate in both sides of the river mouth.

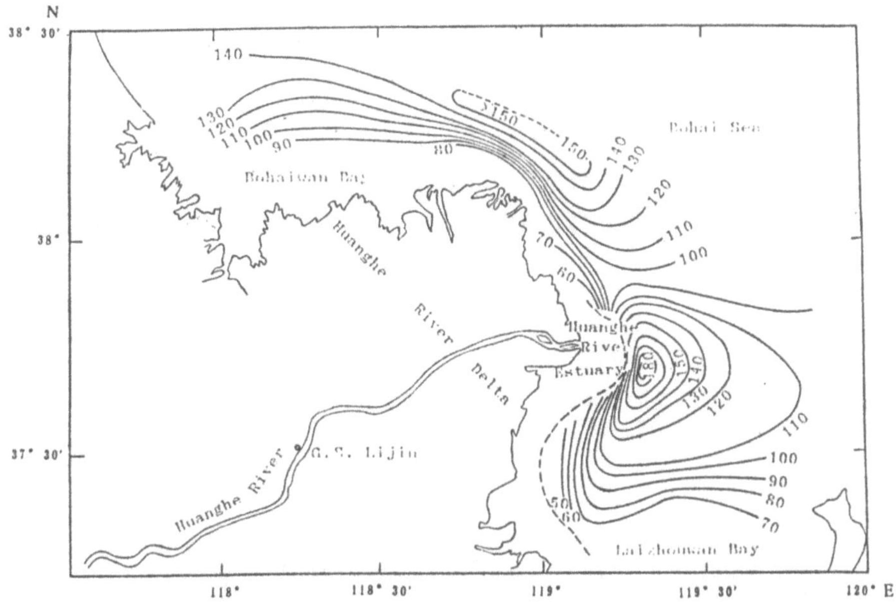

Figure 3: Distribution of the maximum possible tidal current speed (cm/s) around the Huanghe river delta.

As stated by Yalin (1972), when the grain size mobility number, which reflects the influence of specific weight of a grain in fluid and characterizes the ratio of the magnitude of the tractive fluid dynamics force acting on a grain to the grain weight, is "large", then the value of total transport rate must rise with the fourth power of shear velocity (or of any other velocity) in the total-load formula of R. A. Bagnold. Actually, in the case of "large" mobility number, the total-load formula is in agreement with many field observations carried out for flows transporting a heavy suspended load. It is just the case here in the Huanghe River estuary. These results suggest that the tidal zone around the Huanghe River mouth acts as a sediment trap as shown by satellite pictures and hydrographic data.

Based on the above-mentioned 2D numerical model, which has been verified by Le et al. (1993), the Eulerian tide-induced residual circulation is given in Fig. 5. Then the movement of marked water parcels is determined by Eqs. (4) and (5). Shown in Fig. 7 are the labelled water parcel trajectories computed in a Lagrangian sense over 2 periods of tidal constitute M2. Figs. 5 and 6 show the following interesting features.

1. The Eulerian tide-induced residual currents off the Huanghe River Delta basically follow the bathymetric contours. The strongest residual current appears over the shoal south of the Huanghe River mouth, reaching more than 20cm/s and flowing northeastward.

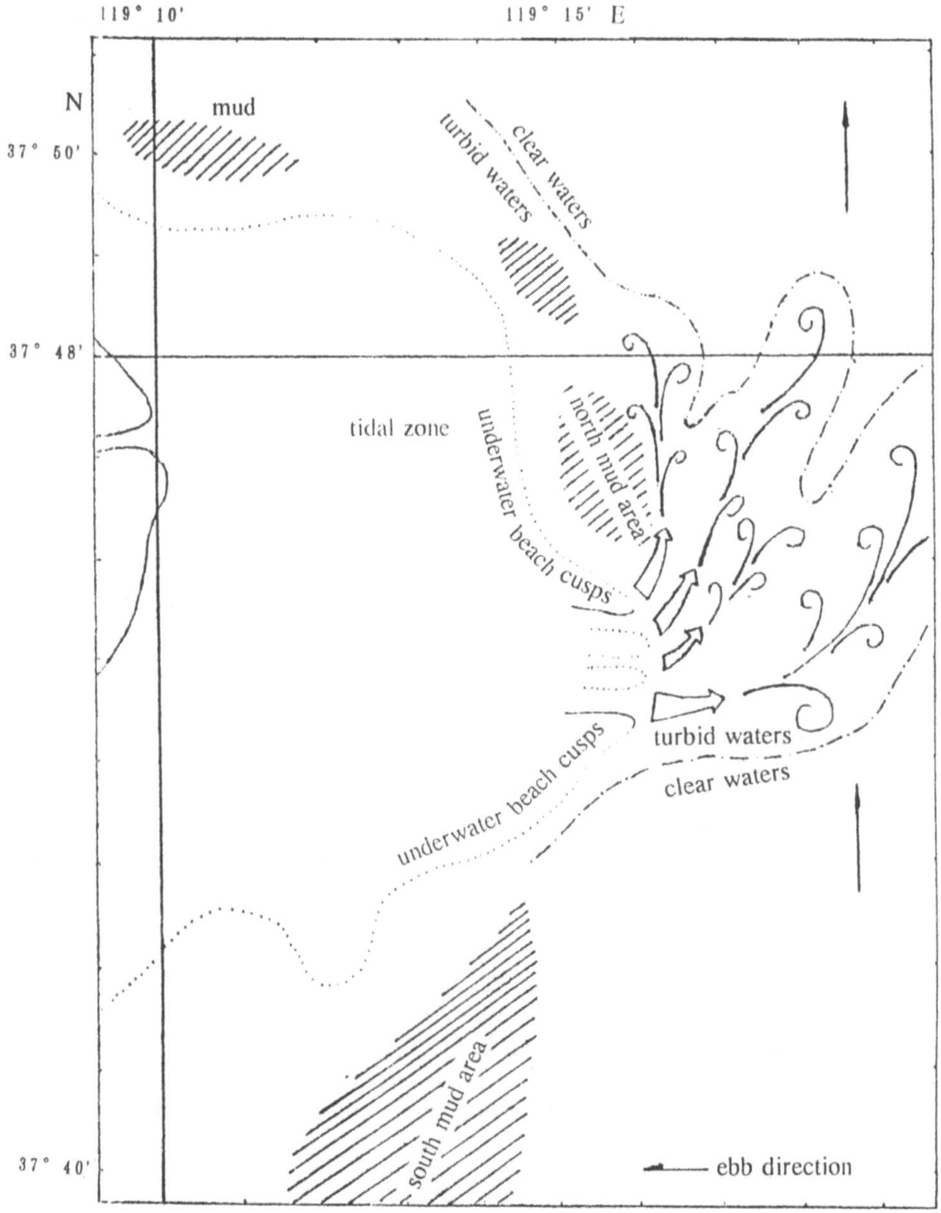

Figure 4a: Distribution of the turbid waters in the ebb period (at 7:30 Sep. 30, 1983)

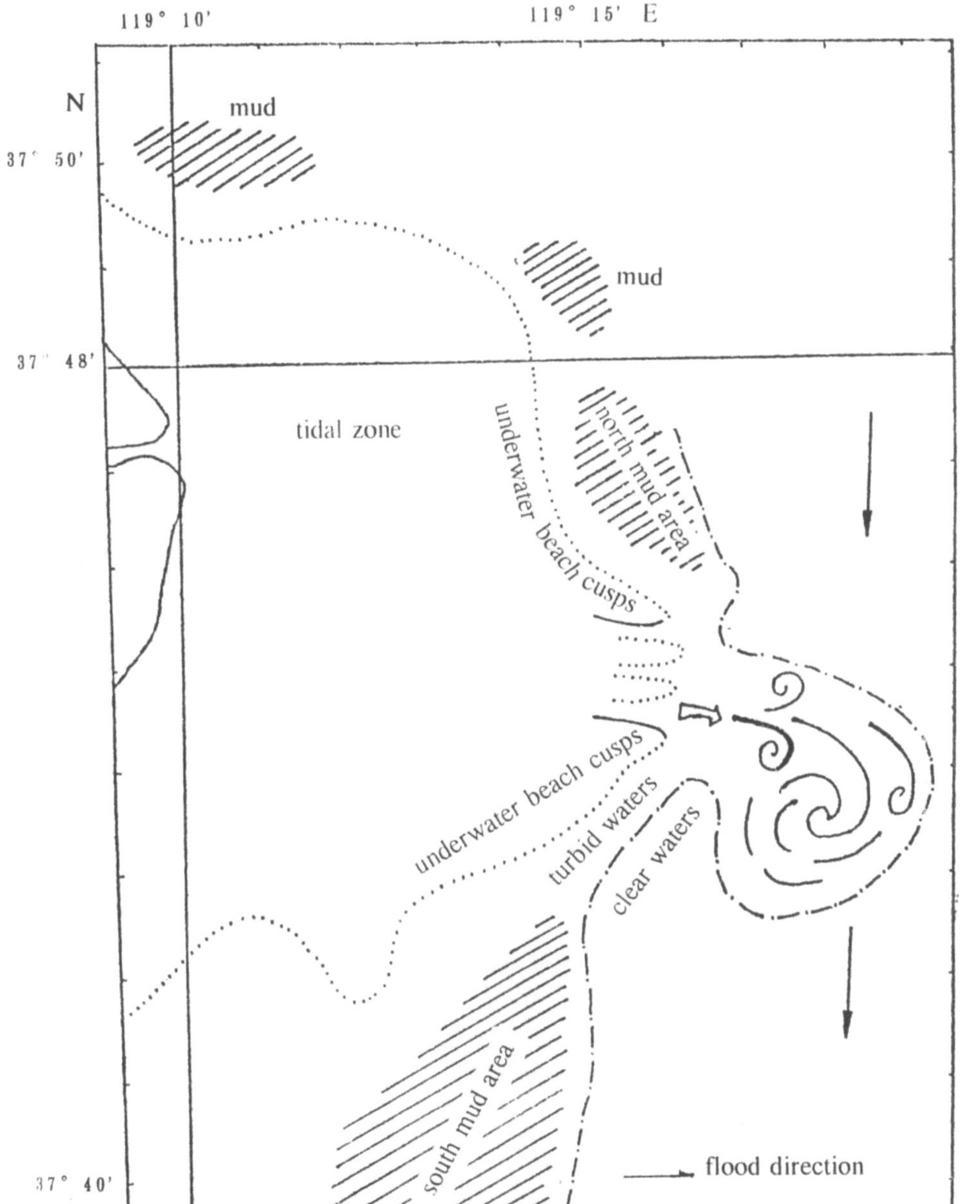

Figure 4b: Distribution of the turbid waters in the flood period (at 6:20, Oct. 9, 1983)

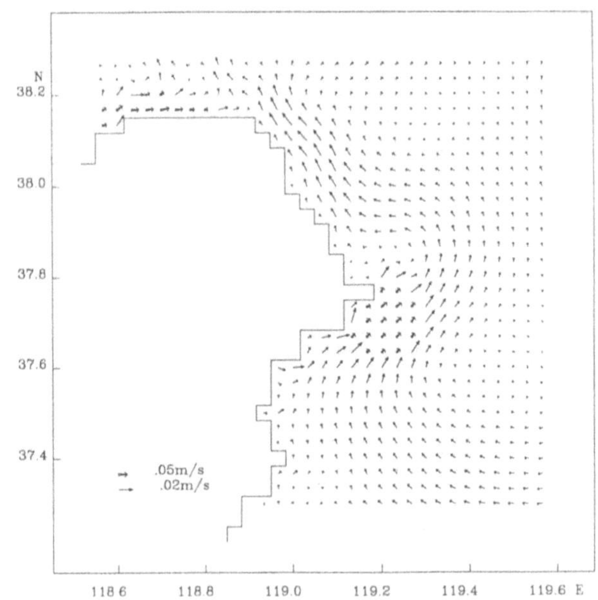

Figure 5: Eulerian tide-induced residual circulation with no discharge of the Huanghe river.

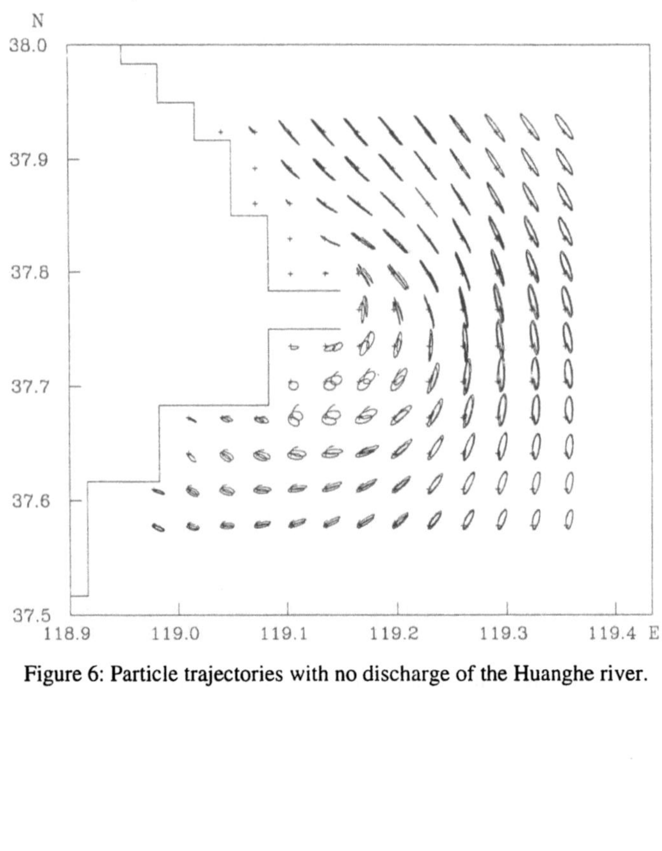

Figure 6: Particle trajectories with no discharge of the Huanghe river.

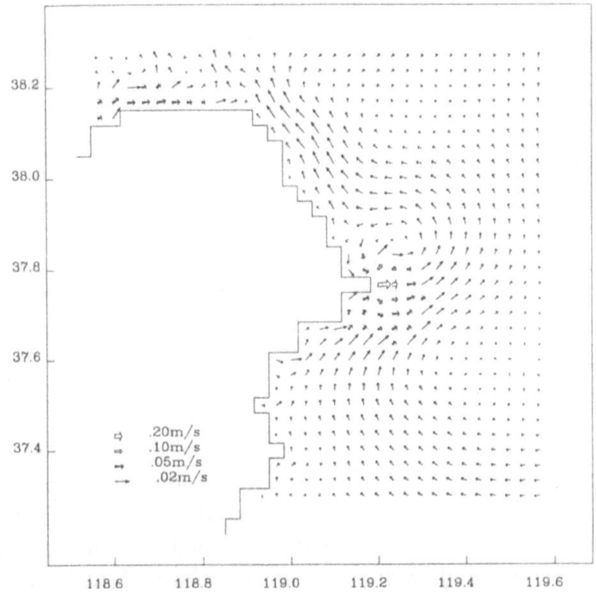

Figure 7a: Eulerian residual circulation in Case 1.

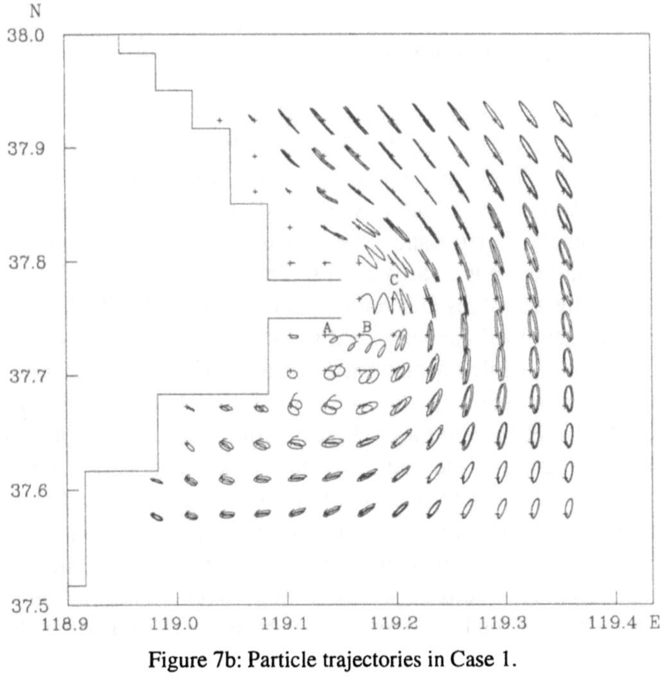

Figure 7b: Particle trajectories in Case 1.

2. The tracer trajectories seem to be classified into three types. In the nearshore region around the Huanghe River mouth, the tracer trajectories take a non-closed looped shape. The tracer trajectories in the offshore area off the Huanghe River Delta display an enclosed ring. The third kind of the tracer trajectories is bidirectional.

3. The distance of the tidal excursion is approximately 1-3 km in most of the fine net area while it is less than 1 km over the tidal zone.

4. The net displacement of water parcel, ranging from decades of meters to hundreds of meters, is smaller by an order of magnitude or much smaller than the tidal excursion in most of the area. While in the tidal zone, it is smaller than the tidal excursion as required, but it is large enough to be the same order of magnitude as the tidal excursion.

As pointed out by Zimmerman (1978), induced residual circulation in a periodic flow is primarily due to the nonuniform spatial distribution of amplitude or phase or of both. In the tidal zone off the Huanghe River Delta the rapid variations of bathymetry result in large spatial variations in tidal amplitude and phase angle. Though residual circulation is induced directly by spatial variation of amplitude and phase of tidal wave, the actual cause of the residual circulation are the nonlinear effects due to large variations in its basin bathymetry.

4.2 Simulation of the Movement of Suspended Load

To model the sedimentation process off the Huanghe River is a very complicated problem. Firstly, the sediment load carried by the Huanghe River has heavy concentration. A statistics shows that the average sediment concentration of the Huanghe River is 37.6kg/m3 (Qian and Dai, 1980). Secondly, the sediment load with heavy concentration differs greatly interannually. For instance, the maximum annual sediment load is 3.91´109 tons (the year 1933), and the minimum annual sediment load is only 488 million tons (the year 1928),as shown by the statistical data (Qian and Dai, 1980). Thirdly, the run-off discharge and the sediment load have seasonal marked variations. To be more precise, the oncoming load in a year is highly concentrated, forming several sediment peaks during the flood season. Fourthly, sometimes the run-off of the Huanghe River tail becomes intermittent because of human activities. These factors result in a complicated source of water and sediment. Therefore the sedimentation process is very complicated off the Huanghe River mouth. The major feature of the sedimentation off the mouth is the existence of the sand bar, north mud area and south mud area around the mouth. To understand this feature of the sedimentation, we take 3 sets of the data of the monthly-averaged river discharge and sediment concentration at Gauging Station Lijin, which is listed in Table 1, as a base of the boundary conditions at the current Huanghe River mouth for modelling.

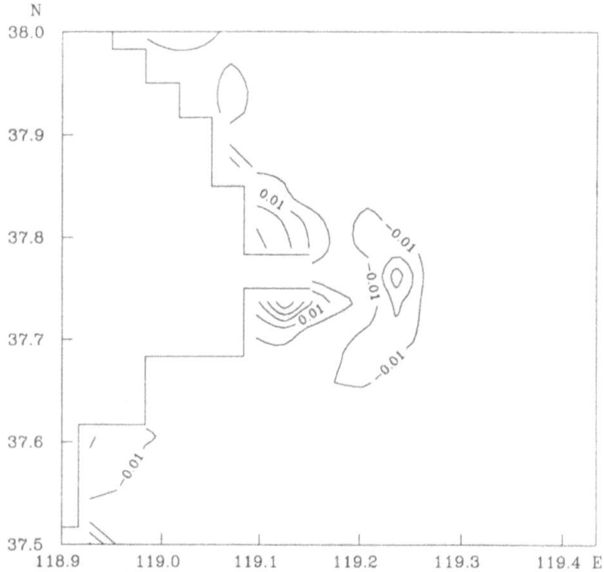

Figure 7c: Deformation of sea bed (η, in m) after 30 days in Case 1.

Table 1. Some typical data at G. S. Lijin for modelling sedimentation off the Huanghe River mouth

Case	1	2	3
Month (in the year 1984)	Dec.	Oct.	Aug.
Mean run-off discharge (m3/s)	941	2614	3808
Mean sediment concentration (kg/m3)	4.8	18.6	31.2

4.2.1 Case 1

Case 1 in Table 1 shows a typical condition in the wintertime Huanghe River. The run-off discharge, sediment load and sediment concentration reach their minima in winter. The modelling results are given in Fig. 7 describing the main features of sedimentation, which are quite consistent with the observed ones in this season.

The Eulerian residual circulation shown in Fig. 7(a) reveals the following influence of little run-off discharge on the circulation. (1) The current speed become stronger in the nearshore region off the river mouth.(2) A cyclonic gyre is formed to the northeast of the river mouth.

As shown in Fig. 7(b), the particle trajectories in the nearshore region around the river mouth have bigger net displacements, which reflect the influence of the run-off discharge.

The distribution of sediment concentration and the deformation of sea-bed in this case are given in Fig. 7(c) respectively, which were obtained after the model equations (6) and (7) had been integrated for 30 days. However, it is found from Fig. 7(c) that the elevation of sea-bed is asymmetric and there exist two silting areas located in both sides of the river, whose locations are consistent with that of the observed mud areas.

To understand the sedimentation process around the river, the evolutions of sedimentation at 5 points marked in Fig. 7(b) were computed. The evolutionary lines at points A, B, C are given in Fig. 8, where abscissa is the integrated time from 10 days on and three ordinates indicate the elevation of sea-bed (plus stands for silting and negative for scouring), the sediment concentration and the sea level in this case, respectively. It is found from Fig. 8 that (1) the variations in z and C are periodic but their phases are just opposite at all 5 points, which are in agreement with the observed ones; (2) the evolution of sedimentation at the 5 points can be classified into 3 types: "linear growth", "periodic growth" and "mixed growth". The evolution of sedimentation at point C belongs to "periodic growth type", although the envelope of its large amplitudes seems to have an increasing tendency, its rate of increase is very little as shown in Fig. 8(c). Fig. 8(a) shows the behavior of "linear growth type" at point A and the same behavior was shown at point E, but the growth rate is larger at point A than that at point E. Even if in case 1, there should exist two mud areas at both sides of the river mouth because of the growth tendency of the siltation at points A and E. The siltation at point D is basically the same as that at point B shown in Fig. 8(b), but the silting rate is much larger at point B than that at point D. All above-mentioned features are consistent with that indicated by the observed data (Le and Huang, 1993).

4.2.2 Case 2

During summer the Huanghe River empties its enormous sediment loads into the Laizhouwan Bay, with which as a boundary condition the sedimentation off the river mouth will be simulated by integrating the model equation here. In this case the open boundary conditions at the river mouth come from the data of case 2 in Table 1. In this case, the large run-off discharge has a significant influence on the current speeds around the river mouth but a little influence on the patterns of the circulations. Even in summer the run-off of the Huanghe River only affect the movements of particles just around the river mouth.

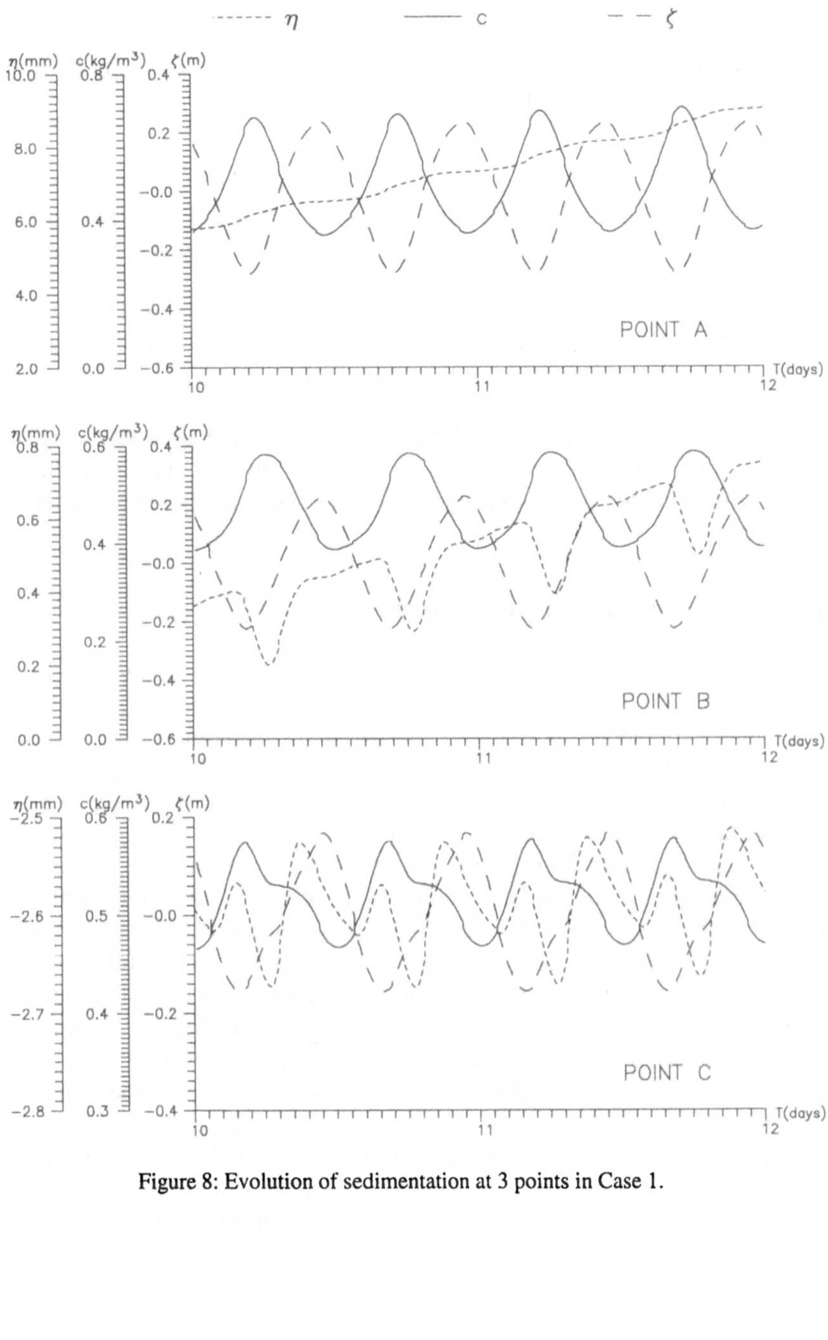

Figure 8: Evolution of sedimentation at 3 points in Case 1.

The sediment loads from the Huanghe River spread widely and quickly, which suggest that the sedimentation process be much quicker and wider in summer than that in winter. And the following marked features are different from that shown in Fig. 7(c). Firstly, the rate of deposition is much greater in summer than that in winter. Secondly, the deformation pattern of sea-bed is asymmetric, which means that the sediment loads deposited to the south of the river mouth are more than that to the north. This explains why the area of the south mud area is larger than that of the north one.

4.2.3 Case 3

The data of case 3 in Table 1 are typical of a sediment peak during the flood season since the 1980's. With these data the simulated results after 10 day's integral are given in Fig. 9.

The main feature of the residual circulation in case 3 is the existence of a little gyre located to the south of the river mouth as shown in Fig. 9(a), which is different from that shown in Figs. 7(a). There exist two high deposition areas in Fig. 9(b), which are located to the northeast and the southeast of the mouth, respectively. The measurements since 1984 show that the tail of the Huanghe River have been developing towards the southeast owing to the siltation in both sides of the river mouth. From Fig. 9(b) it can be expected that this fact will be bound to happen because of the behavior of the sediment deposition.

5. DISCUSSION AND CONCLUDING REMARKS

The main features of the sedimentation process off the Huanghe River mouth are quite well simulated by using the 2D model described above. Since the sedimentation process off the Huanghe River mouth is characterized by large run-off inflow and high sediment concentration, the challenge is how to specify the parameters and boundary conditions. In the sediment model the formulae of sediment transport capacity and settling probability are obtained from laboratory data following Duo (1963) and Duo (1987). Generally speaking, the sediment model given by Duo (1987) can be used in a wider range of current speed. Although the limitation of the model was not discussed by him, the fundamental equation of non-equilibrium transportation of non-uniform suspended load, which does not involve the diffusion-type transport explicitly, is adequate for the present study. However the sensitivity of the sediment model is one of the main problems in modelling sedimentation process. In this study parameters α_j and β_j are not so sensitive because of equation (18). In this case equation (18) is useful for the total sediment concentration. Since the open boundary condition (10) is formulated from the observed data, so the computed tidal current field is well consistent with the observed one and can be applied to predict the tidal current field in the study area quantitatively. However, to predict the development of the Huanghe River Delta the land boundary should be movable with the flood-and-ebb stream. Thus, the relevant parameters involved in the sediment model should be changed properly.

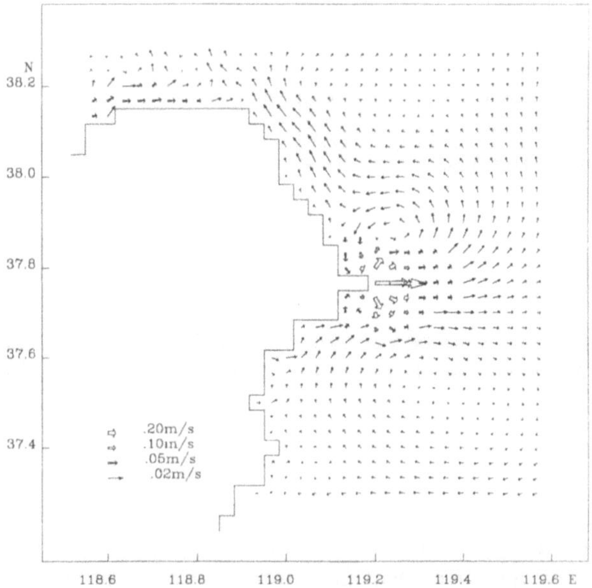

Figure 9a: Eulerian residual circulation in Case 3.

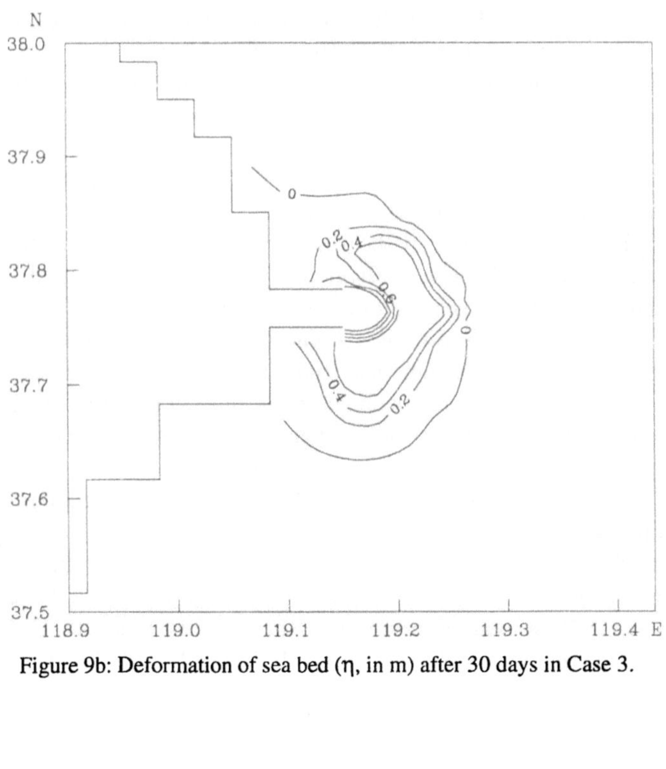

Figure 9b: Deformation of sea bed (η, in m) after 30 days in Case 3.

The present model should be improved in the following aspects to make the system be able to predict the development of the Huanghe River Delta.

(1) Include storm process;
(2) Formulate a three dimensional model system;
(3) Involve moving land boundary;
(4) Improve parameterizations.

ACKNOWLEDGMENT

The work in this paper was supported by the Provincial Natural Science Foundation of Shandong, China. We would like to thank Miss Yu Kangling for her careful typing.

REFERENCES

Burt T.N, 1986. Field settling velocities of estuary mud, in Ashish J. Mehta (Editor): Estuarine Cohesive Sediment Dynamics, Springer-Verlag. pp. 126-150.

Cheng R.T. and V. Casulli, 1982. On Lagrangian residual currents with applications in South San Francisco Bay, California, *Water Research* 18(6):1652-1662.

Dou G., 1963. Computation of suspended load movement and its scouring and silting in tidal flow. *J. Hydraulic Research*, 4. (in Chinese)

Dou G., Zhao S. and Huang Y., 1987. Study on two-dimensional total sediment transport mathematical model. *Journal of Nanjing Hydraulic Research Institute*, No. 2:1-12.(in Chinese)

Fang G. and T. Ichiye, 1983. On the vertical structure of tidal currents in a homogeneous sea. *Geophys. J. R. astr. Soc.* 73:65-82.

Le K., 1993. Influence of changes of the Huanghe River estuary on the Diluted Water distribution of the Huanghe River. *Studia Marina Sinica*. (Submitted - in Chinese)

Le K. and Huang S., 1993. Basic features of changes of the Huanghe River Delta in modern time and their relations to the current field, ibid.(in Chinese)

Le K., Liu X. and Shi J., 1993. A numerical study on influence of changes of the Huanghe River estuary on the tidal wave movement in its adjacent sea. ibid. (in Chinese)

Le K. and Zhou C., 1993. Influence of changes of the Huanghe River estuary on the current field in its adjacent shelf area, ibid. (in Chinese)

Leendertse, J. J., 1967. Aspects of a computational model for long-period water-wave propagation RAND Corp.165 pp.

Pang J. and Shi S., 1979. Fluvial process of the Huanghe River estuary, I. Changes in modern time, *Oceanologia et Limnologia Sinica*, 10 (2): 136-141. (in Chinese)

Pang J. and Shi S., 1980. Fluvial process of the Huanghe River estuary, II Hydrographical character and the region of sediment silting, ibid., 11 (4): 295-305. (in Chinese)

Pang J. and Shi S., 1982. Fluvial process of the Huanghe River estuary,III Influence of the evolution of the Huanghe River estuary on the lower reach, ibid., 13 (3):218-223. (in Chinese)

Partankar, S. V. 1980. *Numerical heat transfer and fluid flow*, McGraw-Hill.

Qian N. and Dai D., 1980. The problems of river sedimentation and the present status of its research in China. In: Proceedings of the International Symposium on River Sedimentation (March 24-29,1980, Beijing) Ed. by the Chinese Society of Hydraulic Engineering Guanghua Press. 1: 3-39.

Yalin M. S.,1972. *Mechanics of Sediment Transport*, Pergamon Press. pp.121-127.

Zhang Z., 1992. Tentative idea of comprehensive control of lower reaches of the Yellow River. *Coastal Engineering* 11(2):1-29 (in Chinese).

Zimmerman, J. T. F., 1978. Topographic generation of residual circulation by oscillatory (tidal) currents. *Astrophys. Fluid Dyn.*, 2: 35-47.

HAZARDS FROM SEA LEVEL RISE AND THEIR IMPACTS ON RESOURCES UTILIZATION IN THE YANGTZE RIVER DELTAIC PLAIN, CHINA

YANG GUISHAN

Nanjing Institute of Geography and Limnology, the Chinese Academy of Sciences. Nanjing 210008, People's Republic of China

Abstract. Sea level rise will cause or exacerbate a series of coastal hazards during the 21st century. The Yangtze River deltaic plain, China will be the region most vulnerable to accelerating sea level rise, in this region, relative sea level may rise 38-66 cm by the year 2050, while the eustatic sea level rises 20-30 cm. Accelerating sea level rise will increase the probabilities of occurrence of storm surge and the intensity of salt water intrusion into the Yangtze River Estuary; exacerbate coastal erosion; impede the drainage of floodwater of coastal lowlands and inundate tidal flat and coastal wetlands. This paper focuses the discussion on exacerbated coastal hazards by sea level rise and their impacts on resources utilization (land, water and living resources) in the Yangtze River deltaic plain.

1. INTRODUCTION

The mankind are living in a "Global Change" era, various natural and man-made hazards have brought about more and more losses of people's life and property. In the coastal area, for the one hand, global warming, caused by the atmospheric buildup of trace gases (CO_2, CH_4, N_2O, CFC_S, etc.) , can raise eustatic sea level (ESL) through thermal expansion of the oceans and the melting of glaciers; for the other hand, land subsidence, as a result of both nature and unreasonable human activities, will also accelerate the rising of relative sea level (RSL) in major river deltas of the world.

Sea level rise will create a series of environmental effects, such as exacerbate coastal hazards (storm surge, coastal erosion, salt water intrusion, etc.) , inundate coastal wetlands and lowlands, impede the drainage of floodwater of coastal lowlands and polluted water of seaside towns and cities, and then endanger socio-economic development and people's daily life. According to the estimates of the Environmental Protection Agency (EPA) , in the United States, the total losses (including the cost of protecting ocean resort communities and developed areas along sheltered waters and the inundated loss of coastal wetlands and undeveloped lowlands) for a one metre sea level

M. I. El-Sabh et al. (eds.), Land-based and Marine Hazards, 179–189.

rise will reach 270-475 billion US dollars, ignoring future development (Titus et al., 1991). In the Nile River delta, the rising of relative sea level during the 21st century will drown 22-26 percent of the habitable land and displace 19-24 percent of the Egyptian population. In Bangladesh, due to higher rate of local surface subsidence, relative sea level rise by the year 2100 can drown 26-34 percent of the land; displace 27-35 percent of the population and cause a loss of 22-31 percent of the Gross Domestic Product (GDP) in the affected area (Milliman et al., 1989). This potential threat will become a major natural disaster which the people living in coastal areas have to deal with before the 21st century.

The Yangtze River deltaic plain in China is densely populated and economically well develope, with the population density nearly 1500 persons per sq. km and mean GDP density over 12.5 million RMB yuan per sq. km (about $2.2 million /sq.km) , these densities are far higher than the average values in all of China. Shanghai, the biggest industrial and harbour city in China, is located in the centre of this region, where the population density exceeds 5000 persons/sq.km and the GDP density reaches 24 million RMB yuan/sq.km (about US$4.2 million/sq.km. Figures above-mentioned are all 1993 (Zhou et al., 1995)).

The topographic relief in the Yangtze River deltaic plain is low-flat, surface elevation (all adopt Wusong Datum Plane) is generally 1.5-4.5 m and the considerable part of habitable lands are below local mean high tidal level (approximately 3.0 m). At present, these coastal lowlands are all protected by sea dykes. Outside the dykes, the coasts belong to muddy marshes, where the typical tidal flat is developed, the area of tidal flat between the dykes and the datum plane of theoretical depth totals 2000 sq.km, making up about 10 percent of the total area of tidal flat in China (Figure 1).

The Yangtze River deltaic plain belongs to the remnant subsidence. In addition, a large amount of groundwater has been overpumped, the mean rate of land subsidence has reached 2-5 mm/a in recent forty years. It has exceeded 7 mm/a at Wusong area of Shanghai (Hu et al., 1990). Due to the effects of various natural and man-made factors, the mean rate of land subsidence can reach 3-6 mm/a during the first half part of the 21st century (Xie, 1992). So, if eustatic sea level rises 20-30 cm by the year 2050, the relative sea level in this region will rise 38-66 cm, doubling ESL rising value.

2. COASTAL HAZARDS EXACERBATED BY SEA LEVEL RISE

2.1 Storm Surge

The Yangtze River deltaic plain is one of the regions which suffer the heaviest casualties and losses from storm surge in China. The probabilities of occurrence of tropical storm is second only to the coastal area of South China. Especially in Shanghai area, a combination of flood, astronomical high tidal level and typhoons will bring about

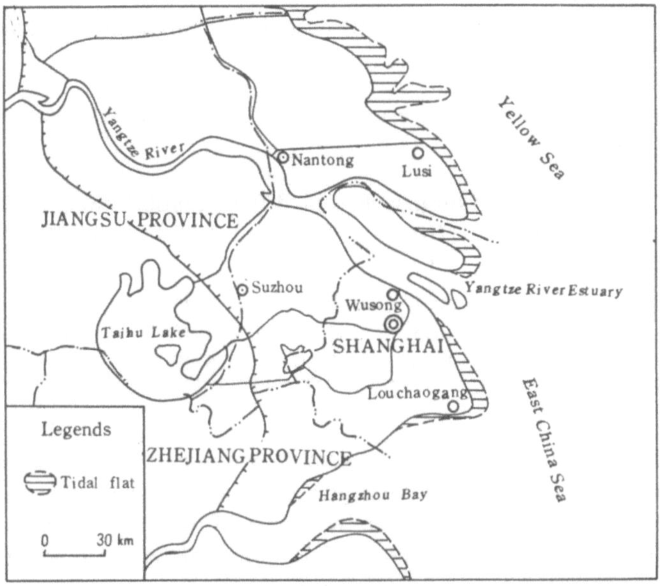

Figure 1 Sketch map showing the Yangtze River deltaic plain

enormous losses. As an example, in August, 1956, an extraordinarily serious storm surge caused 4626 persons to die and over 20 thousand to be injured in Shanghai and its adjacent areas. Yuan (1991) estimated according to the historical records that the losses by storm surge have constituted approximately 70 percent of total ones by natural disasters in Shanghai area.

The observed data indicate that there are obvious rising trends in storm surge level of Shanghai area. At Huangpu Park tidal gauge station, for example, the observed yearly highest sea level was 4.65 m in 1954; 4.76 m in 1962; 4.98 m in 1974 and 5.22 m in 1981. In the 1950's and 1960's, yearly highest sea level over 4.65 m only occurred once in every decade, but it occurred twice and four times in the 1970's and 1980's respectively. The results obtaoined from numerical calculation of tidal current and storm surge in the Yangtze River estuary (Zhu et al., 1995) show that, although sea level rise will cause some slight decrease in surge wave setup (it may on average decrease by 9 cm in the northern coast and 6 cm in the southern coast of the Yangtze River Estuary, while future sea level rises 50 cm) and change the process of the astronomical tide, its effects on the highest and lowest tidal level are very small. So the probabilities of occurrence of the storm surge level at a given height will surely have a great increase due to the rising of initial mean sea level. According to the results of numerical calculation mentioned earlier and the statistical analysis of yearly highest tidal level observed at Wusong (Shanghai) and Lusi (Jiangsu Province) tidal gauge stations, a 50 cm rise in sea level may increase the probability of occurrence of storm surge from 1/100 to 1/10-1/20 in the southern coast and from 1/100 to 1/20 or so in the northern coast of the Yangtze River Estuary (Table 1).

Table 1: The impacts of sea level rise on storm surge in the Yangtze River deltaic plain (Wusong Datum Plane, m)

Coast section	SLR	Storm surge level with different probabilities of occurrence					
	(cm)	1/5	1/10	1/20	1/50	1/100	1/1000
The southern	0	5.02	5.17	5.32	5.54	5.69	6.24
coast	50	5.46	5.61	5.76	5.98	6.13	6.68
The northern	0	4.22	4.45	4.68	4.96	5.17	5.88
coast	50	4.63	4.86	5.09	5.37	5.58	6.29

Presently, sea dykes in this region are designed to withstand a storm surge level with a probability of occurrence of 1/1000 in Shanghai's urban and industrial areas and 1/100 in the rural and agricultural areas. As shown by table 1, if future sea level rises 50 cm, the defensive capacity of the dykes will become 1/100 or so in the urban and industrial areas and 1/20 or so in the rural and agricultural areas respectively.

2.2 Coastal Erosion

Coastal erosion can cause the loss of a large amount of beaches and coastal wetlands, accelerate the deposition of port pools and channels, as well as damage various coastal projects etc.. In the Yangtze River deltaic plain, the length of eroding coast is about 55 km, making up approximately 10 percent of the total length of coastline. Due to the protection of sea dykes, the retreat of coast in this region has effectively been controlled; but the vertical erosion of beaches is very serious. In the coastal area near Lusi, mean elevation of beaches outside sea dykes had declined from 4.5 m to 2.1 m during 1966-1981, the average eroding rate exceed 16 cm/a (Ji et al., 1993).

There are many conditions favourable to coastal accretion, such as the large amount of silt from both the Yangtze River estimated at 480 million tons per year and the undersea delta of Abandoned Yellow River in the North and the protection of huge radiating sand-bars outside the northern coast of the Yangtze River Estuary, etc.. However, the coasts of this region are still subjected to erosion and its intensity is continuously increasing. Undoubtedly, sea level rise has played an important role.

The "Bruun rule" is widely used to predict the coastline response to sea level rise. It states that a typical concave-upward beach profile erodes sand from the beachface and deposits it offshore, so as to maintain constant water depth (Figure 2; Bruun, 1962, 1983). The estimates applying this rule show that, a 1 cm rising in sea level can result in 8.3-9.1m retreat of the coastline in the north and 2.6-16.7 m retreat in the south coasts of the

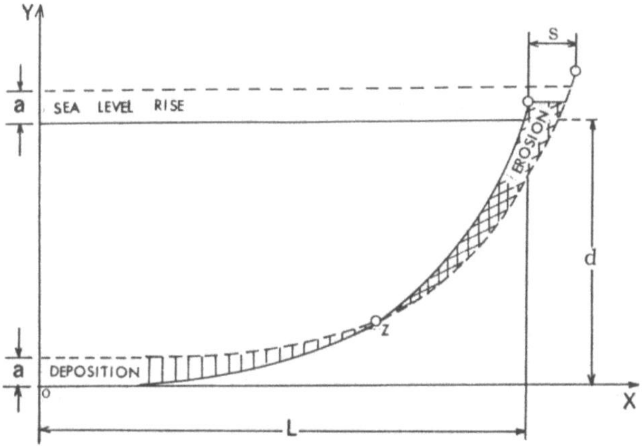

Figure 2 Schematic illustration of the coastline response to sea level rise, according to the "Bruun rule". a=sea level rise; s=total coastline displacement; L=length of beach profile; d=height of profile.

Yangtze River deltaic plain, assuming the coasts are not protected by sea dykes (Ji et al., 1993). In fact, due to the limits of sea dykes, the coastline in this region can retreat no more, coast erosion exacerbated by sea level rise will mainly be vertical erosion of the beaches. If relative sea level rises 50 cm, the contribution of sea level rise to the total impacts of all factors on coastal erosion will increase from 9-10 percent at present to 50-54 percent by 2050. As the rate of sea level rise increases in the future, its contributions to the total coastal erosion will also greatly increase (Ji et al., 1993).

2.3 Salt Water Intrusion into the Yangtze River Estuary

The Yangtze River Estuary is an opening river mouth without tide-control gates. During dry season, the runoff of the Yangtze River is obviously diminished and higher estuarine chlorinity can contaminate water supplies for both urban and rural areas. For example, during the period between winter 1978 and spring of 1979, the intrusion distance of salt water into the Yangtze River reached 170 km due to extreme drought. The whole Chongming Island was surrounded by salt water for more than 150 days. The highest chlorinity at Wusong water intake of Shanghai was nearly 4000×10^{-6}, 16 times over the drinking water standard ($CL<250\times10^{-6}$) and the total time of daily average chlorinity over 250×10^{-6} reached 102 days.

Observed data indicate that the intensity of salt water intrusion into the Yangtze River will increase associated with a decrease of the Yangtze River discharge and a rise of the tidal level in the Estuary, so future sea level rise will surely exacerbate this hazard. The estimates show that, if sea level rises 50 cm, the intrusion distances of the isohalines of 1‰ and 5‰ in dry season will be increased by 6.5 km and 5.3 km respectively, in the dry season of dry year. The two isohalines may respectively reach Gaoqiao (about 70 km from Yinshuichuan station outside the Estuary) and Xiaojiuduan (about 59 km from Yinshuichuan) upstream. Daily average chlorinity at Wusong intake will increase from $131\text{-}171\times10^{-6}$ at present to $206\text{-}276\times10^{-6}$ in the future, approximately 60 percent higher than the current values (Yang 1991, 1992; Yang and Zhou, 1993).

2.4 Episodic Flooding

As stated above, surface elevation in the Yangtze River deltaic plain is generally between 1.5 and 4.5 m. Some lowlands, for example the eastern lowland of the Taihu Lake area, with mean elevation only 3.0 m or less, are easily exposed to floods combined with an unbroken rain in the season of plum rains (the rain season, usually in May and June in the middle and lower reaches of the Yangtze River) or heavy storm rain. During the serious flood of August, 1991, as an example, the cultivated lands of 2100 sq.km in the eastern lowland of the Taihu Lake area were submerged, occupying 32 percent of total cultivated area of the lowland (Liang and Li, 1993).

As a result of sea level rise, the high and low water levels in the tidal river reaches are expected to rise and the streaming action of tidal current intensified. These may result in the reduction of the drainage capacity of lowlands and exacerbating flood and waterlogging disasters. A numerical calculation shows that, a 40 cm rising in relative sea level can decrease by 20 percent of the natural drainage capacity of the eastern lowland of the Taihu Lake area. Under the condition of the current flood-control works, if this region are hit by the floods of 1991-type again, the total submerged area will reach 2700 sq.km and the increased area by sea level rise will exceed 600 sq.km, making up 30 percent of total submerged area (Zhu et al., 1995).

Surface elevation in Shanghai urban district is generally 1.4-1.9 m, the lowest area is below 0.7 m (De, 1991). The current standard of flood-control walls of the urban districts was designed on the basis of a water level of 5.86 m at the Huangpu Park tide gauge station. By taking into account the highest tidal level in historical records and a 50 cm rising in sea level, the storm surge level at the Huangpu Park station will reach 5.72 m or so. Besides the effects of increasing wave height, not only the security of flood-control walls but also the draining efficiency of urban drainage systems will markedly be affected. As a result, the probabilities of Shanghai urban districts hit by floodwater will greatly increase.

2.5 Permanent Inundation

Permanent inundation is a direct impact of sea level rise on the coastal area. The inundated losses will mainly depend on the local gradient, surface elevation and the rising value of relative sea level. Although the habitable land in the Yangtze River deltaic plain is low-lying, its inundated risk is still very little due to the protection of sea dykes. Only the tidal flat and coastal wetlands outside sea dykes may partly be inundated due to sea level rise. This inundated area will be related to the rising rate of relative sea level vs. the rate of vertical accretion or sedimentation of beaches. In the coasts around the Yangtze River Estuary where the rate of accretion of beaches is more than that of relative sea level rise, the existing tidal flat will not be inundated. However, decrease in the rate of accretion due to relative sea level rise can cause the potential losses of tidal flat. Approximately 35 sq.km of tidal flat will be lost, assuming a 50 cm rising in relative sea level by 2050. In the northern and southern coasts with the rate of accretion of tidal flat lower than that of relative sea level rise (including the eroding coasts) , the inundated area will reach nearly 18 sq.km, so if relative sea level rises 50 cm, the total inundated area of tidal flat in this region will reach 53 sq.km (Ji et al., 1994).

The affected extent of sea level rise on coastal wetlands depends on the rates of both loss and development. If the rate of wetlands development can generally keep pace with the rising rate of relative sea level, they will not be lost. In the Yangtze River deltaic plain, development of landward coastal wetlands is impeded by sea dykes, so sea level rise will also cause a loss of considerable part of coastal wetlands. Ji et al. (1994) estimated that a 50 cm rising in relative sea level will cause a loss of approximately 25 sq.km of wetlands, making up 6 percent of total area. The loss types of coastal wetlands will mainly be *Spartina angelica* marshes and reed swamp.

3. IMPACTS ON RESOURCES UTILIZATION

3.1 Land Resources

Land resources in the Yangtze River deltaic plain are in great demand due to densely population and developed industrial and agricultural systems. The total land area is about 22, 800 sq.km, including nearly 2, 000 sq.km of tidal flat, in which only 57% of the land are cultivate. Cultivated area per capita is less than 500 sq.m, far below the 1000 sq.m per person of the average value of all China. 14% of the land are used to both urban and rural residential areas, industrial production and roads; and the other 29% are forest farm, pastureland and inland waters. Cultivated area is decreasing while land for urban and industrial development is rising. So, enclosing tidal flat and coastal wetlands in this region has become the most important means of expanding the cultivated area during the past more than one thousand years or so. For example, in Shanghai, newly increased area through enclosing tideland has reached 670 sq.km since the year 1950, occupying 11% of the total land area.

However, coastal erosion and inundation exacerbated or caused by sea level rise will result in the loss of a large amount of tidal flat and coastal wetlands. Our results show that a 50 cm rising in relative sea level will result in a loss of 80 sq.km of tidal flat in the Yangtze River deltaic plain. This represents 5% of the total area of tidal flat, in which, the loss caused by coastal erosion is about 27 sq.km and that due to inundation and decrease in accretion totals 53 sq.km. Meanwhile, disappearance or decrease in elementary types caused by the loss of the middle and lower tidal flat requires the degeneration of advanced ones to be compensated. So, sea level rise will also result in the adverse succession of coastal ecosystems. The original *Spartina angelica* marshes may degenerate into bare beaches in the northern coast and the original reed swamp may degenerate into *Soirpusx mariqueter* marshes or bare beaches in the southern coast of the Yangtze River deltaic plain. Correspondingly, due to the fact that degenerative wetlands can not be compensated, the area of wetlands, especially the ones with higher use value, will obviously decrease. Ji et al. (1994) estimated that a 50 cm rising in relative sea level may cause a loss of 19 sq.km of *Spartina angelica* marshes and 6 sq.km of reed swamp in this region, so the total loss area of wetlands will reach 25 sq.km, representing 6% of the total area (Table 2).

Table 2 The estimates of the losses of land resources caused by 50 cm of sea level rise in the Yangtze River deltaic plain (According to Ji et al., 1994)

Types	total area (km^2)	loss area (km^2)	loss rate (%)
tidal flat	1550	80	5.2
coastal wetlands	420	25	6.0

In addition, sea level rise will also result in the degradation of land quality owing to the rising of water level and the landward moving fresh-saline water front of coastal aquifers. For example, in the eastern lowland of the Taihu Lake area, about 400 sq.km of cultivated land will be affected by salinization for 80 cm rising in relative sea level. Yearly mean loss of grain yield may reach 50, 000 tons (Mao, 1992).

3.2 Water Resources

The Yangtze River deltaic plain is located in humid climate region of the eastern coast of China. Yearly average precipitation is approximately 120 cm. Although per capita local water resources is not plentiful because of densely population (only 800 m^3 per person, analogous to one-third of average value of the whole China. Yang, 1991) , there is plenty of exterior runoff from the upper and middle reaches of the Yangtze River pass through this region, representing about 40 times the local runoff. Plentiful exterior runoff provides reliable sources of water supply for this region and makes up the inadequacy of local water resources to a certain extent. However, because the variations with time of the external runoff almost keep pace with that of the local runoff, in wet season or year, the external runoff pass through this region is also very plentiful. It may exacerbate local flood and waterlogging, in dry season or year, so that the external runoff is also inadequate and can only play a limited role in mitigating drought damage and salt water intrusion into the Yangtze River Estuary. The shortage of water resources in this region during dry season or year is still serious.

Future sea level rise will impede the drainage of flood water of coastal lowlands and polluted water of seaside towns and cities and exacerbate salt water intrusion into both river mouth and the coastal aquifers. All these will surely aggravate the pressure of water supply. In the Yangtze River deltaic plain, the most obvious impacts of sea level rise on water supply will be the exacerbation of salt water intrusion into the Yangtze River Estuary. At present, the water intakes of Shanghai are mainly distributed in the reaches near Wusong of Huangpujiang River. Associated with the vigorous development of Pudong new district of Shanghai, daily mean value of water supply will increase from 5 million cubic metre today to 10 million cubic metre by 2030. Limited water resources of Huangpujiang River will not satisfy the needs of future development, so the South Branch of the Yangtze River Estuary will be the only reliable water sources of Shanghai. Our research shows that the probabilities of hourly chlorinity over 200×10^{-6} (the upper limit chlorinity for industrial water supply) and 250×10^{-6} (the upper limit of drinking water standards) at Wusong intake during dry season will respectively increase by 73-85% and 83-100%, assuming a 50 cm rising in sea level, during the dry season of dry year, the two probabilities may respectively increase from 36% and 32% today to 88% and 70% in the future (Table 3).

Table 3 Effects of sea level rise on the probabilities of different chlorinity values at Wusong water intake of Shanghai during dry season (According to Yang et al., 1993)

Sea level rise (cm)	Probabilities (%)					
	$CL \geq 200\times10^{-6}$			$CL \geq 250\times10^{-6}$		
	dry year	normal year	wet year	dry year	normal year	wet year
0	> 36	15-24	< 7	> 32	12-20	< 3
50	> 88	26-44	< 5	> 70	22-41	< 5

3.3 Living Resources

Enormous silt and nutrient materials flowing into the sea from the Yangtze River provide suitable conditions for the reproduction and growth of living beings in the coastal zone of the Yangtze River deltaic plain. They enable this region to have not only abundant aquatic resources (such as fish, shrimp, shellfish and algae etc.) but also various resource plant (such as medicinal herbs, perfume plant, fibre plant and forage grass etc.), rare birds and wild animals (red-crowned crane and Davids deer etc.). At present, only existing total volume of animal resources living in the middle and lower tidal flat exceeds 90, 000 tons, and the annual biological yield reaches 57.2 g/m^2.

Future sea level rise will cause marked decrease in the volume of living resources. The estimates according to annual biological yield show that a 50 cm rise in sea level will decrease the existing volume of animal resources by nearly 5, 000 tons annually due to the loss of tidal flats. Furthermore, losses of the coastal wetlands will also cause great losses of living resources. In the Yangtze River deltaic plain, coastal wetlands are mainly used to protect the environment, enclose for cultivated land, produce reeds and breed aquatic product. A 50 cm rising in sea level will not only annually lose 7, 500 tons of reeds and 1300 tons of clamworms but also worsen the habitat of rare birds and wild animals (Zhu et al., 1995).

4. CONCLUSIONS

Sea level rise will be one of the major natural disasters in global coastal areas during the 21st century. Its rate and affecting pattern and extent will be different in each coastal area due to various environmental and socio-economic factors. Therefore, the response strategy to future sea level rise must vary from place to place.

The Yangtze River deltaic plain will be vulnerable to future sea level rise. The rising rate of relative sea level can be double the global mean value owing to rapid natural and man-made land subsidence. Sea level rise will markedly exacerbate storm surge, vertical erosion of beaches and salt water intrusion into the Yangtze River Estuary in this region. It will also aggravate flood damage of coastal lowlands and inundate tidal flat and coastal wetlands outside sea dykes. All these hazards resulting from sea level rise will undoubtedly endanger land, water and living resources and their utilization with a consequent impact on the socio-economic sustainable development and people's daily life in this region.

REFERENCES

Bruun P. (1962) Sea level rise as a cause of shore erosion. American Society Civil Engineers Proceedings, *J. Waterways and Harbours Division*, **88**, 117-130.

Bruun P. (1983) Review of conditions for uses of the Bruun Rule of erosion. *Coastal Engineering*, **7**, 77-89.

De Z.G. (1991) A summary of urban flood prevention of Shanghai City. Yangtze River, **22** (8) , 14-17.

Hu H.M., L.R.Huang, G.H.Yang (1990) Recent crustal movements in the Yangtze River Delta and its adjacent area. In: Studies on Climatic and Sea Level Changes in China (I) , Y.F.Shi, editor, China Ocean Press, Beijing, pp.64-65.

Ji Z.X., Z.X. Jiang, J.W. Zhu and G.S.Yang (1993) Impacts of sea level rise on coastal erosion in the Changjiang Delta and North Jiangsu Coastal Plain. *Acta Geographica Sinica*, **48** (6) , 516-526.

Ji Z.X., Z.X. Jiang, J.W. Zhu and G.S.Yang (1994) Impacts of sea level rise on tidal flat and coastal wetlands in the Changjiang River Delta and its adjacent area. *Oceanologia et Limnologia Sinica*, **25** (6) , 582-590.

Liang L.J., H.Y.Li (1993) 91' Catastrophic Flood in Taihu Lake Basin, 1st Edition, Hehai University Press, Nanjing, 31-37pp.

Mao R. (1992) The effect of sea level rise on drainage system around the Eastern Lowland of Taihu Lake area. In: Studies on Climatic and Sea Level Changes (II) , Y.F. Shi, editor, China Ocean Press, Beijing, pp 88-89.

Millman J.D., J.M. Brodous, F. Gable (1989) Environmental and economic impact of rising sea level and subsiding delta: the Nile and Bengal. *Ambio*, **18**, 340-345.

Titus, J.G., P.A. Park and S.P.Leatherman (1991) Greenhouse effect and sea level rise: the cost of holding back the sea. *Coastal Management*, **19**, 174-204.

Xie Z.R. (1992) A preliminary study on the trends of relative sea level in the Yangtze River Delta and its adjacent area. In: Studies on Climatic and Sea Level Changes (II) , Y.F. SHI, editor, China Ocean Press, Beijing, pp 79-80.

Yang G.S. (1991) The water utilization and its management in the Yangtze River deltaic area. *Natural Resources*, **3**, 41-47.

Yang G.S. (1992) The changes of discharge and tidal level in the future and their impacts on water quality in the Yangtze River Estuary. *Geographical Research*, **11** (2) , 68-76.

Yang G.S., J.W. Zhu (1993) A study of impacts of global sea level rise on salt water intrusion into the Yangtze River Estuary. *Science in China (Series B)* , **36** (11) , 1391-1401.

Yuan Z.L. (1991) Storm surge and its mitigation of Shanghai area. History of the Yangtze River, **4**, 50-64.

Zhu J.W., Z.R.Xie et al. (1995) The trends of sea level change and evaluation on its impacts in the Yangtze River Delta. In:Sea Level Change in China:Past, Present and Future, Y.F.Shi, editor, Shangdong Science and Technology Press, Jinan, pp.249-348.

THE TRAPPING OF OIL AT DAWHAT AD DAFI - AN ANALYSIS OF RESULTS FROM THE 1991 ARABIAN GULF OIL SPILLS

S. VENKATESH
Environment Canada, 4905 Dufferin Street, Downsview, Ontario M3H 5T4, Canada

and

T.S. MURTY
National Tidal Facility, The Flinders University, GPO box 2100, Adelaide 5001, Australia

Abstract. The Gulf war of 1991 resulted in some of the worst oil spills in the Arabian Gulf. While the amount of oil spilled was one of the largest in the world and the damage extensive, the desalination plants near Al-Jubail on the Saudi Arabian coast were not affected by oil contamination. Contamination of these plants would have threatened the water supply not only to the Saudi population but also to the military personnel that were involved in fighting the war. The amount of oil reaching this area of the Arabian Gulf was rather minimal compared to the total quantity of oil that was actaully spilled into the Gulf waters.

Most of the spills originated in the northern portions of the Gulf in mid to late January 1991 and travelled down the west side of the Arabian Gulf affecting its coastline along the way. By mid February the oil had reached the area of Abu Ali island protruding out into the Gulf just north of the desalination plants. The space between the mainland and Abu Ali island was blocked by a causeway. This area, known as Dawhat ad Dafi, would be an ideal catchment area for the oil if the oil could be steered that way.

Through numerical model simulations this paper examines the role of the winds and tides in influencing the accumulation of oil in the Dawhat ad Dafi area. It is found that the changes in the wind direction from their climatological norm of northwesterlies, particularly at the time the oil reached the Ad Dafi area, played a critical role in the accumulation of oil in that area. The numerical simulations did not show any significant contribution from the tidal currents towards this accumulation. Experiments carried out by the authors to determine the impact of the cuaseways blocking the area between the Saudi mainland and Abu Ali island showed that the presence of these causeways did not produce any material change in the tidal current magnitudes, but they did prevent the oil from escaping southward from that area.

1. INTRODUCTION

The 1991 Gulf war resulted in oil spills several times the size of the Exxon Valdez spill where about 11 million gallons of oil was spilled. Estimates of the spill size during the

M. I. El-Sabh et al. (eds.), Land-based and Marine Hazards, 191–202.

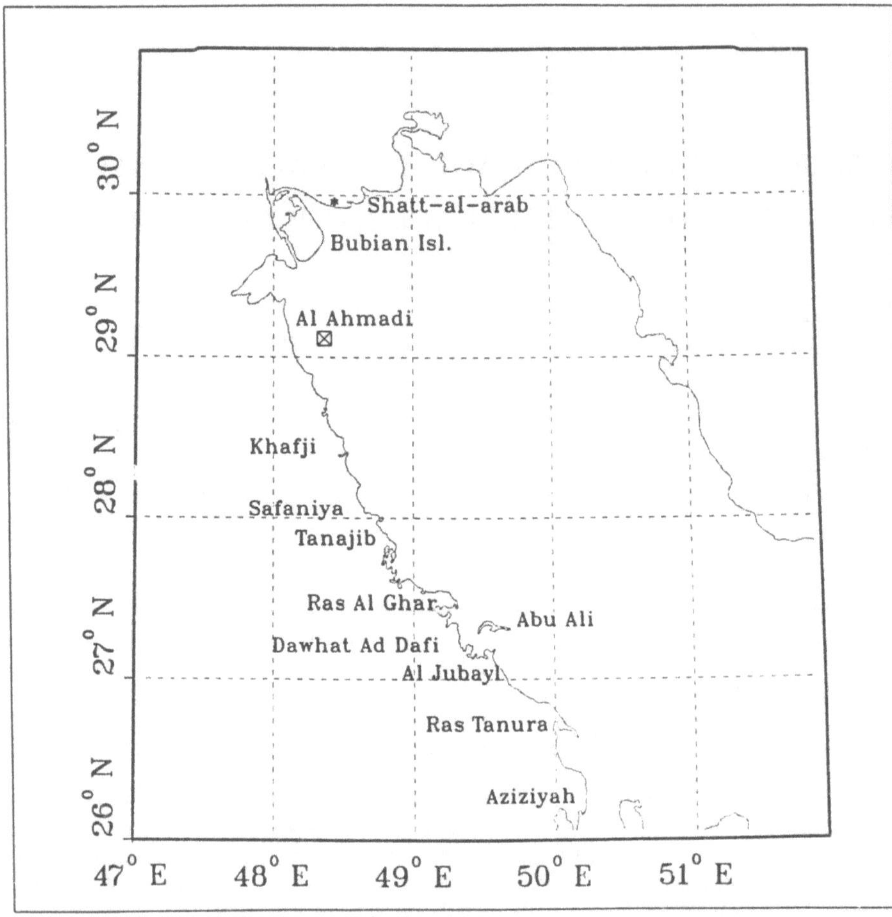

Fig.1. Western coastline of the Arabian Gulf affected by the 1991 oil spill.

1991 conflict range upwards of 100 million gallons (O'Brien and Gallagher, 1993). Beginning on January 20, 1991 oil was spilled at more than one location in the northern part of the Arabian Gulf. A major portion of the oil was moving southward near the coast of Saudi Arabia and by late January was affecting industrial water intakes in the headlands (the word Ra's) of Safaniya and Tanajib (Figure 1). It was feared that the oil would eventually pose a danger to the water intakes of the desalination plants at Al Jubayl, just south of Abu Ali island and at Aziziyah near Dhahran. These deslination plants are a major source of water supply for a large portion of the population in the central and eastern regions of Saudi Arabia.

As luck would have it, not only did very little oil go past the eastern side of Abu Ali but also most of the oil got trapped in the area between the island and mainland Saudi Arabia. This area, known as Dawhat Ad Dafi (the word Dawhat means large tree with a lot of branches), is rich in fishery including shrimp and comprises a shallow estuarine system with mud flats and mangrove stands. The causeways connecting the island and the mainland provided a barrier between the northern and southern portions of the Gulf in that area. Thus this area provided a natural barrier for any oil that got there. A significant amount of oil from the northern spill sources did arrive in this area, with oil thicknesses building up to nearly 2 metres. Such high oil thickness enabled easy recovery by skimmers. O'Brien and Gallaher (1993) give a detailed account of the methods employed in the trapping and removal of the oil arriving in the Dawhat Ad Dafi area.

Climatological data show that during the January-March period the prevailing winds are generally northwesterly. Model simulations carried out with such winds predict the oil to stay offshore away from the Saudi Arabian coast. The question arises as to why a large portion of the spilled oil ended up in the Dawhat Ad Dafi area. O'Brien and Gallagher (1993) suggest the trapping of oil was caused by tidal flow in the area. In a recent study Venkatesh and Murty (1994) found that winds extracted from weather maps during the period of the spill had a significant amount of directional variability. The wind direction, compared to the climatological wind direction of northwesterly, varied from northwesterly through easterly to south-southeasterly. Is it possible that it was the winds that were instrumental in trapping the oil? In this paper we investigate the impact of the winds on the trapping of oil in the Dawhat Ad Dafi area. Attention will be particularly focussed on the middle third of February when the oil reached the Dawhat Ad Dafi area and started accumulating there. We also investigate, through model simualtions, the role of tidal currents as the oil reached the Dawhat Ad Dafi area.

2. CLIMATOLOGICAL VERSUS ANALYSED WINDS

In the area of the Arabian Gulf oil spills the climatological wind speed and direction for January are 5.3 m s^{-1} and 142 degrees (winds blowing towards the southeast or from the northwest). The winds change very slightly for February and March with values of 5.5/132 and 5.0/132 respectively. In general, for any given location, the real-time winds can be quite different from the climatological values for that area. In the case of the Arabian Gulf, given the war-time conditions, it was very difficult for us to obtain information on the real-time observed or forecast winds in the area. The primary source of meteorological data at the time the spill was in progress was the weather charts that were made available through the Canadian Department of National Defence beginning on February 7, 1991. 6-hourly geostrophic winds were manually extracted form these hand-anlysed pressure charts. Figure 2 shows the combination of climatological and manually extracted winds for the period January 18 - March 1. It is quite apparent that there is a considerable amount of variability in the real-time winds compared to the climatological values, particularly in direction. From Figure 2 it can be seen that the transition from climatological winds to the real-time winds on February 6 is rather abrupt. The effect of

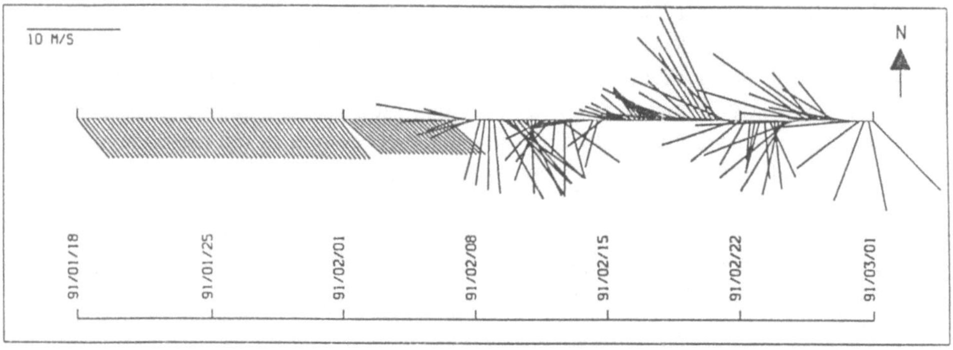

Fig. 2: Winds in the Arabian Gulf. Climatological winds are shown for the period 91/01/18/00 - 91/02/06/18 and analysed winds extracted from weather maps for the rest of the period. Wind direction is from the axis to the tip of the stick.

this on spill movement will be discussed later in this paper. The variability in the winds also raises the question of accuracy of these winds and their impact on the accuracy of spill movement predictions. This aspect is briefly discussed below.

2.1 The Accuracy of Winds

In open waters the expected travel per day of an oil spill is about 30 km. Thus a 1% prediction error (99% accuracy) will mean the prediction of spill location to within 300 m. If wind was the only variable affecting spill motion then we will need a 99.5% accuracy in each of wind speed and direction to achieve a 99% accuracy in spill location (0.995*0.995=0.99). The present state of wind prediction methodologies clearly precludes such accuracies. Even for a 10% error (3 km) in spill location the wind speed and direction need to be specifed with 95% accuracy. If one considers the speed and direction errors in both winds and water currents, i.e., four degrees of freedom, we need ~97.5% accuracy in each parameter to get within a 10% error overall in spill location (0.975**4 = 0.9). These requirements can be relaxed if one or more of the parameters were to be known to an accuracy greater than 97.5%.

In the open ocean environment the wind speed (and direction to some extent) is relatively constant over space scales of the order of 100 km. Beyond these scales the spatial variability of the wind has to be taken into account. The temporal variability of the wind can, of course, be quite high even in the open ocean environment. Under such conditions one can expect wind speeds to be specified to within 1-2 m/s or 10% of the speed, whichever is greater and wind direction to within 10-20 degrees. Over a one day travel of 30 km, a 10 degree error in wind direction alone can result in a location error of over 5 km.

In coastal areas with complex terrain, the winds generally have preferred directions and hence there is less directional variability. In such cases the emphasis can be on wind speed definition, its variability in time being important. One can expect speed accuracy similar to the open ocean case but the direction accuracy can be better. (See Venkatesh (1992) for additional details).

3. MODEL SIMULATIONS AND ANALYSIS OF RESULTS

Venkatesh and Murty (1994) have carried out model simulations of the movement of the oil from two different sources (the Mina Al Ahmadi and Mina Al Bakr) in the Arabian Gulf. The period of the model simulations was from January 20 to March 7, 1991. The model simulations were compared with available observations as given by Al-Rabeh et al. (1992). It was found that the lateral spread of the oil was well simulated with a turbulent dispersion coefficient value of 200,000 cm^2s^{-1}. Therefore in the present study all simulations will be carried out with this value of the turbulent dispersion coefficient. It was also shown in that study that it was the oil from the Mina Al Ahmadi spill that formed the bulk of the oil reaching the Abu Ali area. We will therefore concentrate our attention on movement of the oil from this source.

As stated earlier we shall focus on the middle part of February 1991 when the oil from the spills started to arrive in the Dawhat Ad Dafi area. The different model simulations will be compared to base simulations obtained with climatological winds. All model simulations permit refloatation of oil upon shore contact (for details see Venkatesh and Murty, 1994). Figure 3 shows the model simulated location and spread of the oil from the Al Ahmadi spill source and driven by the tidal currents and climatological winds. The oil bypasses the Saudi Arabian coastline and stays well offshore with no accumulation in the Ad Dafi area. As noted by O'Brien and Gallagher (1993), modelling efforts at the time of the spill also made similar predictions. By February 14 the oil was predicted to reach the northern tip of Bahrain. However the oil was observed to be only at Abu Ali island by this time. It is thus clear that climatological winds could not be used for model predictions of spill movement.

Model simulations were then carried out with winds extracted from weather charts. As seen from Figure 2 the analysed winds were available only from February 7. The model simulations of oil spread and movement using these winds are shown in Figure 4. With only climatological winds available prior to February 7 the model simulations do not show any contact with the shoreline near Safaniyah and Tanajib as noted by O'Brien and Gallagher (1993). However, the winds on subsequent days clearly contribute to the push of the oil closer to the Saudi coastline. Compared to Figure 3 where quite a bit of the oil heads well south of Abu Ali, the oil in Figure 4 is held north of Abu Ali, as was observed, with only a few parcels escaping south of that island by February 14. In Figure 4 it is also clear that there is oil accumulating in the Dawhat Ad Dafi area from February 9 onward.

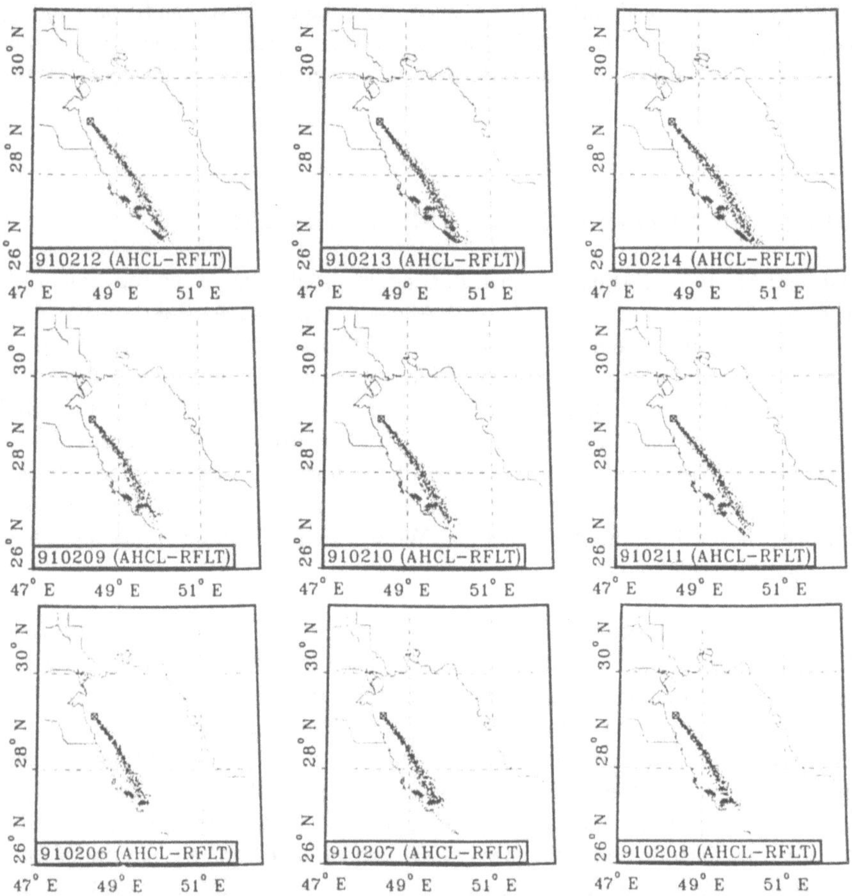

Fig. 3. Model simulations of the movement and spread of oil from the Al Ahmadi oil terminal using only climatological winds. The valid date for each panel in the figure is shown at bottom of panel (e.g., 910206 is 1991 February 6).

In Figure 2 there is an abrupt change in wind direction at the time we started using analysed winds instead of climatological winds (around February 7, 1991). In light of the earlier discussions on accuracy of winds we need to examine the effect of wind errors on spill motion. As such an abrupt change is unlikely, we chose to make a gradual change in the wind direction from the climatological value to that analysed on February 6 at 1800 hrs. In this case we adopted a linear interpolation. The resulting winds are shown in Figure 5. As a result of the interpolation there is an easterly component to the wind for two days prior to February 7. Model simulations using these winds are shown in Figure 6. The figure shows that no oil goes past Abu Ali during the period February 6-14. The oil is

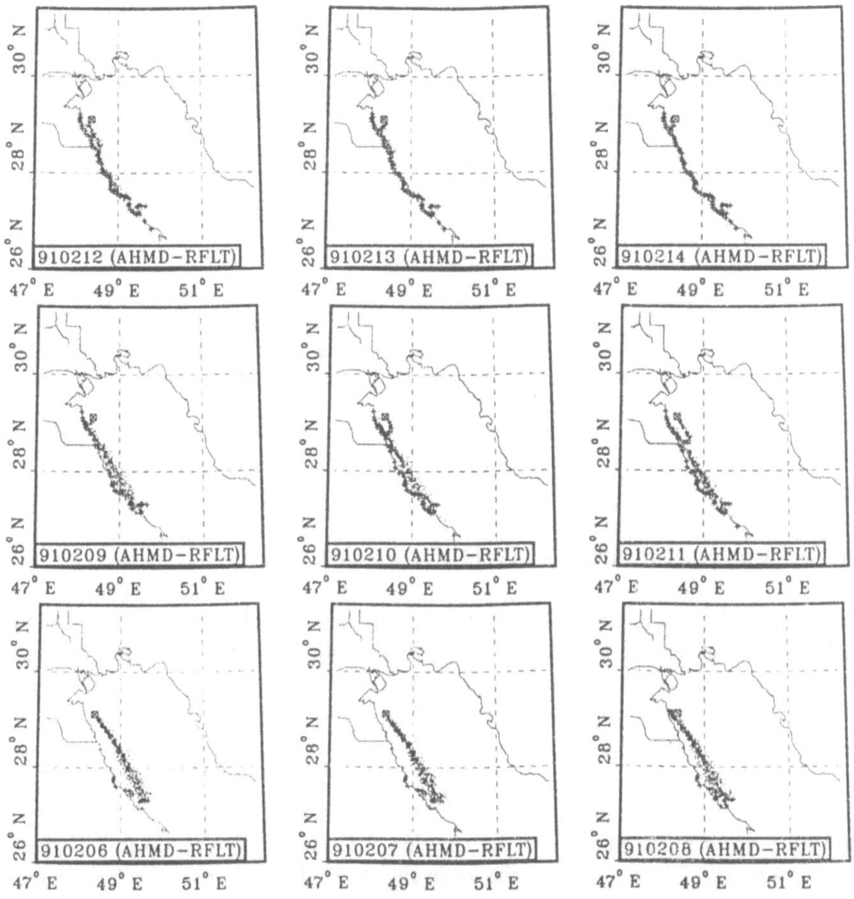

Fig. 4. Same as in Figure 3 except using winds shown in Figure 2.

driven towards the Saudi coastline much sooner than shown in Figure 4 and there is oil entering the Dawhat Ad Dafi area.

Venkatesh and Murty (1994) have compared model simulations of the motion of oil from both the Al Ahmadi and Al Bakr sources with observations that were available for specific times. They show a long residence time for the oil north of Abu Ali. The present study clearly shows the influence of the winds in keeping the oil close to shore as it arrives near Abu Ali island, a condition essential to prevent the oil from escaping around the eastern end of the island. Other related studies can be found in Al-Rabeh et al. (1992) and

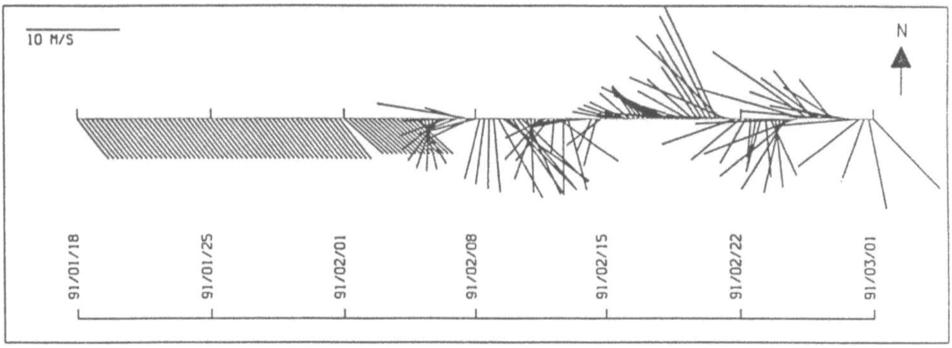

Fig. 5. Same as Figure 2 except wind speed and direction between 91/02/04/18 and 95/02/06/18 are obtained by linear interpolation. Note that the wind speeds at the two times are nearly the same.

Spaulding et al. (1993).

4. TIDAL CURRENTS IN THE DAWHAT AD DAFI AREA

In the above model simulations, although tidal currents are included, we have essentially examined the effect of winds on spill motion since the spill locations are shown only at one day intervals, averaging out the effects of the tides. Figures 7 and 8 show spill motion at 3-hourly intervals using the winds from Figures 5 and with winds set to zero during the 24 hour period from February 6-7, respectively. The tidal model used here has a grid resolution of 4.74 km. The model generated tidal currents are rather weak in the Dawhat Ad Dafi area. Comparing figures 7 and 8 it is seen that the tidal currents have very little effect on spill motion in the vicinity of the Dawhat Ad Dafi. There is not much change in spill location over the 24 hour period shown in Figure 8 The relative effects between the two figures can be seen near the coast around latitude 27°30'N and further north. In their study of the area Spaulding et al. (1993) also found that tidal currents had very little contribution to overall spill motion (see also Al-Rabeh et al., 1992). Thus one can conclude that the wind changes are mainly responsible for spill motion during the period shown in Figure 7. In all above model simulations the causeways between the Saudi mainland and Abu Ali island prevented any flow between the north and south portions of the Gulf in this area.

Experiments were also conducted with the area between the Saudi mainland and Abu Ali island being open. Given the very low water depths in the area (< 5 m) and because of strong bottom frictional effects, the tidal currents were also correspondingly very weak and did not induce any strong flows between the island and the mainland. There would however be some flow of oil through this area in this case.

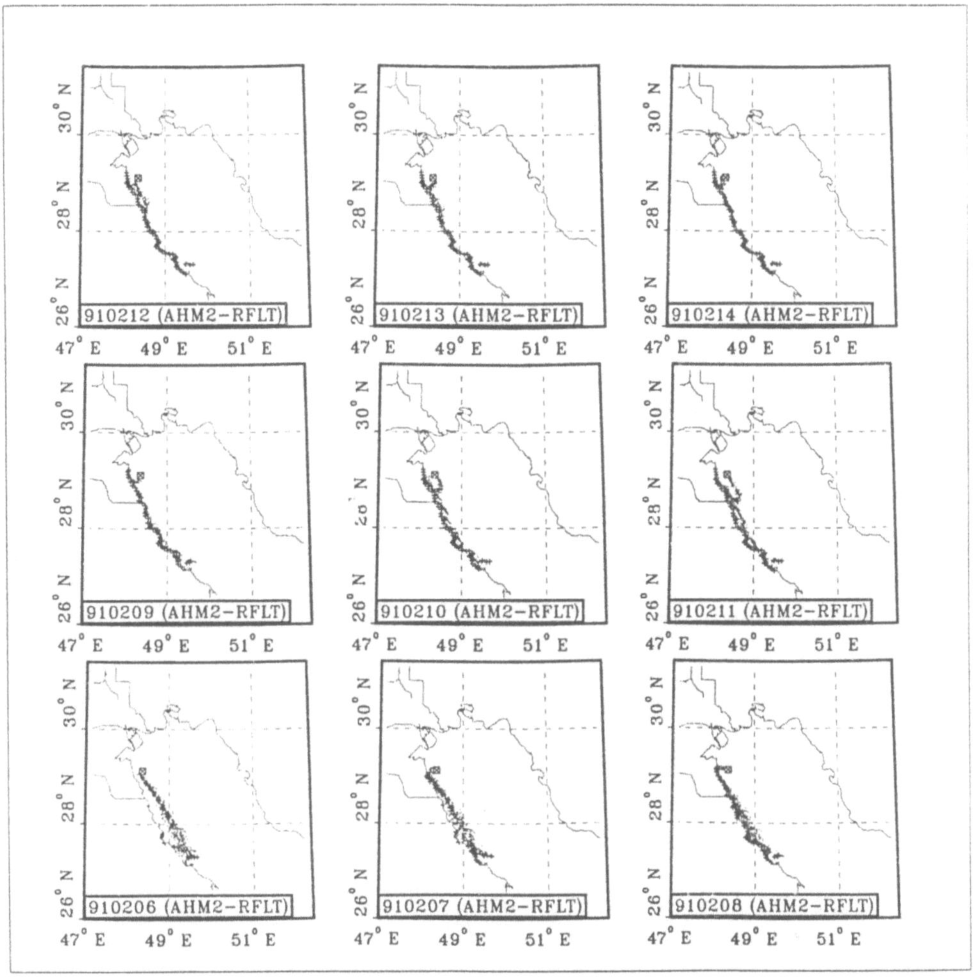

Fig. 6. Model simulations of the movement and spread of oil from the Al Ahmadi oil terminal using winds shown in Figure 5.

5. SUMMARY AND CONCLUSIONS

In this study we have investigated the factors that were responsible for the trapping of oil in the Dawhat Ad Dafi area of the Arabian Gulf. Through model simulations of the movement of oil from the Al Ahmadi source we have shown that the winds, the direction in particular, played a major role in moving the oil towards the Dawhat. These winds were also responsible for keeping the oil north of Abu Ali and in the Dawhat Ad Dafi area

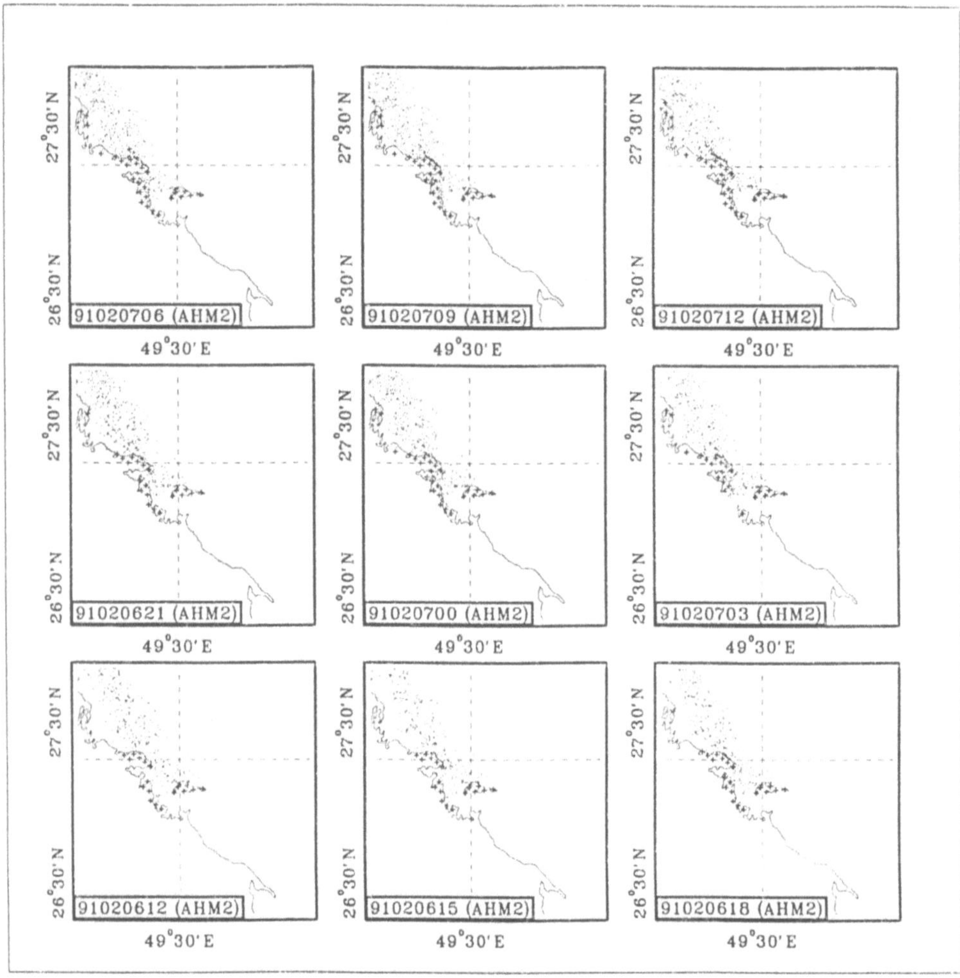

Fig. 7. Model simulated spill position near Abu Ali island and the Dawhat Ad Dafi area at three hour intervals and using winds shown in Figure 5.

for a sustained period of time enabling its removal. The blockage of the area between the Saudi mainland and Abu Ali island by man-made causeways prevented the oil from flowing southward out of the Dawhat Ad Dafi area, once it was trapped there. Detailed model simulations also show that the tidal influences were minimal in the trapping of the oil.

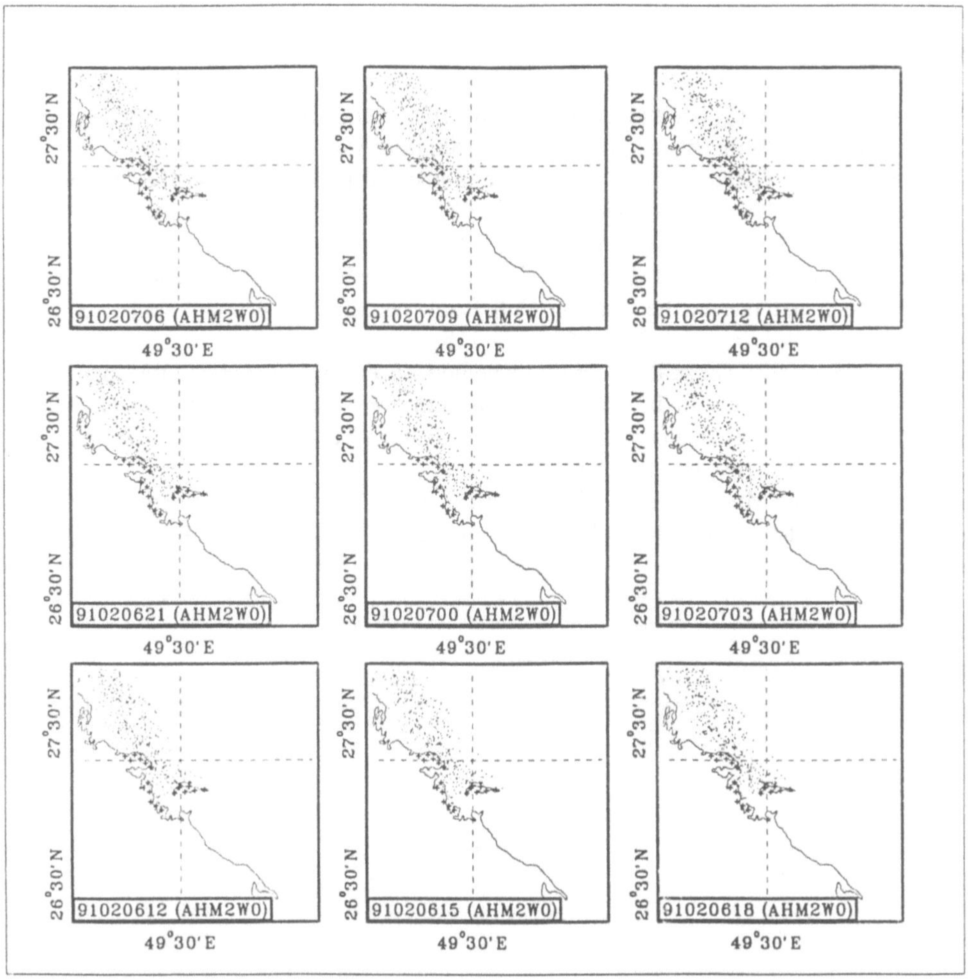

Fig. 8. Same as Figure 7 except winds are set to zero during the period 91/02/06/12 and 91/02/07/12.

ACKNOWLEDGEMENTS

We thank Mr. Do Kyu Lee of the Institute of Ocean Sciences of the Canadian Department of Fisheries and Oceans for his assistance in the operation of the Arabian Gulf tidal model.

REFERENCES

Al-Rabeh, A.H., Gunay, N., and Cekirge, H.M., 1992. Modelling the fate and transport of Al-Ahmadi oil spill. *Water, Air, and Soil Pollution*, Vol. 65, pp 257-279.

O'Brien, J.L., and Gallagher, J.J., 1993. The mother of all oil spills and the Dawhat Ad Dafi. *Proc. 1993 International Oil Spill Conference*, Tampa, USA, March 29-April 1, 1993. US Coast Guard, Washington, D.C.

Spaulding, M.L., Anderson, E.L., Isaji, T., and Howlett E., 1993. Simulation of the oil trajectory and fate in the Arabian Gulf from the Mina Al Ahmadi spill. *Marine Environmental Research*, Vol. 36, pp 79-115.

Venkatesh, S., 1992. Oil spill trajectory modelling - Prediction accuracies for open water spills and oil spills in broken ice, a concern for Canada. *Proc. Ocean Model Workshop*, Halifax, January 15-17, 1992. Canadian Tech. Rep. of Hydrography and Ocean Sci. No. 140, pp 151-153.

Venkatesh, S., and Murty, T.S., 1994. Numerical simulation of the movement of the 1991 oil spills in the Arabian Gulf. *Water, Air, and Soil Pollution*, Vol. 74, pp211-234.

DAMAGE TO FINE STRUCTURE OF GILLS OF MYTILUS SP. DUE TO POLLUTION BY LAS (LINEAR-ALKYL-BENZENE-SULPHONATE) AND CADMIUM

L. DALLA VENEZIA, S. STRADELLA, G. CAMPESAN AND A. MENETTO
Institute of Marine Biology, CNR, Riva 7 Martiri, 1364/A, 30122 Venice, Italy

Abstract: The possible effects of the use of domestic detergents containing LAS (Linear-alkyl-benzene-sulphonate) in the gills of Mytilus sp., in presence and absence of cadmium, are investigated. Preliminary analyses were carried out to determine the trend of LAS concentration with time in the aquaria containing mussels. The experimental conditions were: 6 mg/l LAS; 50 μg/l Cd; 6 mg/l LAS + 50 μg/l Cd. After 15 and 30 day exposure ten mussels were sacrified and their gills observed by Scanning Electron Microscope (SEM). Another group of ten mussels was sacrified and used for Atomic Absorption Spectrophotometry analysis of Cd accumulation. SEM investigations after 30 days showed that gill lamellae were severely damaged by LAS and LAS + Cd. Cadmium alone does not cause any damage, which can be seen by SEM.

1. INTRODUCTION

The global environment is formed of biotic and abiotic components which, if changed by man, cause alterations in the natural balance.

The growing interest in mussels as a subject of studies in physiological ecology and as possible indicators of various types of pollution is mainly due to their great resistance to many different environmental conditions, their importance as a human food, and the low level they occupy in the food chain (Goldberg, 1975).

In filter-feeding animals such as mussel, the gills are an organ with many important and different functions. They are the site for gas exchange, excretion, probably osmotic regulation and, because of the presence of cilia, also food filtering.

It is useful to recall briefly how the filtering process occurs in the body of mussel: water enters the internal cavity of the shell through the inhalant siphon and carries food particles, which are conveyed by the cilia covering the W-shaped gills to the marginal groove (Fig. 1) and hence to the mouth. The water then penetrates through the filaments into the suprabranchial cavity, and leaves the body of the mussel through the exhalant siphon. The cilia covering the external side of the gills play an important role in the nutrition process (Bayne et al., 1976a; 1976b).

M. I. El-Sabh et al. (eds.), Land-based and Marine Hazards, 203–214.

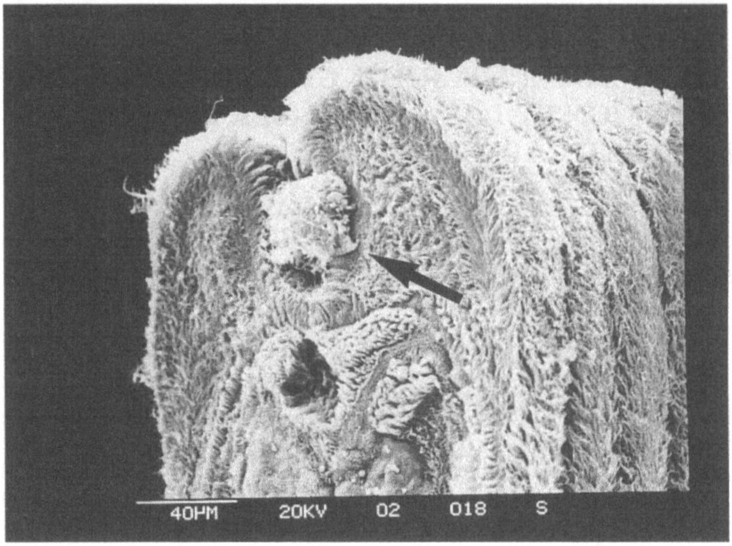

Fig.1. *Control mussel.* Narrow marginal groove of gill in mussel fed on Marine Liquifry in laboratory. Note food particle, which is too large to pass.

The present paper deals with the latter aspect and shows the possible effects of the use of domestic detergents containing LAS (linear-alkyl-benzene-sulphonate) on the gills of Mytilus sp., in the presence and absence of cadmium.

LAS and cadmium were chosen as pollutants because they are widespread in coastal waters throughout the world. LAS is in fact the major component of all laundry compounds and its concentration in the marine environment ranges from 1 µg/l to 1 mg/l (Majori et al., 1969; Cosovic et al., 1985). Every year in western Europe, Japan and the USA, 3-4 million cubic meters of synthetic detergents are produced, 28% of which is made up of LAS (Breen et al.,1992). In coastal waters, where this pollutant is carried by waste, 30% of LAS is irreversibly trapped by sediments (Marin et al.,1991); the rest is biodegraded by bacteria (Seassaro, 1973).

Cadmium is frequently used instead of chromium for plating the metal parts of engines and other devices, and its concentration in the marine environment is less than 1 µg/l (Boyle et al., 1976; Scarponi et al., 1989). The world consumption of cadmium is about 15,000 tons /year (Sames, 1971). The effects of Cd on mussel gills have been extensively studied by Sunila (1986).

2. MATERIALS AND METHODS

Mussels were collected at a station in the Northern Adriatic 8 miles off the Lagoon of Venice, where salinity is 34±1 PSU. After an acclimatization period, they were transferred to 18-l aquaria containing sea water, at the same salinity, dosed with: 1) 6 mg/l LAS; 2) 50 µg/l Cd ; 3) 6 mg/l LAS + 50 µg/l Cd. There was also one aquarium as a control. Two experiments were done of 15 and 30 days. The concentration of LAS in the aquaria was checked in a preliminary experiment by HPLC analysis (Matthijs and De Henau, 1987; Cavalli et al., 1987) to follow its degradation with time. On the basis of the results, reported in table 1, the water plus pollutants was completely renewed every second day in all aquaria. Mussels were fed with Marine Liquifry.

At the end of each experiment, ten mussels were sacrified for Scanning Electron Microscope (SEM) observations, and ten for Atomic Absorption Spectrophotometry (AAS) analysis of Cd.

Gills were carefully cut and treated for SEM observations (Sciala and Pasquinelli, 1987). AAS analyses were carried out on pooled samples of gills (Fossato et al., 1989).

3. RESULTS

Table 1 presents preliminary analyses of LAS concentrations, determined by HPLC, in the aquaria with LAS and LAS+Cd, in the presence of mussels.

Table 1. Decrease with time of LAS concentration (expressed in mg/l) in water of aquaria with mussels. Preliminary analyses.

TIME	TREATMENT	
(Hours)	LAS	LAS + Cd
1	6.59	4.88
3	1.99	1.96
5	1.46	1.37
21	0.85	0.72
24	0.86	1.04
48	0.54	0.24
72	0.24	0.11

Table 2 shows Cd concentrations in the gills of mussels after 15 and 30 days of exposure to Cd and LAS+Cd respectively.

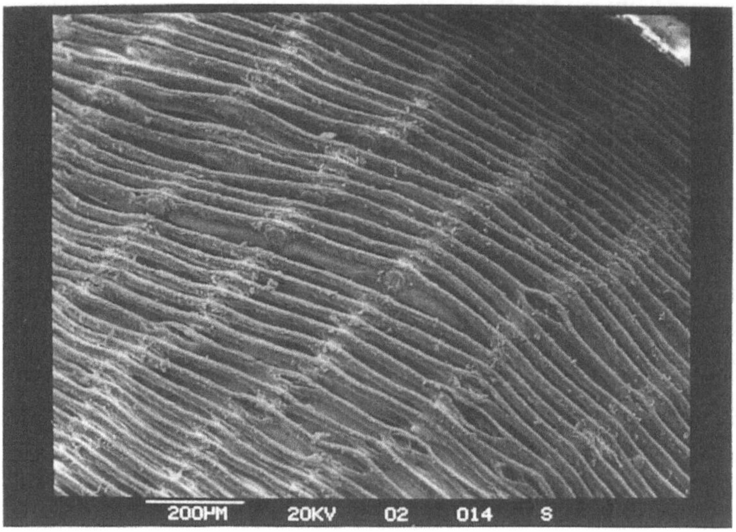

Fig.2. *Control mussel.* Internal side of lamella, with clearly visible "ciliated discs".

Table 2. Accumulation of Cd in gills of Mytilus sp. (expressed in µg/g wet weight).

TIME (Days)	TREATMENT Controls	Cd	Cd+LAS
0	0.36	0.36	0.36
15	0.50	3.73	3.88
30	0.37	5.44	4.51

A complete SEM investigation showed that mussel gills are made up of about 1500 narrow filaments (Fig.2), attached to each other by means of "ciliated discs" (Fig.3). Instead, the descending and ascending lamellae are connected to each other by "interlamellar junctions" through which haemolymph circulates (Fig.4). The difference in the U-shaped marginal groove of mussel gills fed with Marine Liquifry and wild mussels is due to the size of the food particles: in the former the marginal groove is almost closed, because the food is nearly liquid (Figs. 5 and 6).

However, the most interesting results are summarized in the following pictures, which, beside the "normal" lamellae of control gills, show the damaged lamellae of gills exposed to LAS or LAS+Cd.

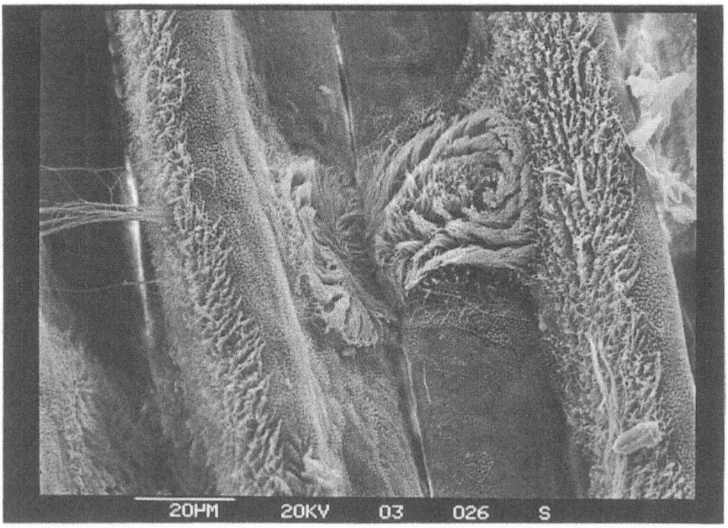

Fig.3. *Control mussel.* Internal side of two filaments and two corresponding "ciliated discs", like press-studs.

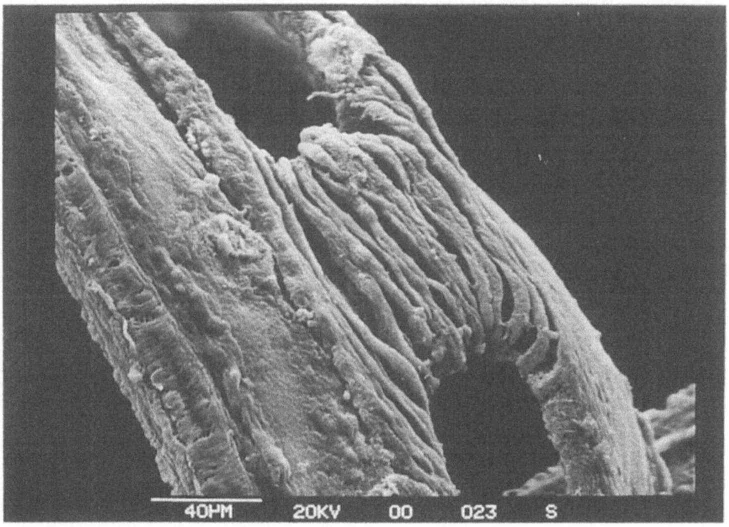

Fig.4. *Control mussel.* Infolded "interlamellar junction".

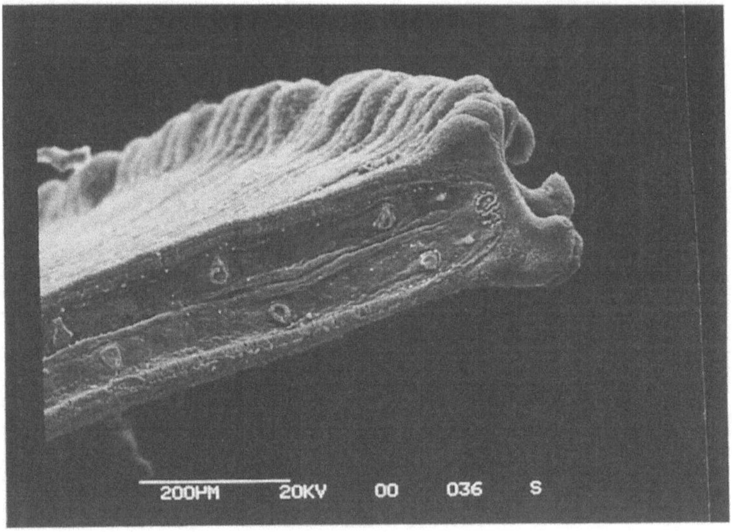

Fig.5. *Wild mussel.* Gill with very large U-shaped marginal groove.

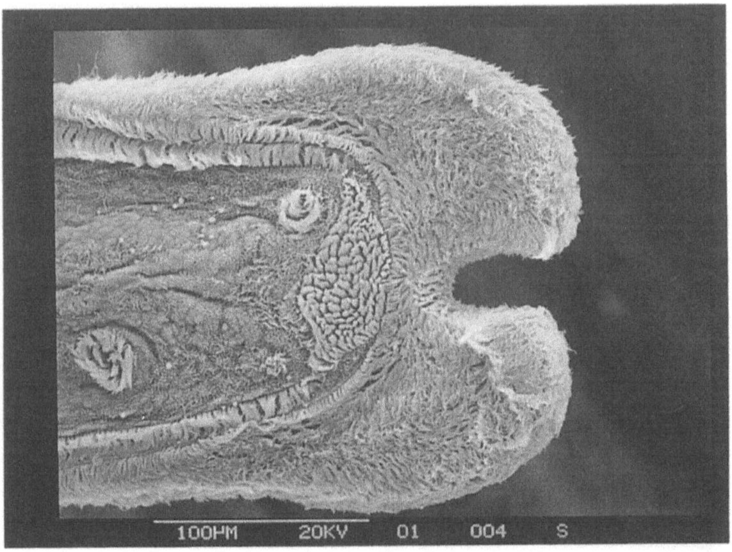

Fig.6. *Control mussel.* Narrow U-shaped marginal groove of mussel fed on Marine Liquifry .

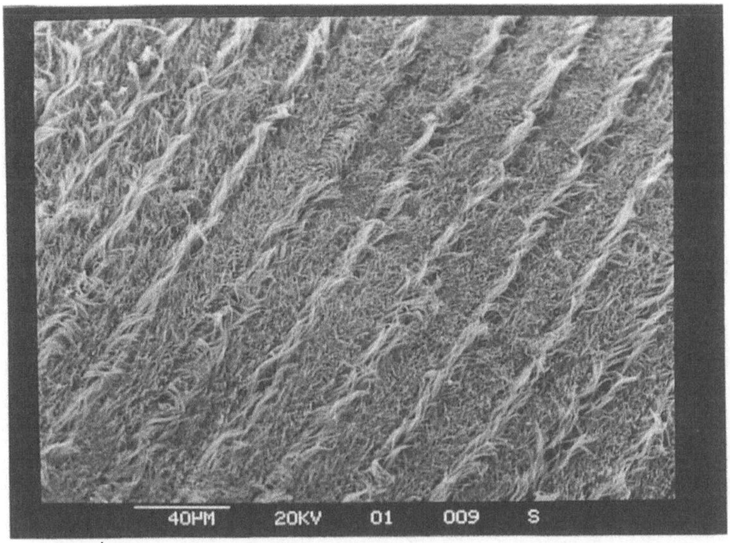

Figure 7. *Control Mussel.* External side of gill lamella, normally covered with cilia.

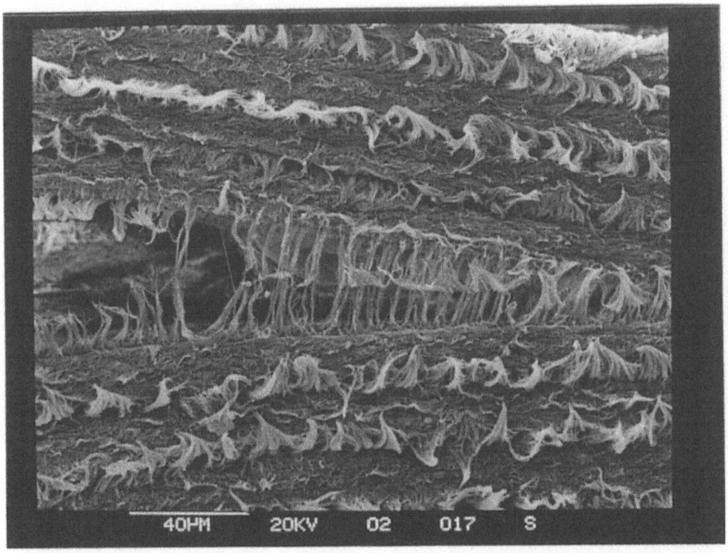

Fig.8. *Treated mussel.* At same magnification as Fig.7, external side of gill lamella after 30 day-exposure to LAS: cilia appear frayed.

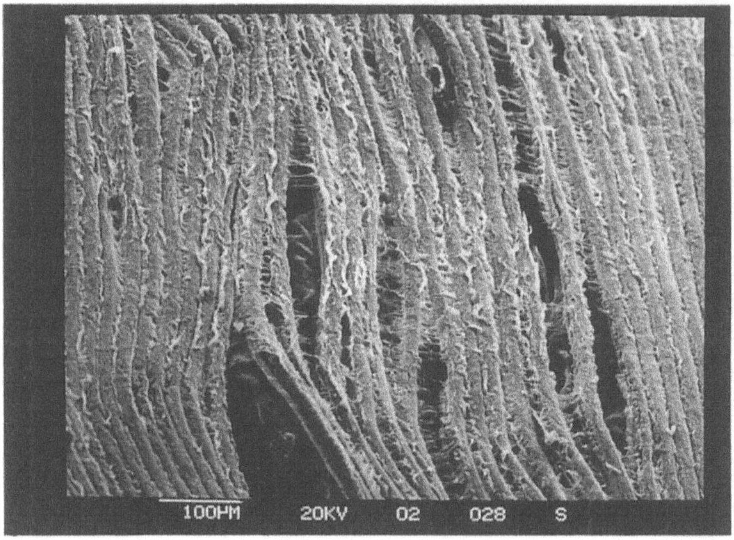

Fig.9. *Treated mussel.* Wider view of external side of gill lamella after 30-day exposure to LAS: detergent has dissolved components of cilia membranes and consequently filaments are unhooked.

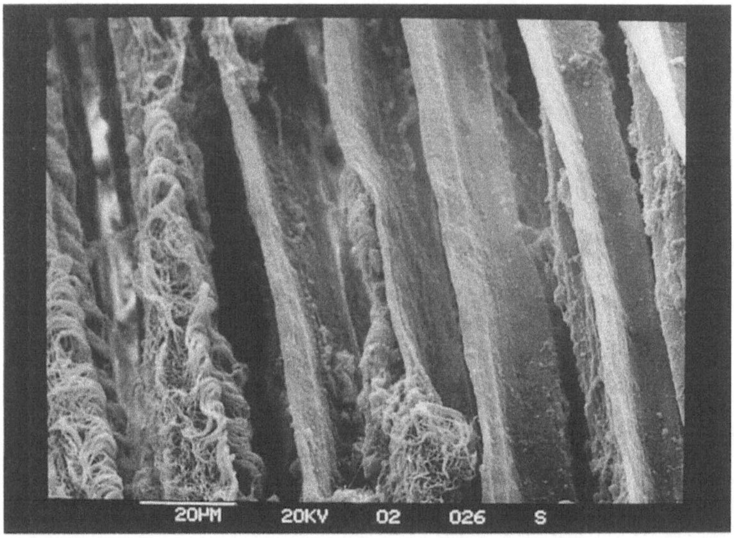

Fig.10. *Treated mussel.* External side of gill lamella after 30-day exposure to LAS: some filaments appear completely destitute of cilia.

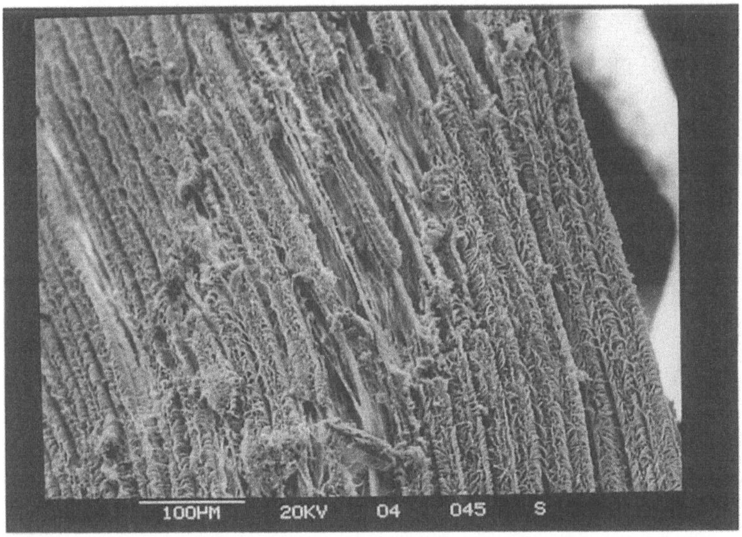

Fig.11. Treated mussel. Extensive damage due to 30-day exposure to LAS+Cd.

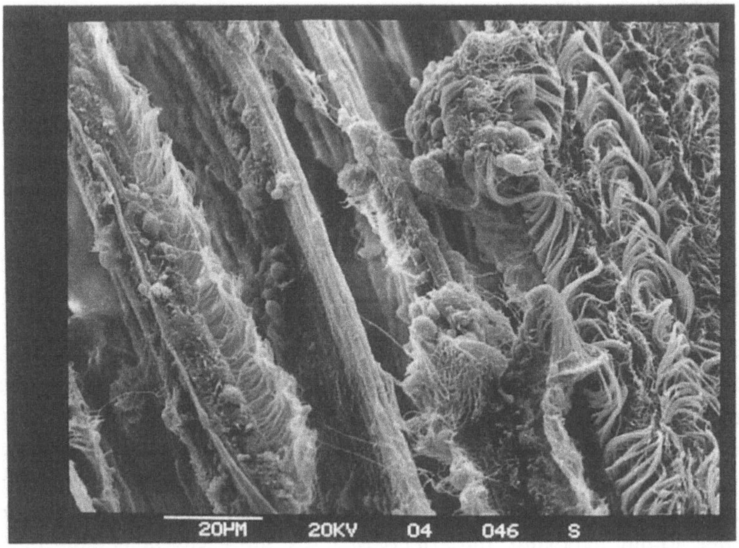

Fig.12. *Treated mussel.* Same gill as in Fig.11 at higher magnification: again, filaments appear destitute of cilia.

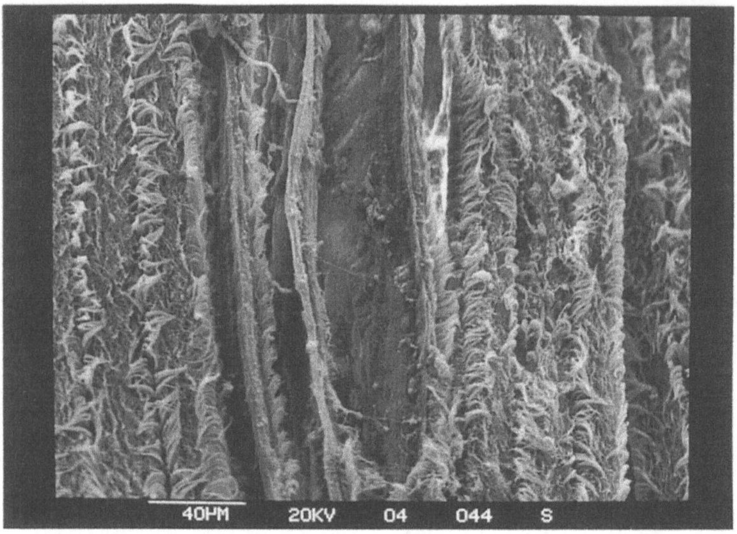

Fig. 13. Treated mussel. Damage similar to "torn window", through which food particles can enter suprabranchial cavity, after 30-day exposure to LAS+Cd.

Fig. 7 shows the external side of a control lamella, covered with normally preserved cilia along the filaments.

Figs. 8, 9 and 10 show the effects on cilia of 30-day exposure to 6 mg/l LAS, at various magnifications. Figs. 11, 12 and 13 show the effects of 30-day exposure to 6 mg/l LAS + 50 µg/l Cd.

4. DISCUSSION

The results of SEM analysis did not revealed damage to the gills of individuals exposed only to cadmium. On the contrary, in the gills of the same individuals histological analysis showed some alterations such as hypertrophy and swelling of cells (Da Ros et al. , 1995). The analyses of this element, as shown in Table 2, demonstrate that, after 30-day exposure to 50 µg/l Cd, this tissue has accumulated a quantity of Cd about fourteen times higher than the initial value. The presence of LAS does not increase the accumulation of Cd in the gills. After 30-day exposure, the effects of LAS, are easily detectable by SEM. The cilia, which in control animals are arranged along the filaments and play the important role of filtering food particles, are completely destroyed in the experimental animals. Damage similar to "torn windows" appears on the external side of the gill lamellae. This is not surprising, if we recall the results obtained by Kirkpatrick et al. (1974) and Nakanishi et al. (1986), who observed that detergents may dissolve animal membrane components such as proteins, lipids and phospholipids.

The obvious consequence of this damage is that food particles are allowed to enter the suprabranchial chamber instead of the stomach, thus causing disease.

This result may be considered in correlation with the disappearance of mussels, the resistance of which is well-known, from many sites severely polluted by urban waste, such as the internal canals of the city of Venice.

REFERENCES

Bayne B.L., Thompson, R.J. and Widdows, J., (1976a). Physiology I. In: Marine mussels: their ecology and physiology, B.L. Bayne, editor, Cambridge University Press, Cambridge, pp. 121-206.

Bayne B.L., Widdows, J. and Thompson, R.J., (1976b). Physiology II. In: Marine mussels: their ecology and physiology, B.L. Bayne, editor, Cambridge University Press, Cambridge, pp. 207-260.

Boyle E.A., Sclater F.R. and Edmond J.M., (1976). On the marine chemistry of cadmium. *Nature*, 263, 42-44.

Breen A., Jimenez L., Sayler L. and Federle G.S., (1992). Plasmid incidence and linear alkylbenzene sulfonate biodegradation in wastewater and pristine pond ecosystems. *Journal of Industrial Microbiology*, 9, 37-44.

Cavalli L. and Lazzarin M., (1987). Applicazione della cromatografia liquida su colonna nel campo dei tensioattivi. La rivista delle sostanze grasse, 64, 263-267.

Cosovic B., Zutic V., Vojvodic V. and Plese T., (1985). Determination of surfactant activity and anionic detergents in sea water and sea surface microlayer in the Mediterranean. *Marine Chemistry*, 17, 127-139.

Da Ros L., Nasci C., Campesan G. Menetto A. and Stocco G., (1995). Cadmium accumulation and biological effects of Cadmium and LAS on gills of Mytilus sp. Società Italiana di Ecologia, Atti 6° Congresso. 16, 635-637.

Fossato V.U., Campesan, G., Craboledda L. and Stocco G., (1989). Trends in chlorinated hydrocarbons and heavy metals in organisms from the Gulf of Venice. *Archivio di Oceanografia e Limnologia*, 21, 179-190.

Goldberg E.D., (1975). The mussel watch. A first step in global marine monitoring. *Marine Pollution Bulletin*, 6, 111.

Kirkpatrick F.H., Gordesky S.E. and Marinetti G.V., (1974). Differential solubilization of proteins, phospholipids and cholesterol of erythrocyte membranes by detergents. *Biochemica Biophysica Acta*, 345, 154-161.

Majori L., Rausa G., Morelli M.L., Diana L. and Gasparini V., (1969). L'inquinamento delle acque del mare nell'Alto Adriatico. *Igiene moderna*, 62, 715-749.

Marin M.G., Bressan M. and Brunetti R., (1991). Effect of linear alkylbenzene sulphonate (LAS) on two marine benthic organisms. *Aquatic Toxicology*, 19, 241-248.

Matthijs E. and De Henau H., (1987). Determination of LAS. *Tenside Surfactants Detergents*, 24, 193-198.

Nakanishi H., Takebayashi J., Tsuda T., Aoki S., Fukui S. and Hirayama T., (1986). Comparative studies on changes in lipid composition and some histological responses of gills from carp by exposure to acute levels of surfactants. *Comparative Biochemistry and Physiology*, 85, 375-379.

Sames C.W., (1971). Die Zukunft der Metalle, Suhrkamp-Wissen 17, Frankfurt, 183 pp.

Scarponi G., Capodaglio G. and Cescon P., (1989). Determinazione di zinco, cadmio, piombo e rame nell'acqua sub-superficiale del mare Adriatico mediante voltammetria di ridissoluzione anodica. *Bollettino di Oceanologia Teorica e Applicata, n.s.*, 233-240.

Sciala C. and Pasquinelli A., (1987). Microscopia elettronica a scansione in biologia, CLEUB, Bologna, 30 pp.

Seassaro E., (1973). La degradazione microbica degli alchilbenzensulfonati lineari. Inquinamento, 4, 27-30.

Sunila I., (1986). Chronic histopathological effects of short-term copper and cadmium exposure on the gill of the mussel, Mytilus edulis. *Journal of Invertebrate Pathology*, 47, 125-142.

PART II: MANAGEMENT ISSU

NATURAL AND TECHNOLOGICAL DISASTER MANAGEMENT: UNCERTAINTY IN TECHNICAL, SOCIOPOLITICAL AND SCIENTIFIC ISSUES

H. DENIS

Department of Industrial Engineering, Ecole Polytechnique, University of Montreal, Montreal, Quebec H3C 3A7, CANADA

Abstract. A certain number of natural and technological disasters in Canada and around the world are presented to illustrate the role that complexity and uncertainty play in the issues pertaining to many catastrophes: technological, sociopolitical and, sometimes, scientific issues. In order to deal with this uncertainty, the interaction between organizational structure, culture and technology, as applied to developing countries, sheds new light on disaster management.

1. INTRODUCTION

This paper deals first with the response to a disaster. In order to do this, a typology of issues faced by disaster managers must be considered (technical, sociopolitical and scientific issues), as well as the relationships both within and between issues that render disaster management more complex. The uncertainty which may or may not underlie each of these issues then allows a comparison to be made between natural and technological disasters. Finally, the disaster management problems in developing countries are better grasped by differentiating between organizational structure, culture and technology in the response.

2. METHODOLOGY

The typology of issues was developed as part of a study on the management of a fire at a PCB warehouse in St-Basile-le-Grand, Canada, in August 1988 (Denis, 1990a). At this time, it offered a better understanding of the importance of scientific issues, and of their relationship to the technical and sociopolitical issues (Denis, 1991). Because of its usefulness, the typology was then applied to a research study on disaster management in Canada. Some results of this research are presented here, together with data on other disasters around the world.

M. I. El-Sabh et al. (eds.), Land-based and Marine Hazards, 217–228.

The research concentrates on the response to disasters. Even though no two disasters are quite alike, comparisons of the ways in which natural and technological disasters are managed were undertaken, based on the hypothesis that very often there is no clear distinction between these two types of disasters. This is in part because industrialization tends to implicate technology in many "acts of God": for example, the Mississippi River floods of 1993 involved dangerous goods, as did the Edmonton tornado of 1987 (Showalter and Myers, 1992).

The present paper is based on a particular number (N=77) of disasters. The data (Table 1)

Table 1: Sample of disasters by type and geographic distribution [a]

TYPES OF DISASTERS		GEOGRAPHIC DISTRIBUTION						TOTAL
		Americas						
		Canada	USA	Other	Asia	Europe	Africa/ Other	
Natural	Earthquakes	1	3	2	1	1	1	9
	Landslides	2						2
	Volcanic -eruptions		1	2	2			5
	Floods	4			1	3		8
	Forest fires	7					1	8
	Wind storms	6	2		2			10
	Snow storms	2						2
		22	6	4	6	4	2	44
Techno-logical	Transportation							
	- planes	3	1			2		6
	- ships	2	1			2		5
	- trains	3						3
	- satellites	3						3
	Chemical							
	- fires	3	1	2	1	2		9
	- leaks	3						3
	Nuclear	1 [b]	1			1		3
	Structures		1					1
		18	5	2	1	7	--	33
TOTAL		40	11	6	7	11	2	77

[a] This list is far from being exhaustive, depending on the availability of data on the subject.

[b] Does not include 2 cases of nuclear material in satellites (included above).

come first from documentary material, like information reports written by various government agencies during and after a disaster, and newspapers and scientific articles; second, from in-depth interviews conducted with first responders, elected representatives and a few media people, all of whom played a major role in the response to disasters in

Quebec which occurred after 1988, more precisely after the PCB fire in Saint-Basile-le-Grand. Finally, participant observation was used for two disasters, namely a tornado and a fire at a used-tire dump.

The methodology must also define what is under the microscope, so to speak. Operationally, a disaster is viewed in the research as a sudden, low-probability event with such severe consequences in terms of loss for a given community (human, material, financial, etc.) that it causes tensions in its social fabric (Denis, 1993). This definition therefore excludes long-term environmental disasters. It also excludes human-engineered disasters such as mass-murders, to the extent that they involve security problems (unless their impact is felt physically by a whole community, as in the Lockerbie airplane crash). Finally, the research is limited to the response phase of a disaster. Needless to say, all these choices are strictly operational and in no way involve value judgements as to the importance of any other alternative.

3. TECHNICAL, SOCIOPOLITICAL AND SCIENTIFIC ISSUES: COMPLEXITY AND UNCERTAINTY

3.1. The Issues

Emergency response generally involves at least two types of issues: technical and sociopolitical, to which a third is sometimes added, scientific issues. An issue is a broad category of activities which surrounds a problem to be solved. Technical issues deal with the source of the threat and include activities like search-and-rescue, firefighting techniques, safety of the site, etc. Sociopolitical issues are related to the needs of the community affected: evacuation, shelter, communication, health care, etc. and relations with elected representatives. Scientific issues are related to science: sampling, data analysis and interpretation of a past or future event (predicting, for example, a volcanic eruption (Nevado del Ruiz) or earthquake aftershocks (Mexico, Saguenay)).

3.2. Issues and Complexity

Disasters are complex because most of the time the event (or the response, as will be seen later) presents certain difficulties that have to be solved. These difficulties can arise because of the sheer number of elements contained in the problem, or because of the relationships among these elements (Perrow, 1984). Unlike routine response, the problem is difficult for a human being to grasp, i.e. difficult to understand in terms of pre-established responses or patterns of responses. For example, the 1988 Saguenay earthquake in Canada was a relatively simple event to manage, apart from a chemical leak. Had the leak been a more serious one, however, requiring evacuation (an added element to manage), the situation would have been more complex, as was the case in the 1985 Mexico earthquake (Dynes et al, 1990).

Complexity can originate in the relationships among elements within and between issues. Within technical issues, there can be a relationship between a lack of knowledge concerning a source of danger and firefighting techniques, as shown by the 1979 Mississauga train derailment in Canada (Scanlon, 1989). Within sociopolitical issues, there can be relationships among elements like evacuation, shelter, communications, etc., as in the case of the forest fires which occurred in Western Canada in 1989: for some Amerindian people, evacuation meant that for the first time in their lives they had to face problems of language, lifestyle and communication with their leaders (groups were scattered) all at once. Finally, within scientific issues, sampling, analysis and the interpretation of results can be interrelated. We all know that sampling can influence the results, as shown by Three Mile Island (Thornburgh, 1987).

Complexity can also result from the relationships between categories of issues: between scientific and sociopolitical issues, when scientific research about the level of toxicity in the St-Basile fire transformed what was expected to be a 24-hour evacuation into an 18-day evacuation (Denis, 1990a). At Three Mile Island, communications about sampling results precipitated a spontaneous evacuation, in a kind of panic. If science can influence sociopolitical issues, the reverse is also true: Chernobyl fallout around the world has illustrated how the scientific issue of an acceptable radiation norm can be subjected to political pressures which, in fact, may vary from country to country (Czada, 1991). Scientific and technical issues are also interrelated, as in train derailments, where, for example, the knowledge about the behavior of chemicals can influence firefighting techniques.

In the past, only the two first types of issues had to be dealt with, most of the time. The activities in technical and sociopolitical issues were generally well-known to emergency preparedness officers, even when the response was not routine (there have always been cases where the magnitude of the disaster made its management very complex). However, scientific issues are tending more and more to become an integral part of disaster management with, in many cases, not only complexity but uncertainty as well.

3.3. Issues and Uncertainty

Uncertainty stems from a serious incapacity of disaster respondents to solve a problem. With complexity, the problem can be solved within a reasonable length of time, although with some difficulty. With uncertainty, the problem can also be solved, but the time delay will be much longer and it will require much greater resources. The research on the PCB fire has shown two different levels of uncertainty. On one level, uncertainty is subjective (Duncan, 1972): it comes from the perception, individual or organizational, of a relative incapacity to respond appropriately to a given situation (Crozier, 1963; Turner, 1992). On another level, uncertainty arises from the existence of two or more conflicting "certainties", that is, from a lack of consensus among respondents, as long as a decision has not been reached on the appropriate solution to the problem.

An example of the first level of uncertainty would be a refusal by hospital administrators to accept patients following a disaster, based on what they perceive as a lack of sufficient resources to face the problem (sociopolitical uncertainty). This perception is subjective, because other hospital managers may have evaluated exactly the same situation quite differently. An example of the second level of uncertainty is provided by the choice of the most appropriate firefighting technique to use at a used-tire dump fire in St-Amable, Canada, where there was no consensus on the best way to extinguish the fire: foam, water, sand or laisser-faire (technical uncertainty). Uncertainty can also be explained, at both levels, by a lack of scientific knowledge *per se* within the scientific community: What are the effects on health of a PCB fire, a particular virus, or the combined effects of two or more chemicals as, for example, in a fire following a train derailment?

4. UNCERTAINTY IN NATURAL AND TECHNOLOGICAL DISASTERS

Using this typology of issues and its characterization by uncertainty, we can now look at some differences or similarities between technological and natural disasters. This field has been explored in the past by Quarantelli (1990), who concluded that there was not much difference between the two types of disaster. Table 2 presents a few cases (in a very brief form, assuming that the reader is familiar with the literature on disaster management). It does not claim to be exhaustive, nor does it differentiate between cases that were only natural and technological, and those which were both, i.e. natural with a technological component (Showalter and Myers, 1992). What can we observe from this table?

1. Both technological and natural disasters may present uncertainty related to firefighting techniques. Technical uncertainty in natural disasters is often related to dangerous goods (Tierney, 1990), though not always, as shown by the ash problem in the Mt. St. Helens explosion (Drabek et al, 1981).

2. Natural disasters seem to be easier to perceive, and their danger easier for the public to interpret (Denis, 1990a). Nevertheless, certain natural disasters can also constitute an invisible threat, as in the 1987 Edmonton tornado in Canada, where dangerous goods were involved (Anonymous, 1988).

3. Because they are "acts of God", natural disasters tend to be more accepted by the victims (Quarantelli, 1990). This could mean less confrontation with political representatives (Maskinongé tornado, Canada, 1991), compared with technological disasters where, most of the time, governments are blamed (Rosenthal, 1988).

Table 2: Uncertainty in issues for natural and technological disasters

TYPES OF DISASTERS	UNCERTAINTY		
	TECHNICAL ISSUES	SOCIOPOLITICAL ISSUES	SCIENTIFIC ISSUES

TECHNOLO-GICAL DISASTERS	• knowledge of the source (Seveso, '76; Bale, '86; Chernobyl, '86; St-Basile, Qué., '88)
	• firefighting techniques (St-Amable, Qué., '90)
	• cleansing techniques (Seveso, '76; Binghampton, '81; Exxon-Valdez, '89)
	• toxic ashes problem: (St-Basile, Qué., '88)
	• dangerous goods: (Mississauga, Ont., '79)
		• evacuation
		  • spontaneous (T.Mile Island, '80)
		  • voluntary (St-Amable, Qué., '90)
		  • refusal
		• shelter (St-Basile, '88)
		• political convergence (St-Basile, '88)
		• public information (T.M.I., '80; Chernobyl, '86)
		• security of homes (St-Léonard-d'Aston, '89)
			• samples (T.M.I., '80)
			• measurement indicators (Chernobyl fallout)
			• effects on health (Seveso, '76; Bhopal, '84; Chernobyl; St-Basile)

NATURAL DISASTERS	• firefighting techniques • day / night (forest fires, Canada, '89) • ash problem (Mt. St-Helens, '80) • levees' behavior (Mississippi floods, '93) • chemical leaks (Saguenay, earthquake, '88) • dangerous goods (Edmonton tornado, '87)	• relocation refusal (landslide, St-Fabien, Canada, '76) • communication problems (Perth Andover floods, Canada, '87) • sociological problems with shelter: language, culture, etc. (forest fires, Canada, '89) • insurance problems, e.g. with mortgage (St-Bonaventure tornado, Canada, '75) • refusal of expert judgment concerning structures (Philippines earthquake, '90)	• fire behavior (forest fires, Qué, Canada, '89) • smoke toxicity (forest fires, Western Canada, '89) • volcanic predictions (Nevado del Ruiz, Colombia, '85; Mt. Pinatubo, Philippines, '91) • aftershock predictions (Saguenay earthquake, Canada, '88)

4. In both types of disasters, shelter can be a source of uncertainty, particularly when different communities are involved. In the St-Basile fire, for example, there were three municipalities with different average incomes and different shelter patterns, at least at the beginning of the evacuation: public shelters, hotels, etc. Probably because someone is considered responsible for the event in the case of a technological disaster, the victims tend to ask for more comfortable evacuation conditions (at least in North America).

5. In the two types of disasters, financial aspects can cause uncertainty, mostly with regards to the insurance question. Technological disaster assistance can also present uncertainty when an owner acts in an unethical way, for example by leaving the government to cope with the problem, as in the St-Basile PCB fire. Uncertainty can also be present, even if the owner does act in a responsible way, if there is disagreement concerning the amount of compensation, as in Bhopal (Shrivastava, 1987). The difficulty stems in part from the fact that victims are less easily identifiable in technological disasters, not only in relation to the previous point about visible and invisible threats, but also in relation to their effects on health (Clarke, 1989; Couch and Kroll-Smith, 1990).

6. Communications present no clear pattern of uncertainty, probably because there always seems to be uncertainty in communications. Here again, communications can engender more uncertainty in technological disasters, because people will tend to question the content of information coming from disaster managers, experts, political representatives and owners, all of whom are perceived as having been unable to prevent the disaster (Lagadec, 1981; Otway et al, 1988).

7. Contrary to what we hypothesized a few years ago, which is that scientific uncertainty is present only in technological disasters, scientific uncertainty is also present in natural disasters (Podesta and Olson, 1988; Voight, 1990).

From this, it follows that there is as much uncertainty in natural as in technological disasters. The only difference stems from the perception of the event, more or less visible, and from its degree of acceptance by the victims. Of course, many other variables, like familiarity with the event, alert time, etc. (Quarantelli, 1990) can come into play, but it cannot be said, looking at the uncertainties in the various categories of issues, that one type of disaster involves more uncertainty than the other. And this also applies, contrary to our first hypothesis, to scientific uncertainty.

5. THE RESPONSE IN DEVELOPING COUNTRIES: TECHNOLOGY, STRUCTURE AND CULTURE.

Up to now, the focus has been on the scene of the disaster, the issues in fact showing the specificities of an event which require a response. This response must in turn be considered. In order to do so, three concepts are now added to the typology of issues: the structure, technology and culture of respondents. These three interrelated concepts, theoretically developed elsewhere (Denis, 1990b), are important to consider in order to gain an understanding of the nature of disaster management, particularly in developing countries.

Structure refers to the division and corresponding integration of responsibilities among the responders (Duncan, 1972; Galbraith, 1977; Mintzberg, 1979). Technology (or technical support) refers to the tools, in the broadest sense of the term, that are used in the response action (Gillespie and Mileti, 1977; Gille, 1978; Zuboff, 1985). Culture means the collective values, frames of reference, management styles, etc. of the participants (Turner, 1991; Gephart, 1984; Pidgeon et al, 1991; Denis, 1993). As in the case of the issues, there may or may not be complexity or uncertainty surrounding these response dimensions.

The classification presented here is aimed at gaining an understanding of some of the characteristics of disaster management in developing countries. Our sample is not large due to the regrettable lack of information coming from these countries. Nevertheless, the following elements would only like to give a few clues as to how some of the differences can be explained between developed and developing countries concerning the response to disasters.

First, and this is not new, technical support is clearly lacking in developing countries, particularly sophisticated equipment, as shown by the absence of detection equipment for earthquakes or volcanic eruptions (Voight, 1990). Because such tools can help to save lives, it is crucial that they be obtained by every means possible. However, there are difficulties inherent in the process.

Clashes occurring among responders may also complicate the process of managing the disaster and result in the need, often in the midst of the event, to create harmony among the various responders in terms of the structures and values underlying the technology. This was so in the case of the Armenia earthquake of 1988 (Krimgold, 1989). Other difficulties stem from the fact that developing countries must import, along with a technology, an underlying structure or culture (or both). For example, every search-and-rescue team, as illustrated by the San Salvador earthquake (Comfort, 1989), carries a given structure and culture with its technology. Technology is never only technological. This is why it is important to bear in mind that in the transfer of emergency technology, as in the transfer of other technologies, the recipient of the technical support receives much more than just equipment, procedures, etc.

Another point is that response structures may differ from one country to another. We are thinking here not only of structures based on the classical preference for either centralized or decentralized models, but of the differences in the formal acceptance or otherwise of volunteers. In some countries, for example, the role of the volunteer has become an integral part of the national response plan in a time of disaster (in Switzerland, for example, although volunteers here are trained and not "spontaneous"), while in others volunteers are excluded, either officially (1989 Canadian forest fires) or unofficially (downgrading of the role of volunteers by officials in the 1985 Mexico earthquake). The lack of formal training sometimes explains why the assistance of volunteers may be refused in the case of technical issues, but accepted in the case of sociopolitical issues. What is important to remember is that response structures, like technology, are not neutral. They involve an underlying culture which includes values, styles, behaviors, etc.

These structures and cultures correspond not only to a single organization participating in the disaster (army, ministry of the environment or health and welfare, etc.), but to the disaster mega-organization (Denis, 1993, 1995) as well. The mega-organization is an umbrella coordinating group bringing together the various responding organizations in a disaster. Obtaining a consensus, which is already a difficult process in a country when the response involves various levels of government, becomes highly complex when international assistance comes into play (Anisuzzman, 1990) in a national mega-organization.

What is more, not only are international experts from the various countries providing assistance involved in this dynamics, but international organizations as well, each with its own response culture (Ali,1992). In addition, pre-disaster alliances or conflicts among helping countries can increase the complexity of disaster management in developing countries.

This very brief section is aimed at pinpointing some of the factors influencing the response in developing countries. While it is exploratory, and must be treated as such, it can nevertheless be helpful when the question of receiving or giving help comes into play.

6. CONCLUSION: TAMING THE UNCERTAINTY

In conclusion, a central question arises: What can be done, in very concrete terms, about uncertainty? Both typologies presented above provide a means for positioning a potential disaster, first in terms of the various categories of issues it involves, and their complexity or uncertainty, and, second, in terms of the dimensions of structure, culture and technology in response to an event, each with, in some cases, its own uncertainty.

What is important is to apply these typologies *before* the disaster occurs. It is imperative, in this exercise, to consider the intersections between one element of a typology and the whole. It is these crossing points that constitute the crux of disaster management. At the same time, it is important to remember that these are open systems we are dealing with, and not programmable mechanisms, hence the need for flexibility in the response. This is why we prefer to coin the expression "taming the uncertainty", rather than using others which emphasize mastery or control, or even overcoming uncertainty.

This does not mean to imply that disaster management is a process in which everyone does as he or she wishes. On the contrary, clearly established coordination in the mega-organization constitutes the basis for the success of the operation. But this coordination must not be a closed system either. In some cases, it is not even possible to think in terms of closed-system coordination, when the event is so complex and involves so much uncertainty that it is uncontrollable. Hence the necessity to tame the uncertainty.

ACKNOWLEDGMENTS

Financial support for this research was provided by a grant from the Social Sciences and Humanities Research Council of Canada (SSHRC) in 1991. Emergency Preparedness Canada and la Sécurité civile du Québec both collaborated generously with their time and data. The Saint-Basile-le-Grand study was sponsored by the "Comité interministériel de protection civile du Québec" in 1988. The author also wishes to thank the evaluators, whose suggestions were greatly appreciated.

REFERENCES

Ali, L. (1992). Symbolic planning and disaster preparedness in developing countries: The Presbyterian Church in Vanuatu, *Internat J. of Mass Emergencies and Disasters*, **10**, 293-314.

Anisuzzman, M. (1990). Managing flood disaster in Bangladesh: Implications for public administration, *Asian Review of Public Administration*, **II**, 12-28.

Anonymous (1988). *Tornado: A Report - Edomonton and Strathcona County, July 31st 1987*, Alberta Public Safety Services, Edmonton.

Clarke, L. (1989). *Acceptable Risk? Making decisions in a toxic environment*, University of California Press, Berkeley, 229 pp.

Comfort, L.K. (1989). The San Salvador earthquake. In: *Coping with Crises*, U. Rosenthal, M.T.Charles and P.T. Hart editors, Ch.C.Thomas, Springfield, pp. 323-340.

Couch, S.R. and J.S. Kroll-Smith (1990). Patterns of victimization and the chronic technological disaster. In: *The Victimology Handbook*, E.C. Viano, editor, Garland Publ. Inc., New York and London, pp. 159-176.

Crozier, M. (1963). *Le phénomène bureaucratique*, Seuil, Paris, 413 pp.

Czada, R.M. (1991). Muddling through a nuclear-political emergency: Multi-level crisis management after radioactive fallout from Chernobyl, *Industrial Crisis Quarterly*, **5**, 293-322.

Denis, H. (1990a). *La gestion de catastrophe: Le cas d'un incendie dans un entrepôt de BPC à Saint-Basile-le-Grand*, Les Publications du Québec, Québec, 115 pp.

Denis, H. (1990b). *Stratégies d'entreprise et incertitude environnementale,* Agence d'Arc, Montreal and Économica, Paris, 328 pp.

Denis, H. (1991). The complexity of technological disaster management: Technical, sociopolitical, and scientific issues, *Industrial Crisis Quarterly*, **5**, 1-18.

Denis, H. (1993). *Gérer les catastrophes: L'incertitude à apprivoiser*, Les Presses de l'Université de Montréal, Montréal, 248 pp.

Denis, H. (1995). Coordination in a governmental disaster mega-organization, *International Journal of Mass Emergencies and Disasters*, **13** (1), 25-43.

Drabek, T.E., H.L.Tamminga, T.S. Kilijanek and C.R. Adams (1981). *Managing Multiorganizational Emergency Responses: Emergent Search and Rescue Networks in Natural Disaster and Remote Area Settings*, University of Colorado Institute of Behavioral Science, Boulder, Co.

Duncan, R.B. (1972). Characteristics of organizational environments and perceived environmental uncertainty, *Administrative Science Quarterly*, **17**, 313-327.

Dynes, R.R., E.L. Quarantelli and D.Wenger (1990). *Individual and Organizational Response to the 1985 Earthquake in Mexico City, Mexico*, Disaster Research Center, University of Delaware, Newark, Del., 135 pp.

Galbraith, J.R. (1977). *Organization Design*, Addison-Wesley, Reading, Mass., 426 pp

Gephart, R.P. (1984). Making sense of organizationaly based environmental disasters, *Journal of Management*, **10**, 205-225.

Gille, B. (1978) *Histoire des techniques*, Gallimard, Paris, 1652 pp.

Gillespie, D.F. and D.S. Mileti (1977). Technology and the study of organizations: An overview and appraisal, *Academy of Management Review*, **2**, 7-16.

Krimgold, F.(1989). Search and rescue, *Earthquake Spectra*, Special Supplement, Armenia Earthquake Reconnaissance Report, 136-149.

Lagadec, P. (1981). *États d'urgence: Défaillances technologiques et déstabilisation sociale*, Seuil, Paris, 406 pp.

Mintzberg, H. (1989). *The Structuring of Organizations*, Prentice-Hall, Englewood Cliffs, N.J., 512 pp.

Otway, H.P., P. Haastrup, W. Cannell, G. Gianitsopoulos, M. Paruccini (1988). Risk communication in Europe after Chernobyl: A media analysis of seven countries, *Industrial Crisis Quarterly*, **2**, 3-15

Perrow, C. (1984). *Normal Accidents: Living with High-Risk Technologies*, Basic Books, New York, 386 pp.

Pidgeon, N., B.A. Turner, B.Toft and D.Blockley (1991). Hazard management and safety culture. In: *Hazard Management and Emergency Planning: Perspectives on Britain*, D. Parker and J. Handmer, editors, James & James Science Publishers Ltd, London, pp. 243-254.

Podesta, B. and R.S. Olson (1988). Science and the State in Latin America: Decisionmaking in uncertainty. In: *Managing Disaster*, L.K.Comfort, editor, Duke University Press, Durham and London, pp. 296-314.

Quarantelli, E.L. (1990). *Similarities and Differences in Institutional Responses to Natural and Technological Disasters*, Disaster Research Center, Preliminary Paper #147, University of Delaware, Newark, Del., 41 pp.

Rosenthal, U. (1988). Disaster management in the Netherlands: Planning for real event. In: L.K. Comfort editor, *Managing Disaster*, Duke University Press, Durham and London, pp. 274-295.

Scanlon, T.J. (1989). Toxic chemicals and emergency management: The evacuation of Mississauga, Ontario, Canada. In: *Coping with Crises*, U. Rosenthal, M.T.Charles and P.T. Hart, editors, Ch.C.Thomas, Springfield, pp. 303-322.

Showalter, P.S. and M.F. Myers (1992). *Natural Disasters as the Cause of Technological Emergencies: A Review of the Decade 1980-1989.* Natural Hazards Research and Applications Information Center, University of Colorado, Boulder, Co., 42 pp.

Shrivastava, P. (1987). *Bhopal: Anatomy of a Crisis*, Ballinger, Cambridge, Mass., 308 pp.

Thornburgh, R. (1987). The Three Mile Island experience: Ten lessons in emergency management, *Industrial Crisis Quarterly*, **1**, 5-13.

Tierney, K.J., with R.C. Anderson (1990). *Risk of Hazardous Materials Release Following an Earthquake*, Disaster Research Center, University of Delaware, Newark, Del., 118 pp.

Turner, B.A. (1991). The development of a safety culture, *Chemistry & Industry*, april, 241-243.

Turner, B.A. (1992). *Stepping into the Same River Twice: Learning to Handle Unique Management Problems*, Middlesex University Inaugural Lectures, London, 18 pp.

Voight, B. (1990). The 1985 Nevado del Ruiz volcano catastrophe: anatomy and retrospection, *Journal of Volcanology and Geothermal Research*, **42**, 151-188.

Zuboff, S. (1986). Automate/informate: The two faces of intelligent technology, *Organizational Dynamics*, Fall, 4-18.

GOVERNMENTAL MEASURES TO MITIGATE EARTHQUAKE IMPACTS IN ALGERIA

DJILLALI BENOUAR
USTHB, Civil Engineering Institute, BP 32 El-Alia, Bab Ezzouar, Alger, ALGERIA

Abstract: This paper presents the different earthquake disaster prevention and post-disaster measures taken by the government of Algeria between 1980 and 1995. It details the measures taken at the country level as well as those involving the co-operation of international organizations. It discusses post-disaster measures including the relief of victims and the methodology of the assessment of the serviceability of damaged structures after the disaster, which represents one of the most important duty of the emergency staff, are also discussed. It examines the range of needed, possible and practical pre-disaster measures which the government of Algeria should undertake and actively encourage. This study briefly presents scientific and non-scientific Algerian organizations.

1. INTRODUCTION

In Algeria, as amazing as it may appear, before the last earthquake disaster of El-Asnam (actually Cheliff) which occurred on the 10th of October 1980, the government and the population as a whole did not consider earthquakes as a threat.

The destructive El-Asnam earthquake of 10 October 1980 occurred in the Central Cheliff Valley affecting a rather densely populated region where about 900,000 people lived within 8,000 square kilometres. The surface-wave magnitude of this event was recalculated at Ms 7.5 (±0.3). The main shock (12h 25mn 24s (U.T)) and its largest aftershock Ms 6.1 (15h 39mn 09s (U.T)) caused 3,000 deaths, injuring more than 8,500 persons and they made about 400,000 homelesses; by destroying or seriously damaging at least 60,000 housing units in 24 towns and villages. This earthquake seriously affected all levels of the regional economic development. Also, the social and economic impacts were felt throughout the entire country. One important conclusion that can be drawn from this event is poor earthquake-resistance performance of the Algerian buildings and other facilities. Whether modern constructions or traditional gourbis (local house), buildings are potentially vulnerable to the recurrence of destructive earthquakes (Benouar, 1993 & 1994). It has shown that large earthquakes, particularly in northern part of the country, could be very destructive and also it has proved that additional loss-reduction measures are needed to mitigate potential earthquake-induced losses. This is not the first

M. I. El-Sabh et al. (eds.), Land-based and Marine Hazards, 229–237.

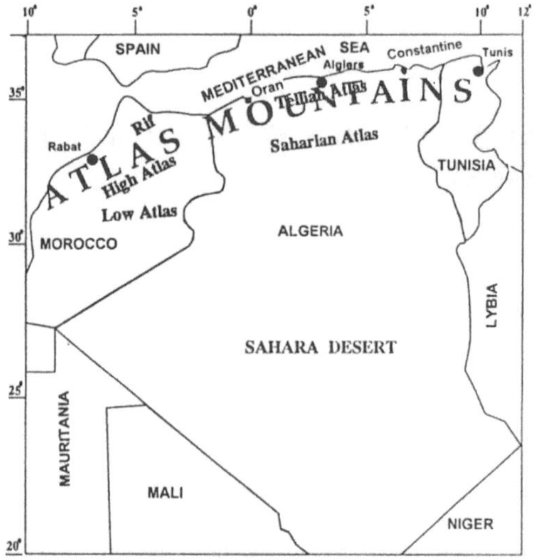

Figure 1: Map of Algeria and adjacent regions

catastrophe of its kind to have struck Algeria and there is no knowledge of when and where a similar disaster may strike again. Thus, there should be implementation of effective disaster mitigation measures to cope with such emergencies.

It is of great importance to mention that Algeria, particularly its coastal part, the seismic activity presents a real threat for the population and their properties. Algeria is in the border which separates the Eurasian and the African tectonic plates. In the case of the African plate, it is in the process of colliding on its northern boundary in the vicinity of the Mediterranean Sea with and being overthrust by the Eurasian plate. In fact, we record everyday a seismic event in the region; however, these earthquakes are of low magnitudes, often not felt by the people but it should be kept in mind that devastating earthquakes may occur again in the region. In recent history, strong earthquakes have occurred in northern Algeria in Algiers (1716), Oran (1790) and Cheliff (1867, 1873, 1922, 1934, 1954 and 1980) (Figure 1). A detailed study and analysis of the seismicity of Algeria and adjacent regions during the twentieth century is presented in Benouar (1993). The study in Benouar (1993) shows that during the twentieth century only, earthquakes have occurred with increasing magnitudes, from 5 in 1922 and 1934 to 6.7 in 1954 to 7.5 in 1980.

2. DISASTER PREVENTION MEASURES

The occurrence of natural hazards cannot be avoided, at least at the present state of our knowledge and ability; however, the impact of such events on humans and their properties can be mitigated. The analyses of the effects of all disastrous earthquakes which have occurred in Algeria during the twentieth century (Benouar, 1993) have shown that the scope of the disaster could be mitigated with a great amount if appropriate governmental predisaster measures were adopted and implemented. Petrovski (1981) has also shown the necessity to implement predisaster measures.

The 1980 El-Asnam earthquake created a serious awareness of the high short-term seismic risk of coastal Algeria (Rapport CTC, 1981). A number of factors, including environmental concerns and an increased public and officials awareness of earthquake hazards, the El-Asnam 1980 seismic event has led to a rapid increase in seismic risk mitigation. Following that earthquake, numerous measures have been taken since to reduce the seismic risk in Algeria and the most important are:

* 1980: New mission is given to the Centre of research in Astronomy, astrophysics and Geophysics (CRAAG), by the decree 2 February 1981; the CRAAG became responsible for the management of the entire Algerian seismological network.

* 1981: Appearance of the first Algerian seismic code for building design and construction (RPA 81).

* 1982: Meeting of Arab Ministers of Construction in June 1982 in Algiers where a programme for the evaluation and the reduction of the seismic risk in the Arab World (PAMERAR) was recommended, with the co-operation of the UN specialized agencies.

* 1982 - 1984: realization of a seismic microzoning study in the Cheliff region (El-Asnam previously named).

* 1983: Appearance of a new version of the seismic code for building design and construction (RPA 1983) where only minor recommendations were added to the RPA 81 seismic code.

*1985: The Algerian Parliament adopted a law 85-231 of August 25, 1985 which establishes disaster relief and civil defence organizations.

*1985: Appearance of the first recommendations fro strengthening, reinforcement and repair of damaged earthquake constructions.

*1985: With the co-operation of UNESCO, UNDRO, UNDP and AFESD, the National Centre for Applied Research in Earthquake Engineering (CGS) was created and located at Algiers.

1985: The Ministry of Construction has appointed a special governmental permanent commission to update the regulations for earthquake resistant design and construction.

*1988: Appearance of a new seismic code for building design and construction (RPA 88) where minor modifications were added to the previous ones.

*1988: Participation of Algeria to the programme for seismic risk reduction in the Mediterranean region (SEISMED).

*1990: New mission to CRAAG to seismically monitor the whole Algerian territory and to pass under the authority of the Ministry of the Interior by the decree of 17 April 1990.

*1990-1993: Revision and analysis of the seismicity of Algeria and adjacent regions during the twentieth century by Benouar (1993). This has led to a complete, homogeneous and as accurate as possible earthquake catalogue for the region limited by 20°N and 38°N and 10°W and 12°E. It also presents the first calibration of historical earthquakes as well as intensity-attenuation models for the Atlas zone and Algeria.

*1994: Publication of the earthquake Maghreb catalogue during the period 1990 - 1990 (Benouar, 1994).

1995: Earthquake hazard mapping of the Maghreb region (Algeria, Morocco, Tunisia) by Benouar (1995). This has led to the construction of seismic hazard maps for the Maghreb region in terms of absolute peak ground acceleration (PGA), expected peak ground acceleration for a probability of 10% of exceedance for economic life time of 50, 100 and 250 years and return period.

*1981-1995: Numerous seminars, symposia, conferences, training courses and workshops have been taking place in Algeria in order to update the knowledge of Algerian scientists and engineers engaged in earthquake engineering.

3. POST-DISASTER MEASURES

The assessment of the serviceability of different damaged structures after the disaster is one of the most important duties of the emergency staff. This task is always accomplished by the specialized engineers, which are familiar with the problems of statics and dynamics of structures, of the Organisme de Controle Technique des Constructions (CTC), a governmental body under the authority of the Ministry of Urban Planning and Construction. The main task of these engineers is to use the same criteria and methodology to classify the damaged constructions. Regarding the considerable stake of this task (safety of the population and safe-keeping of the building patrimony) and the difficulties to conciliate the social and economic imperatives (recuperation of the maximum number of constructions and the quick return to normal life), the CTC (Rapport CTC, 1981) has proposed a methodology to evaluate the damage caused to the constructions which gives a certain garantee to the validity of the classification of damage. To this aim and since the evaluation of damage should be carried out uniformly to allow the engineers involved in this duty to present a homogeneous assessment, a guide with approximately sixty informations to the field engineers. Therefore, all the constructions in the stricken zone are marked with one of the following colors:

* Green: Constructions slightly damaged or not damaged at all and which may be used.

* Orange: Constructions which are damaged and need a second or more detailed expertise to decide whether or not these buildings may be retrieved in taking into account the importance of damage and the cost of repair.

* Red: Heavily damaged constructions which should be destroyed.

This methodology by means of three colors was used in the El-Asnam 1980, Constantine 1985 and Chenoua-Tipaza 1989 earthquakes. However, the evaluation of damage from the point of view of serviceability is a very responsible and complex task since nobody is able to predict if other strong shocks could occur in the near future. In many cases, within six months after the main shock, aftershocks occur sometimes as large as the main shock further damaging the constructions already affected. Therefore, the decision about the serviceability of the constructions should be taken with a great deal of caution, keeping in mind the probability of occurrence of aftershock.

Civil protection stations have been installed across the country in all administrative units Baladia (City), Daira (Borough or County) and Wilaya (District) and also in major industrial plants. The most important role of these units of the civil protection after a disaster is to save as many buried victims from the building ruins, to treat and transfer victims to hospitals and send homelesses to temporary shelters, to give first medical aid, fight fires and make sure that electricity, gas and water systems in the zone affected are shut as quickly as possible after the disaster.

The Algerian Red Crescent (ARC) is also present in all major cities across the country. The main responsibility of the ARC is to give first medical aid, install and manage the shelters (tents), provide victims with food, water, clothes and transportation.

The army, which has also units well distributed across the country, has its own organization for emergency response and management. They provide food, shelters, medical aid, transportation and security to the victims and protect their damaged and non damaged properties from looting.

Although there is no governmental or non-governmental organization which uses the local people in an emergency; El-Asnam 1980 and Chenoua-Tipaza 1989 earthquakes have shown that the help of volunteers from the residents of the stricken area is of great importance during the first hours after the disaster. In these last two destructive earthquakes, many people were rescued and fed by the local population before any official relief team arrives to the site affected.

4. NEED IN OTHER PREDISASTER MEASURES

The range of needed, possible and practical predisaster measures which the government of Algeria can (or may undertake) and actively encourage are as follows:

* Identification of the probable natural hazards that may occur in Algeria

* Identification of high risk areas

* Soil occupation management

* Enforcement of the seismic building code

* Policies to assure the safety of lifelines, critical facilities, essential utilities, emergency services and the vital economic sectors such as agriculture, industry, banking, etc.

* Make earthquake engineers aware of the continuous need for emergency issues in their ongoing programmes

* Design of procedures for emergency management including warning, mobilisation, search, rescue, emergency health actions, sheltering of victims, protection of vital supplies and telecommunications

* Reinforcement of the emergency skill training in the crisis and essential services such as police, army, health sector, public works, etc.

* Involving citizen organizations, local communities, uniformed youth groups (scouts, youth organizations, etc..), schools, the private sector and non-governmental organizations (Mosques, ...) in emergency management

* Public information and education must be an essential component of disaster preparedness and response

* Education programmes should be supplemented by campaigns through television, radio and the press

* Development of internal training programmes for all the civil servants and, employees of the industrial and agriculture sectors

* Monitoring of food and relief supplies and the maintenance of adequate quantities

* Examination of the relevant laws which could be activated under an emergency in order to prevent any possible civil disorder

* Encouraging volunteer organizations to specialises in disaster management.

5. SCIENTIFIC ORGANISMS RELATED TO SEISMIC RISK REDUCTION

Actually, there are essentially, in Algeria, furthermore to the major universities in the country, two research centres which deal directly with seismic risk reduction.

1. CRAAG (Centre of Research in Astronomy, Astrophysics and Geophysics) which is located at the Observatory of Bouzareah (Algiers). For a better use, this centre was transferred from the authority of the Ministry of Higher Education to the authority of the Ministry of the Interior by decree of 17 April 1990 (JO, 1990).

*** Missions of the centre**

- Research and study in the fields of astronomy, astrophysics and geophysics and transmittal of the results obtained to the concerned services of the government authorities and the operational structures

- Insure a permanent monitoring of natural geophysical phenomena

- Develop and Extend the seismological, geomagnetic and gravimetric networks to the whole country

- Insure the use of the results obtained, the management and the maintenance of the equipment

- Establish the earthquake catalogue and seismic hazard maps, collect and keep all historic documents of the seismicity including those in foreign countries to contribute for the constitution of an Algerian seismic history data bank

2. The CGS (National Centre for Applied Research in Earthquake Engineering) was established and attached to the Ministry of Constructions by decree 85-71 of 13 April 1985 (JO 1985). This centre is a governmental body which is supposed to establish the procedures by which the analysis of the seismic resistance of buildings should be carried out.

*** Missions of the Centre**

- Seismic macrozoning and microzoning of Algerian high seismic risk regions
- Seismic risk and hazard assessment
- Research to improve earthquake resistant buildings and constructions
- Update the regulations for seismic resistant building design and construction
- Vulnerability and seismic risk reduction
- Improve significantly the scientific and technical knowledge of Algerian engineers, architects and planners in the field of earthquake engineering.
- Monitoring the national strong ground motion network.

6. SEISMOLOGICAL SURVEY NETWORKS

There are two seismological networks to survey the seismic activity in Algeria: the first one is the National Telemetered Seismological Network (NTSN) which is managed by the CRAAG and the second one is the Accelerograph National Network (ANN) which is managed by the CGS.

The NTSN whose programme started in 1982 under the supervision of the CRAAG. This national network is subdivided into 4 sub-networks located at Oran, Cheliff, Algiers and Constantine (Rapports internes CRAAG). Each one of these sub-networks has the following configuration:

- Three component recording stations: one vertical and two horizontals
- Seven stations with one vertical component radio linked to the recording central station in Algiers.

The four sub-networks are linked to the central recording station in Algiers-Bouzareah (Within CRAAG) by specialized telephone lines. A total of 32 recording stations are supposed to cover the entire northern part of Algeria.

The ANN, which is managed by the CGS, started in 1983 with 90 accelerographs type SMA-1 installed over Northern Algeria with the co-operation of the Ministry of Constructions represented by the CTC (Controle Technique des Constructions). Actually, the CGS has bought 120 accelerographs to install over northern Algeria. The maintenance of the equipment and the use of the results are assured by the CGS (Rapports internes CGS).

It is of interest to mention that dams are also equipped with seismological equipment which are run by the National Dam Agency.

7. CONCLUSION

This paper shows the interest the Algerian government is giving in seismic risk reduction, emergency management. This strategy has given good results during the Constantine 1985 and Chenoua-Tipaza 1989 earthquakes. During these two earthquakes, the different emergency and essential services such police, army, civil protection, Algerian red crescent and even the local communities co-ordinated adequately their efforts to save and help the victims. The two research centres also play an important role in mitigating the seismic risk by communicating their findings to the concerned services of the government charged with seismic risk reduction. However, despite all the efforts made by the government since 1980, there are still some needed and necessary measures to be taken and implemented before the next seismic disaster.

8. REFERENCES

Benouar, D.(1993) The seismicity of Algeria and adjacent regions during the twentieth century, Ph.D thesis, Imperial College, University of London, pp712.

Benouar, D.(1994) Materials for the investigations of the seismicity of Algeria and adjacent regions during the twentieth century. *Annali di Geofisica*, Vol. XXXVII, No.4, July.

Journal Officiel de la République Algérienne Démocratique et Populaire of 15 May 1990, 25 March 1985 and 12 June 1985.

Petrovski, J.(1981) Earthquake engineering aspects of El-Asnam 1980 earthquake, studies and actions for seismic risk reduction. Rapport Habitat Alger, Skopje, June.

Rapport général sur le séisme d'El-Asnam du 10 Octobre 1980, CTC, Alger, 1981.

Rapports internes du CRAAG. 1980 - 1995.

Rapports internes du CGS. 1985 - 1994.

UNIFIED NATURAL HAZARD MANAGEMENT: LESSONS FROM FLOODING

FRANK H. THOMAS[1]
Federal Emergency Management Agency/MIT, 500 C Street, SW, Washington, D.C. 20472, U.S.A.

Abstract. In the United States, authority for flood loss reduction programs is fragmented among thousands of federal, state, local and private sector entities. Experience of the National Flood Insurance Program, has demonstrated the merit of pursuing a unified national program to coordinate the activities of these entities. This experience is applicable to the mitigation of earthquake and hurricane hazards which are often geographically coincident with flood hazards. Lessons from flood experience suggest that the coordination of standards and application of technology are insufficient to achieve unified natural hazards management. In addition, national goals are essential to provide focus and cultural values must be explicitly recognized and accommodated or modified.

1. INTRODUCTION

In the United States, methods to reduce losses of human life and property from floods, earthquakes, hurricanes and other natural hazards have developed independent of one another. The strategies and tools available to reduce losses are basically the same for each hazard. Application of the tools encounters the same problems of societal resistance. In some instances the application of tools to one hazard increases the risk of loss from another hazard. The need for an effective unified national program for natural hazard loss reduction is recognized. How can the nation achieve a unified natural hazard mitigation program wherein the application of mitigation tools to one hazard complements and supports reduction of risk for other natural hazards?

This presentation provides guidance for a unified program by drawing upon lessons from national experience with floodplain management programs, particularly that of the National Flood Insurance Program. Floodplain management experience is summarized and then three sets of lessons learned are examined.

---

[1] Present Address: 110 Tanasi Court, Loudon, TN 37774-3130, U.S.A.

M. I. El-Sabh et al. (eds.), Land-based and Marine Hazards, 239–244.

2. FLOODPLAIN MANAGEMENT IN THE UNITED STATES

Management of the nation's floodplains is conditioned by over 200 years of water-focused development and individual property rights values. Individual property rights are preserved within the Federal and the State constitutions which also partition governing authority among the Federal, State and local governments. Consequently, authority for hazard mitigation tools is spread among 27 federal agencies, numerous agencies in each of the 50 states, 20 thousand local governments and millions of private land owners. These entities may be competitive, redundant, or joined in cooperative ventures. The management of all natural hazards is affected by this partitioned authority.

Federal responsibility for dealing with flooding began to develop in the early 1900's in both the natural and the human resource program areas. Conceptually this development is analogous to a dendritic stream pattern wherein many small tributaries join to form two major streams and then form a main stem.

Prior to the 1960's four major Federal human resource programs focused on the strategy of controlling flood waters by constructing systems of reservoirs and levees such as that used on the Mississippi River system. Little priority was placed on nonstructural alternatives. Federal natural resource programs focused on the conservation and preservation of floodplain environments.

The 1960's experienced a rapid growth in recognition of nonstructural strategies and the acceptance of environmental floodplain management objectives. The importance of community property rights increased relative to individual property rights.

For example, new state and local floodplain land use regulations designed to prohibit individual property owners from actions transferring flood risk to neighboring or downstream properties were upheld by the courts. Also, new water quality regulations designed to protect water supply and fish and wildlife resources were upheld by the courts.

A Congressional call for "A Unified National Program for Floodplain Management" resulted the definition of floodplain management as a decision-making process. This process seeks land use compatible with the risk to floodplain environmental and human resources. Compatibility is achieved through application of the "best mix" of loss reduction measures applied to any given floodplain.

Related to the call for unification, the National Flood Insurance Program (NFIP) was created to transfer costs of flood losses to floodplain occupants. The benefit of flood insurance is provided to citizens in return for local government regulation of land use to mitigate future flood risk. The NFIP is a package of nonstructural loss reduction measures including land use management standards, hazard identification, floodproofing, flood damaged property purchase [open space], and insurance (prepaid disaster assistance and rate incentives for mitigation). The NFIP has become the leading force in national floodplain management. National interest in floodplain management has never been greater.

The first comprehensive national assessment of the status of floodplain management was published in 1992. Recent severe coastal flood damages have resulted from four major hurricanes and record 1993 flood damages along the Mississippi River system. Intense scrutiny is focused on floodplain management programs. At the same time, forecasts of severe West Coast and Midwestern earthquakes have led to proposals for multi-hazard mitigation.

3. FLOOD LESSONS: IMPLICATIONS FOR MULTI-HAZARD MANAGEMENT

After 25 years experience addressing directly or indirectly most structural and nonstructural floodplain management tools, the National Flood Insurance Program offers multi-hazard management lessons dealing with loss reduction tools, behavioral responses, and long range national goals.

3.1 Loss Reduction Tools

The first lesson from loss reduction tools is that the government's responsibility to protect the life, health and safety of its citizens provides the strongest argument to support hazard mitigation. Because mitigation tools require expenditure of additional dollars, strong political pressure is exerted for reduction or exemption from mitigation requirements. However, few persons will publicly take a position against the protection of life, health and safety.

Second, a national risk standard understood by the general public is prerequisite to a successful hazard mitigation program. The national base flood standard of one percent chance flood is not perfect, but it is nationally accepted and applied. The widespread adoption of this standard stands in contrast to current problems of standards for other hazards or wetlands.

Third, loss reduction decisions have a limited life expectancy. Flood control projects built 40 years ago for a single purpose now are expected to perform additional purposes such as recreation and water supply. Urbanization of watersheds has reduced the protection level of flood control projects. Floodplain managers and the public must understand that historically sound mitigation decisions may have to be modified to recognize new conditions. Recognition of the need to reevaluate flood control projects was a major problem following the 1993 Mississippi River floods.

The fourth lesson is that State and local land use regulations designed to reduce flood risk are highly effective but require a decade or more before significant benefits can be realized. Public acceptance of land use regulations is a slow process of cultural assimilation. Success requires frequent contacts with local officials, careful record keeping, a fair and equitable system of incentives to enforce obligations, and an appeals process.

3.2 Behavioral Responses to Mitigation Tools

New policies, programs, and technologies to effect changes in flood hazard or the risk to life and property frequently have been enacted or implemented without being sensitive to values held by floodplain stakeholders. In such instances, the unnecessary result has been outright rejection as unfair or insufficient priority. Behavioral responses have too often been overlooked.

The first lesson is that human behavior responds to benefits and penalties. The NFIP is a trade-off of insurance benefits for regulation of floodplain development. Initially, insurance alone proved an insufficient enticement for many communities to adopt unpopular floodplain land use regulations. Participation in the program was stimulated significantly by making the purchase of insurance an eligibility requirement for other federal financial benefits such as federally backed mortgages and federal flood disaster assistance. Careful balancing of benefits and penalties is essential.

Second, the values of individual floodplain occupants are hard to anticipate. One recent mitigation initiative to compensate property owners with 110 percent of their property value for relocating their structures prior to damage or destruction by coastal erosion had a surprisingly low acceptance rate. Non-economic values such as "family summer home" were more important. Behavioral projections must be improved and utilized to anticipate acceptance of mitigation practices.

Third, the values of institutions are hard to change. Government agencies, such as those managing disaster relief programs, want to continue to provide rapid financial assistance without deference to opportunities for long term reduction of flood risk. The result often is perpetuation of bad floodplain location and construction decisions. Persistent effort is needed to bring about change.

The final behavioral lesson is that regulatory programs stimulate creativity. Floodplain land use regulations have created a tremendous incentive for property owners to minimize the impact of the rules. Many property owners and local governments have developed highly effective, innovative loss reduction techniques. Incentives such as reduced insurance rates are needed to encourage compliance and harness this creativity to improve floodplain management, not evade it.

3.3 Long-Term National Goals

From its inception, the " Unified National Program for Floodplain Management" focused on programs, projects, and technical standards to achieve coordination among the programs of different levels of government and the private sector. Technical progress has been achieved but a broad program of national uniformity has been unattainable. In 1994, a new strategy for attaining a more unified national program focused on establishment of four national goals toward which all governmental and nongovernmental floodplain management decision makers can strive over a period of years. The goals are as follows:

1. Goal: Establish a National Goal-Setting and Monitoring System. A lead Federal agency needs to be charged by statute with convening floodplain constituency representatives for systematically setting, monitoring, and revising national floodplain management goals. A national forum on floodplain management goals should be convened to formalize the national goal process.

2. Goal: Establish Positive Public Attitudes Toward Floodplain Management. This requires a national leadership strategy to encourage acceptance of responsibility and mitigation initiatives by floodplain decision-makers. New financial incentives also are needed to credit the integration and application of floodplain management tools. The public needs to understand that mitigation is required when proposed floodplain development adversely impacts public or private property or natural floodplain functions such as water supply.

3. Goal: Establish In-House Floodplain Management Capability Nationwide. Comprehensive floodplain management training and professional certification need to be made available to assure resident professional floodplain management capability at all governmental levels. Professional floodplain management services need to be provided to previously unserved non-metropolitan areas.

4. Goal: Reduce by one-half the number of high risk floodplain properties and high risk environmental functions of the nations floodplains. All existing floodplain structures and environmental functions need to be inventoried preparatory to initiating a mitigation program to eliminate the most serious risks within 25 years.

4. CONCLUSION

Floodplain management experience affirms the need for a unified national hazard management program. Experience also shows that most loss reduction strategies and tools are common to all natural hazards, but, reliance upon coordination methods and technical standards is insufficient to achieve a unified program. Common national goals acceptable to all hazard area decision-makers are necessary and these must address both human and natural resources. We have also learned the following:

1. Publicly accepted national risk standards for each hazard are prerequisite to achieving consistent, effective mitigation.

2. Hazard specific application of individual standards must be compatible with risk from other hazards [building codes & infrastructure standards].

3. The costs of mitigation are met with individual and governmental resistance because the benefits of mitigation are long term and the public believes the next disaster will not affect them.

Reflecting on this experience, it may be concluded that because achieving acceptance of mitigation practices is a slow process, the ultimate prerequisite to successful multi-hazard mitigation is patience.

5. SELECTED REFERENCES

Federal Emergency management Agency (1994). A Unified National Program for Floodplain Management, FEMA-248, Washington, D.C., 43 pp.

Federal Emergency management Agency (1992). Floodplain Management in the United States: An Assessment Report, Vol. I, FIA-17, Vol. 2, FIA-18, Washington, D.C., 70 pp. and 501 pp.

Federal Emergency management Agency (1993). Building Performance: Hurricane Iniki in Hawaii: Observations, Recommendations, and Technical Guidance, FIA-23, Washington, D.C., 100 pp.

Federal Emergency management Agency (1993). Building Performance: Hurricane Andrew in Florida: Observations, Recommendations, and Technical Guidance, FIA-22, Washington, D.C., 93 pp.

ON THE NATURAL DISASTER REDUCTION POLICY AND MANAGEMENT SYSTEM IN CHINA

HUATING YANG
China Institute for Marine Development Strategy, State Oceanic Administration, No 8, Dahuisi Str., Haidian District, 100081 Beijing, China

ABSTRACT. China is one of the world's most disaster prone countries, having been hit over the centuries with famine, flood, earthquakes, to give only three examples. China has fought these disasters with massive works projects and with relief for hundreds if not thousands of years. This paper deals with the historical investigation of natural disaster reduction in China and the analysis of its three phases: the empirical, semi-empirical and the development of an integrated disaster response policy under the theory of system engineering. In order to suit this development the Chinese Government will have to adjust its disaster reduction policy "give top priority to the disaster's prevention and to combine it with the relief operation for an integrated system of response". A proposal is made to reform the existing management system of disaster reduction and to draw up a national law of integrated disaster reduction management.

1. ON FORMATION OF DISASTER REDUCTION POLICY

1.1 Main Experience

China is a country with vast territory, large population and frequently hit by almost all kinds of natural disasters except volcanic eruptions. China has fought those disasters with massive works projects and with relief for hundreds if not thousands of years. Therefore, the history of the Chinese nation is a history of combating natural disasters and a history of disaster reduction. For example, China has created great water conservancy engineering projects of very long standing such as the Dujiangyan Water Division and Irrigation Projects of Sichuan Province and Sea Dike on the coast of Hangzhou Bay in Zhejiang Province. During the long period of struggle practice, Chinese people have accumulated rich experience in reducing damages and losses of disaster. The disaster reduction that refer to the national economy and the public livelihood is almost a national policy in every dynasty throughout history.

The area along the lower reaches of Yellow River is the main national economic zone of past dynasties, but the situation of "hung river" is very unfavourable for this area, so that here becomes "the Misery of China". After Zhou Dynasty (1100-256 B.C.), every feudal

M. I. El-Sabh et al. (eds.), Land-based and Marine Hazards, 245–256.

dynasty almost had organization of water-control administration, which was responsible for "construct levees and canals, link up rivers and waters, rush down flood and waterlog, reservoir water to use in risk period". The Qing Dynasty (1616-1911 A.D.), was titled as governor of river, and the number of river soldiers exceeded that of river worker, so that the river protection system became specialized flood protection army. The cost in Yellow River flood prevention is amazing, for example, in the year of Daoguang 24 (1845 A.D.), Qing Dynasty, the cost used in blocking up was more than 11.9 million liang silver (1 liang=1.10 oz.), which was about 1/4 of the state revenue at that time! Ancient policy to Yellow River flood reflected the spirit of restraining damages to the lowest limit spent. This represents a stability of political power and safety of the society. It also represents a contribution to the economy of the country, e.g., setting up the flood detention basins and the constructions of flood control is a kind of measure that give the benefits of undeveloped area in exchange for the safety of the important economic areas (Zhou, 1990). Futhermore, most of the governors paid more attention to relief measures following disasters in order to avoid peasant uprising.

Because of the limitation of society conditions and of science and technology related to disaster reduction at that time various kinds of nature disaster continued to occur until modern times. Since the founding of the People's Republic of China in 1949, while going all out for economic construction, the Chinese Government have led the Chinese people to exert their efforts in mitigating natural disasters, and have allocated huge financial and manpower resources for this purpose. A series of strategies and policies were adopted that conform to the realities in modern China.

In the early period following the founding of the People's Republic, with the Ministry of Water Conservancy and Ministry of Interior as the core, the Chinese government set up the Central Commission on Natural Disaster Relief to take direct charge of the management of natural disasters. And, in the light of the prevailing conditions of frequent floods and droughts, the Commission formulated the disaster relief policy of "Self-help through production, practicing economy to tide over starvation, expanding mutual help, providing work as a form of relief and supplemented by the necessary relief funds"; and gigantic projects were implemented to harness the historically flood-causing rivers, such as the Yangtze River (Changjiang River), Yellow River (Huanghe River), Huaihe River and Haihe River.

After tiding over the economic recovery period (1950-52) during which successive severe natural disasters have occurred, the Chinese Government initiated natural disaster reduction activities across the country, including massive afforestation drive and large scale efforts to prevent and control crop and forest diseases and insect pests. In the mean time, the Government set up the Central Meteorological Administration (currently called China Meteorological Administration), State Oceanic Administration and State Seismological Bureau, to meet the demands of disaster management in China. In the 1970s, the Chinese Government further set up relevant departments, professional and research institutions to deal with the mitigation of natural disasters in the fields of flood, drought, meteorological, marine and coastal, seismic, geological disasters, as well as agricultural and forest disasters. The capability of preventing and mitigation natural disasters was thereby further improved.

Since the advent of the 1980s, China has persisted in her guiding principles of reforms and opening up. With the progress in socio-economic development, the nation's awareness of the damaging harm of natural hazards has been raised extensively. The existing different types of disaster reduction systems have been gradually modernized. Satellite technology, remote sensing technology, modern communication, computer system, artificial intelligence and various up-to-date monitoring and forecasting equipment have been increasingly used in disasters monitoring, forecast, prevention and rescue, thus providing China with an initial capability to apply modern science and technology to mitigate natural disasters. At the same time, we formed and carried out the disaster reduction policy of "give top priority to the disaster's prevention and to combine it with the resistance and relief operations" in the works of disaster reduction. That is: "in working out the economic development plan, the Government, according to relative laws and regulations, combine closely the constructions of the disaster reduction facilities with the production development enhances the measures and capabilities of preventing disasters; whenever a disaster happens, the emergency rescue actions will be taken to minimize the disaster effects by mobilizing the people and using the effective measures. Afterwards, the Government works out a specific principle on relief operations, that is, "rely on the masses and provide for and help themselves through production, mutual help and mutual aid, supplemented by necessary state relief and supports" (Chen, 1991; Li, 1991).

Since 1989, China's disaster reduction policy has entered a new phase. The Central Government and a number of local governments have established integrated disaster reduction organizations to conduct researches into policy measures on disaster reduction, to expand disaster reduction management and planning, to enhance co-operation with international community, and to further strengthen publicity and education in disaster prevention and mitigation. Meanwhile a large number of disaster reduction projects have been screened, formulated and implemented and activities relating to natural disaster reduction are being progressively incorporated into the nation's overall programme of socio-economic development.

1.2. From Empirical to Semi-theoretical Disaster Reduction

Having analyzed the history of disaster reduction in China, we can say that the disaster reduction policy almost lasted two periods, which are empirical and semiempirical-semitheoretical disaster reduction respectively. The weakness of disaster management policy during these periods is the fact that it was not integrated; each type of disaster was dealt with separately. Now we are creating conditions to go to the third period of disaster reduction based on the theory of system engineering.

The empirical disaster reduction period lasted a long time in China, almost from the time of Dayu, Xia Dynasty (2200-1700 B.C.), who dug canals, and divert waters into the ocean, to the 60th and 70th decades of this century, when we set up and carried out the system that "the Central Government integrate the policy decisions, the Government Departments co-ordinate closely with each other to implement the tasks assigned in charge of disaster reduction respectively according to the integrated policy decisions and their own functions" (Chen, 1991; Li, 1991). This policy is based on the principle that knowledge is mainly stemmed from disaster prevention and reduction experience of social practice, that is to say: when we face a problem, we solve it. Facing natural disaster is almost at the stage of passive management. The theory of disaster prevention during this period, such as the theory of paying attention to agriculture, the theory of reservation, the theory of cultivation, the theory of loan providing, the theory of economy, etc., was combined with the theory of system engineering in disaster's prevention and reduction, but all were at the single original stage. After 1949, more attention was paid to disaster reduction than before. The input in disaster reduction was incomparable in Chinese history, but in generally, it was a passive situation that "when water comes, deal with water; when locust comes, deal with locust". The slogan proposed in this stage, such as "no starve to death" in relieves, "people can beat natural" in disaster reduction, all have much subjectivity. On the other hand, in disaster reduction policy , we only have dealt with each single disaster separately, and no nation-wide integrated all-disasters reduction policy was available..

The semi-theoretical disaster reduction stage has generally been formed step by step with the formation of departments with multiple branches of disaster reduction in China. Its main character is: recognize the development of each disaster by means of the method of modern scientific research. The basis of the government and society disaster reduction activities embraces two parts of experiment and scientific theory. A more clear classification for disaster counter-measures and reduction policies was formed. The achievement of disaster reduction activities is very decentralized, but the disaster reduction policy as "to give top priority to the disaster's prevention and to combine it with the resistance and relief operations" has almost been formed across the country. The importance of disaster prevention and mitigation through science and technology is emphasized. The system is established and carried out so that central and local government could integrate the policy decision and that government departments (include the local government departments) co-ordinate closely with each other to implement the tasks assigned according to their own duty and functions.

1.3. On Formation of New Policy of Disaster Reduction

The effectiveness of the Chinese policy of disaster reduction can be illustrated by the following:

- In the course of economic development, the Chinese Government has allocated the necessary financial resources to disaster reduction. According to preliminary statistics, China investigated, since 1949, 300 billion yuan approximately $40 Billion U.S. through various channels into disaster reduction. The economic benefits have been estimated to be double that value, and the social benefits have also been remarkable.

- Regarding the construction of disaster reduction engineering projects. Over 40 years, China has cumulatively built over 100 billion yuan's (about $12 Billion U.S.) worth of fixed assets in facilities relating to flood prevention, drought combating and water conservancy. Moreover, the nation has completed reinforcement projects against earthquake in 14 main railways, 90 key power plants, 6 major petroleum pipelines, 20 large oil refineries, some super-large ethylene projects, key iron and steel plants, and large-scale reservoirs. Furthermore, China has intensified efforts at afforestation and construction of shelter forests, so that the forest coverage has been raised from 8.6% in the 1960s to the current 13.92% level. In addition, in the last ten years, 1.5 hundred million ha of land areas affected by pests and other ecological disasters have been brought under control annually, and in recent years, disaster prevention bases have been built up in the nation's pastoral areas or grazing land.

- Regarding establishment of monitoring and early warning systems. So far, China has set up a meteorological monitoring and forecast network consisting of more than 2600 weather stations, a hydrological monitoring system encompassing 8000 precipitation and water level monitoring stations, a pre-seismic observatory network comprising over 1000 stations, a monitoring and forecast system consisting of 3000 plant disease and insect pests monitoring stations, a predicting and early-warning system on grassland insect pests and rodent, encompassing 240 stations, a monitoring and early-warning network on marine and coastal disasters consisting of 280 ocean-watch and tidal stations, in addition to the forest and grassland fire monitoring systems and geological disaster surveying and warning systems, etc. Furthermore, telephones, radio communication networks, television and local radio networks have been extensively used in the timely transmission of disaster-related information.

- Regarding organization of disaster relief operations. In an average year (i.e. in the absence of severe natural disasters), the Central Government annually allocates 1.4 billion yuan (is 175 Million U.S.) and more than 1.5 million tons of grains as well as a considerable amount of materials and daily necessities for disaster relief operation; and these are duly distributed, thereby satisfying the basic needs of disaster stricken people. In addition, natural disaster insurance has also been expanded.

- Regarding rescue operations. With the establishment of national networks for epidemic disease prevention and medical rescue, emergency medical treatment has been given to a great number of the sick and the injured. As a result, epidemic diseases have been controlled or eliminated in the seriously hit disaster areas, particularly in the aftermath of disastrous floods.

- Regarding relevant science and technology. China has strengthened disaster reduction through the application of relevant modern science and technology. To date, over 100 specialized institutions have been set up and are engaged in disaster reduction research across the country. And the State has allocated an additional input of one hundred million yuan ($13 Million U.S.) for the scientific and technological research on disaster reduction in the 8th Five-Year Plan period (1991-95).

- Regarding publicity and public education. Extensive publicity and education in disaster reduction has been conducted, and legislation relating to disaster reduction has also been strengthened.

However, the weakness of our disaster reduction system is also clear, For example: (1) Disaster reduction in China is a complex system engineering, and lacks an integrated strategy. In developing the "vertical" disaster reduction system among the departments concerned with water conservancy, meteorology, oceanology, seismology, agriculture disaster, forestry disaster, geology disaster, relief after disaster, etc., of the State Council scarcely considers all hazards approach and connection of disaster reduction systems, so that a strongly united integrated disaster reduction system has not been formed. More duplication and squandering make ratio of input-output of disaster reduction is very poor, the economic efficiency of disaster reduction are also very low. (2) Disaster reduction lacks the link between theory, experimentation and decision-making. (3) The laws and regulations about disaster reduction are very poor, or have no law to depend on. (4) Lack a disaster reduction public education and training program, each type of personnel do not know what to do when disaster comes, so that large damage occurs.

Based on these considerations our natural disaster reduction policy should emphasize the character of integrated disaster reduction, i.e., it should be adjusted into new policy as "to give top priority to the disaster's prevention and to combine it with the relief operations and the integrated disaster reduction". We should consider the following to elaborate programs, plans and measures for disaster reduction:

- Viewing disaster reduction as an important aspect of promoting social progress and economic development. China has incorporated disaster reduction into the overall national socio-economic development programme.

- The guiding principle for disaster reduction, formulated by the Chinese government is "Give top priority to disaster prevention and integrate this with disaster-combating and relief operations".

- China's major policy measures improving the quality on disaster reduction are: (1) Accelerating the establishment and strengthening and monitoring and forecast systems on natural hazards, stepping up construction of disaster reduction engineering projects, and integrating these with economic development so as to raise the nation's overall capability of disaster prevention; (2) After occurrence of disasters, mobilizing all the necessary forces and resources to minimize loss in life and property to guarantee the livelihood of disaster-stricken people and to mitigate the disaster effects; (3) Further improving the related work, including improvement of relevant laws and regulations on disaster reduction, and strengthening publicity and education to raise the awareness of the whole nation; (4) Giving full attention to the role of modern science and technology in disaster reduction.

2. ON THE SETTING-UP OF INTEGRATED DISASTER REDUCTION MANAGEMENT SYSTEM

2.1 Analysis of the Natural Disaster Management System in China

In the above section, we have introduced the formation process of the natural disaster management system in China. As we have seen it finally results in a "vertical" management system that is responsible for the empirical and semi-empirical natural disaster reduction stages. We integrate these management systems according to the principle of disaster reduction system engineering. It is considered that the disaster reduction system include the following three aspects.

The first is information system, including 1) the monitoring and observing of the process and phenomena which lead to damage and other related factors, etc.; 2) the collection of data and information in real-time or semi-real-time transmission to these centres by means of advanced telecommunication; 3) the real-time processing of data, analysis and disaster assessment in every level of these centres, 4) issuing disaster warning and forecast; 5) disaster warning dissemination, etc. We can also regard 1), 2), 3) as the system of collection and analysis of disaster's information, and 4), 5) as the disaster warning system.

The second aspect is emergency affairs, including: 1)command, management and organization of disaster prevention, resistance and relief; 2) disaster recovery and modification after disaster, etc.

The third aspect is disaster reduction engineering, including disaster prevention engineering, disaster control and land conservation, etc.

Regarding disasters with global or regional scales (an example of the regional is the Northwest Pacific Ocean), such as meteorological disaster, marine disaster, etc., there are also global or regional co-operation organizations and programmes: e.g. WWW, GTS of

WMO, Global Ocean Observing System (GOOS), Pacific Tsunami Warning System (PTWS) of IOC/UNESCO, Global Maritime Search and Rescue System (GMSRS) of the International Maritime Organization (IMO), etc. Such co-operation should not be limited to the area of exchange of science and technology, but also should include the global and regional system of operational disaster reduction as a part of the national disaster reduction operations for each country.

In fact, there are five aspects in the disaster reduction management system of China, i.e. the acquirement and analysis of disaster information, disaster warning and dissemination, disaster emergency measures, disaster reduction engineering and international co-operation.

For more than forty years, the natural disaster management system gradually elaborated and developed in China have resisted many enormous natural disasters (especially floods), and earthquakes reduced the damages, prevented various medium and small disasters that occur frequently, even controlled the occurrence of some disasters, and contributed to social and economic benefits, and also prompted the zeal of every profession in developing disaster reduction. The disaster reduction science and technology for each disaster, e.g. monitor station network, forecast system, theory and method of research, etc., development is relatively rapidly. Some of them are advanced in other parts of the world. However, this kind of decentralized disaster management system has many faults. It is clear that an integrated system of disaster response in China is needed.

2.2. Necessity of Setting-up of Integrated Natural Disaster Reduction Management System

The main weakness of the present natural disaster reduction management system in China is decentralizing. Which cannot adopt the highly all-hazards character and approach. Such weaknesses can be illustrated by the following:

1) The separately disaster reduction systems set up without regulation and co-ordination, so that duplication and waste of energy and money cannot be avoided;

2) Single disaster information systems are usually difficult to form integrated disaster response due to the fact that each disaster is connected with others in system;

3) The Government cannot deal with the enormous and sudden disasters as such as the Tangshan Great Earthquake in 1976, and the 75.8 (on August 1975) flood in Henan Province. It should be stated that the flood disaster occurred in basins of Yangtze River, Huaihe River and Taihu Lake in 1991 is not a sudden one.

4) With the speedy development of national society and economy in modern China, the risk has become bigger and bigger, so that the life and economic losses associated with each natural disaster rise rapidly;

5) Because this system is not based on strict engineering principle, the effects of science and technology cannot be utilized efficiently in disaster response. Some larger disaster reduction projects refer to various professions, (e.g. Disaster Satellite Program, national integrated disaster warning telecommunication system, etc.) are difficult to carry out, let alone the establishment of united national emergency management system (e.g., FEMA in the United States);

6) There exist many laws and regulations which involve disaster reduction (some contradict each other clearly). However, disaster response still has no national laws to depend on!

After analyzing the strong and weak points of the actual Chinese disaster reduction management system, and in order to implement the policy of "give top priority to the disaster's prevention and to combine it with the relief operation and the integrated disaster reduction" efficiently, we believe that the actual separate disaster reduction management system should be transformed gradually into integrated one.

Such integrated system of disaster response will lead to the reduction of various damages in a large country like China which is one of the world's most disaster prone countries. Only with an integrated system will we be able to better utilize modern science and technology related to disaster management; ensure that the national economy developed rapidly, damage and loss of lives associated with natural disasters will not rise. It is also clear that only if the integrated system of disaster response is established, we will really be able to mitigate the negative effects of natural disasters by adopting new legislations and laws..

In the process of reforming old system into new one, we believe that we may use the experience of other countries for reference, that is legislate firstly, and then use reform construction according to the laws. That is to say, the national integrated disaster reduction law should be adopted firstly, and then reform the management system accordingly.

3. THE PROPOSAL OF INTEGRATED DISASTER REDUCTION MANAGEMENT SYSTEM IN CHINA

According to above discussion, our proposed scheme as part of China's response to the IDNDR is an integrated management system at the three levels: National, Provincial and Prefectural, which is defined by law or by work control regulations. As a system engineering for disaster reduction, this scheme consists of two parts: disaster information subsystem (right part in Fig. 1) and management subsystem (left part in Fig. 1) for disaster reduction affairs .

The main parts of disaster information subsystem are: 1) the full-time operational disaster monitoring networks under the management of the Departments concerned each disaster, 2) the disaster's data centres dealing with real-time observation data management, data editing and quality control; 3) the analysis and forecast centres responsible for real-time analysis, hindcast, modelling, forecast and warning of all disasters. As a new part of this scheme, the disaster data centres are the integration and hub of information subsystem,. and should provided with the necessary software, that means some data centres of single kind of disaster will be responsible for their function.

The disaster reduction affair subsystem consists of: 1) the disaster reduction centres responsible for disaster assessment, auxiliary decision of disaster's affairs and headquarters of disaster prevention, resistance and relief, 2) the resources and man power needed for disaster prevention: resistance, assistance, and relief, 3) the emergency monitor and watch for comprehensive disasters, and 4) the implementation and management of disaster engineering.

The State Council, provincial and prefectural offices for disaster reduction affairs, that should set up and lead disaster affairs in the system, are designated administrative governmental organs. Those organs, including their centres for disaster reduction (the operational division), may elaborate the way to integrate with disaster reduction system for single hazard.

The National Disaster Reduction Committee, whose members are the heads of Ministries, Commissions and Bureau, will co-ordinate the comprehensive disaster reduction management. The provinces would take the same way to co-ordinate the disaster reduction affairs.

REFERENCES

Chen, H. (1991) On disaster reduction strategy of China. *Natural Disaster Reduction in China*, 1 (1), 32-37.

Li, Z. (1991) Review and prospect for natural disaster reduction activities in China. *Natural Disaster Reduction in China*, 1 (1), 38-44.

Yang, H. (1990) Marine and coastal natural disasters and prevention strategy in China. In: Research on Disaster Reduction in China(Proceedings of the Symposium on

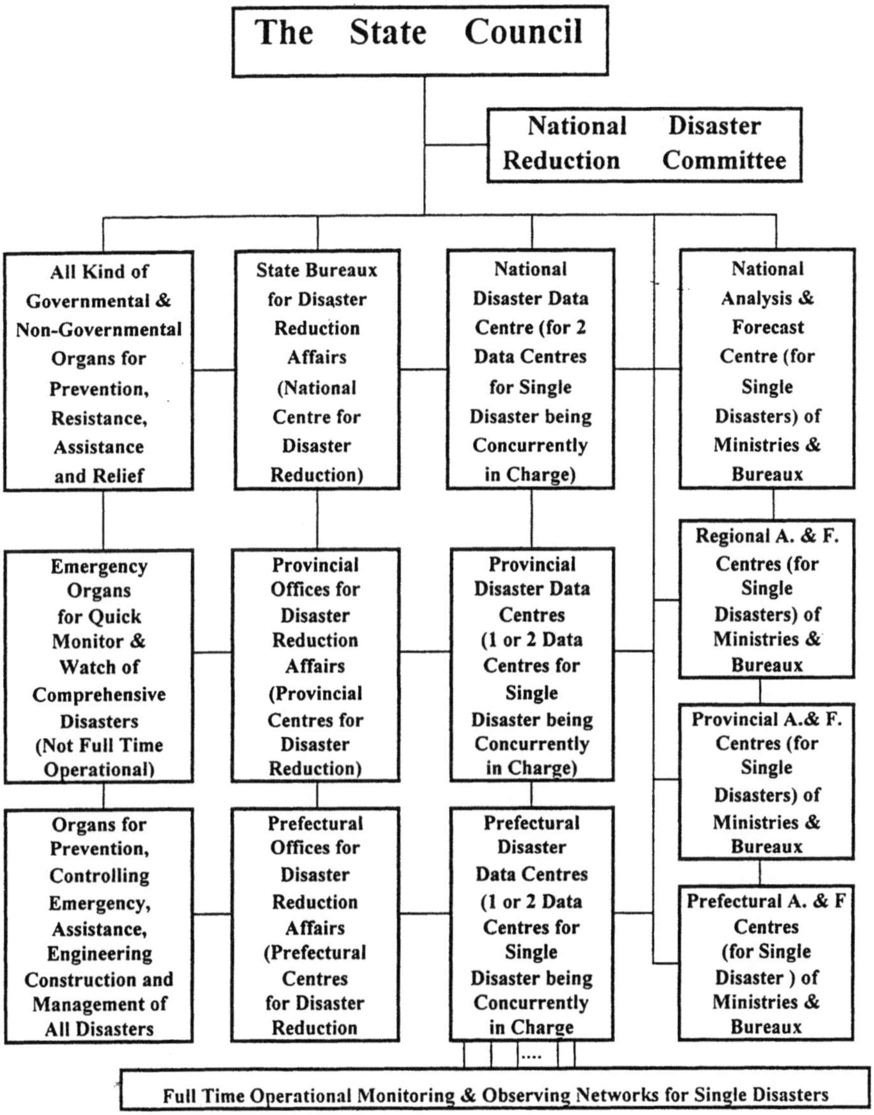

Fig. 1 Scheme of Proposed Integrated Disaster Reduction Management System in China

Natural Disaster Reduction (in Chinese), China Association for Science and Technology, editor, China S & T Press, Beijing, 463-497 .

Zhou, K. (1990) History and situation of disaster reduction in reaches of Yellow River, In: Research on Disaster Reduction in China(Proceedings of the Symposium on Natural Disaster Reduction (in Chinese), China Association for Science and Technology, editor, China S & T Press, Beijing, 85-90 .

DISASTER MITIGATION AND ECONOMIC DEVELOPMENT IN SHANDONG PROVINCE, CHINA

JIANGONG WANG
Vice Governor of shandong Province, Senior Environmentalist, 250011 Jinan city, P. R. China

and

MINGCHUAN JI
Senior Environmentalist, Station of Agriculture Environment Protection and Monitoring of Shandong Province, 17 Minziqian Road, Jinan city 250100, P. R. China

Abstract: Shandong is one of the largest coastal provinces of China. It is confronted with the coordination of economic development and disaster mitigation. Disasters in Shandong are characterized by their severity, frequency, variety and extensiveness. The disasters, which are abnormally active in recent years, constitute great menace to the economic development of Shandong. It is a feasible and wise choice for the province to coordinate disaster mitigation and economic development. This principle is the basis for the strategic goals guiding ideology, and plans for disaster mitigation and economic development.

1. TO COORDINATE DISASTER MITIGATION AND ECONOMIC DEVELOPMENT IS THE ONLY CHOICE FOR SHANDONG IN THE FUTURE

1.1 The Development of the world Economy Offers Shandong More Opportunities

Since the 1980s, the world economy, promoted by new technology, has been changing greatly. The Asian Pacific area, especially the area along the west coast of the Pacific, is very active in economic growth and is becoming the most active and potential area in the world (Anon. 1992a, 1992b). It is predicted that the 21st century will be a Pacific century. China is the largest country in this area. Therefore the development of its economy is a symbol of the international economic development, and Shandong (Figure 1), one of the most advanced coastal provinces in East China, is very noticeable. Shandong lies on the lower reaches of the yellow River between the Yellow Sea and Bohai Sea. It has rich resources, beautiful environment and strong agricultural and industrial bases. It has economic and trading relations with more than 150 countries. The opening coastal area in Shandong is the largest in China and its economic development is

M. I. El-Sabh et al. (eds.), Land-based and Marine Hazards, 257–263.

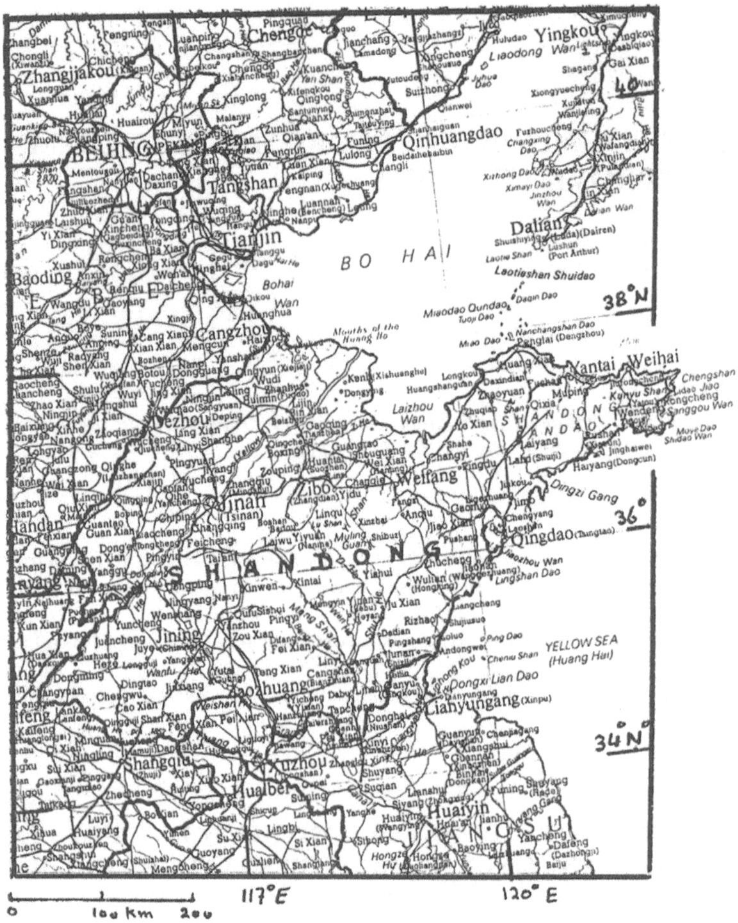

Figure 1: East China and Shandong Province.

also in the lead. The overall strategy of China's economic development could be divided into three phases. The first is the development and opening of Zhujiang Delta in the 1980s with the establishment of Shenzhen Special Economic Zone as an example, and the GNP doubled during this decade. The second is the development and opening of the Changjiang Delta in the 1990s with the establishment of Pudong Economic Development Zone, and the GNP doubled again. And the third, by the middle of next century, China will be a half-developed country. The rise of the economy in the Yellow River Delta and the establishment of the economic zone surrounding the Bohai Sea will be a symbol of the realization of the third phase (Li (1992), Ji (1995)). Both political and academic circles in China shared this conclusion.

1.2 The Economic Development of Shandong is Faced with various disasters

In the next century, due to the changing nature of the lithosphere, the hydrosphere, the atmosphere and the biosphere it is expected that natural disasters will be more frequent than ever (Anon. (1990), Gao and Ma (1990), Gao (1991)). It is predicted that during the 21st century sea level will rise by 40 cm (Luo, 1993). Most large cities and nearly half of the people in the world are living in coastal areas exposed to disasters . This is the case with Shandong which has a coastal line of over 3000km and is often hit by marine disasters (Ji, 1992).

From 1949 to 1987, windstorms and disastrous tides occurred 26 times in Laizhou Bay. The highest tidal level reached 6. 74m. The frequency of windstorm and storm surge induced sea level of over 5. 5m occur every three years and those above 6m occur every ten years (Ji, 1993). In the past ten years 730km^2 of land has been intruded by sea water and great harm has been done to the ecosystem, society and economy (Ji and Geng, 1992). The direct loss caused by disasters is US dollars 250, 000, 000 a year . The disasters also constitute great menace to the life and property of the people and the continuous development of economy. Therefore a better coordination between economic development and disaster management in Shandong is urgently needed before the end of this century (Wang, 1991).

2. CHARACTERISTICS AND FREQUENCIES OF DISASTERS IN SHANDONG

Shandong is exposed to widespread and severe disasters including storm surge, seawater intrusion, typhoon, heavy rainfall, cold wave, frost, gale, sea fog, hail, land slide, earth subsidence, earthquake, drought, waterlogging, soil erosion and salinization, plant diseases, insect pests, and so on. They cover many counties or drainage areas. Natural disasters done to agriculture can serve as an example. Since 1949, the average disaster area is more than 8 million acres a year, making up 30. 5% of the sown area. And the disaster-stricken population amounts to 17 million. Windstorms and storm surges can serve as another example. In the past 2, 000 years, when a windstorm and storm surge occurred, it would reach 20 to 30 kilometres inland, sometimes over 60 kilometres. Great losses are usually caused by disasters, especially those associated with typhoon, windstorm storm surge and flood. They not only destroy infrastructures, but also cause many injuries and deaths. Now the seawater intruded area has amounted to more than 730km^2. Since the mid 1970s, the annual reduction of grain amount to 300 to 500 thousand tons, the annual reduction of industrial and agricultural output is about US dollars 45, 000, 000. Nearly half a million people are short of drinking water, and nearly as many people suffers from endemic diseases.

Historical data shows the interval of disasters to be short. In the past 500 years perennial droughts occurred 80 times and season droughts 401 times. On average, serious droughts occurred once every 9 years and season droughts occurred 8 times every 10 years. During this time, perennial waterloggings occurred 78 times and seasonal ones 362 times. On average, serious waterloggings occurred once every 8 years and seasonal once 7 times every 10 years. Droughts occurred constantly from 1976 to 1989. In the past 3700 years, large earthquakes occurred more than 600 times, averaging once every 6 years. Disasters occurred frequently within the same year. For example it hailed 43 times in 1965 alone.

Natural disasters in Shandong are also interrelated. They usually occur during a certain period of time and certain area. For example, since mid 1970s, multi-disasters such as drought, typhoon, windstorm, storm surges, seawater intrusion, plant diseases and insect pests, occurred in the area around Laizhou Bay (Wang, 1992). Another example is the earthquake of 7. 4 magnitude which occurred 60km to the northeast of the mouth of the Yellow River on July 18 and 19, 1969, then a cleft appeared far to the south of the mouth and past of the surface of Bohai sea rose 15 to 22cm. Droughts along the coast often causes seawater intrusion which in turn will cause shortage of fresh water. People in the area have to drink polluted water and suffer from endemic diseases. Shortage of fresh water will also influence social production.

3. STRATEGIC PLANS FOR DISASTER MITIGATION AND DEVELOPMENT IN SHANDONG

It is clear that the next two decades will be a critical period for Shandong to carry out its developmental strategies. In accordance with the International Decade for Natural Disaster Reduction (IDNDR) and the reality of Shandong, the general strategic goals for disaster mitigation are as follows:

- to prevent and mitigate disasters comprehensively,
- to coordinate disaster mitigation and resource exploitation and environment protection,
- to change the vicious circle of resources, environment and biosystem into a good one,
- to build the coastal area into a model base of disaster mitigation, and
- to realize the aims of "Shandong on land" and "Shandong on the sea".

Detailed goals and requirements are as follows:

1. During the "International Decade of Disaster Mitigation ", loss caused by natural disasters should be reduced by 30%. And by the year 2020, the loss will be reduced by 50%.
2. From 1991 to 2000, the main task is to reduce losses caused by seawater intrusion, windstorm and tide, droughts and so on, to strengthen the opening to

and the exchange with foreign countries and to establish the model area of comprehensive disaster mitigation in the area around Laizhou Bay.

3. Resources should be exploited scientifically and rationally. The resource advantage will be turned into economic and regional advantages. The economic strength to mitigate disasters will be improved.
4. The rate of population growth will be controlled within 10%
5. The people's consciousness of disaster mitigation and the technology to mitigate disasters will be greatly enhanced.
6. A network to monitor, forecast and manage disasters should be set up.
7. A rational system of projects and non-projects to prevent and fight disasters and to provide disaster relief should be set up.
8. The percentage of forest cover should be raised and the ecological environment greatly improved.

In order to realize the above-stated goals, our guiding principles for disaster response, based on ideology and our experience during the last forty years, are as follows:

1. Attention should be paid mainly to disaster protection. In the areas exposed to disasters shelters should be built and disaster monitoring and forecast should be strengthened. When disasters occur all possible measures should be used and coordinated in order to reduce the loss as much as possible.
2. Disaster mitigation, resource exploitation and environment protection should be coordinated and managed comprehensively and included in an integrated system for disaster response.
3. Disaster mitigation and social economic development should be integrated and coordinated.
4. Disaster response plans should be prepared for present and the future.
5. Attention should be paid first to some disaster prone districts.
6. Disaster mitigation should rely not only on large investment, but also on modern science and technology. The technology, consciousness and ability to mitigate disasters should be improved.
7. Self reliance should be adhered to and international cooperation should be enhanced. Based on self reliance, international aid and support of disaster mitigation should be strived for.

Our strategic measures are:

1. Disaster mitigation concerns everyone. All people should be educated and all social powers mobilized. Proper policies elaborated by policy makers and the voluntary participation of the population are the key elements to disaster mitigation.
2. A network to monitor, forecast, and manage disasters comprehensively should be established to monitor and forecast disasters precisely and in real-time.

3. Research on disaster mitigation should be strengthened and scientific achievements and experience of disaster mitigation should be publicized.
4. Installations to mitigate disasters should be reinforced. On the plains, combined with the comprehensive exploitation of agriculture, comprehensive management should be adopted to mitigate disasters such as drought, waterlogging, wind, sand and salinization. In the Yellow River Delta, engineering and biological measures should be used to reduce the agricultural loss caused by natural disasters. In the mountainous area in Central Shandong and along the coast, more trees should be planted to control soil erosion and to improve the ecological environment. In the area around the Laizhou Bay, effective measures should be adopted to control seawater immersion and windstorms. In the area prone to earthquakes, forecasting measures should be improved and the constrictions should be reinforced.
5. Legal systems should be strengthened. First, local legislation should be speeded. Legislation systems for disaster mitigation suitable for Shandong should be set up as soon as possible. Secondly, the laws should be strictly enforced. Laws issued by the state, such as, "Environment Protection Law", "Water Law ", "Earth Administration Law", "Forest Law", "Water and Soil Preservation Law" and "City Planning Law", should be strictly executed. Local laws should also be strictly enforced and disaster mitigation managed legally.
6. Funds for disaster mitigation should be established. Disaster mitigation needs investment. To ensure the supply of fund and to carry out disaster mitigation smoothly, funds for disaster mitigation should be established. Disaster mitigation is a kind of public welfare and a component part of the national economic construction. Therefore, funds for disaster mitigation, except the part provided by governmental finance, can be collected from individuals and units based on the principles that "those who invest will be benefited and those who invest more will be more benefited ".
7. Disaster mitigation is an immense systematic project. To carry out the strategies for disaster mitigation, a ten-year plan should be prepared by the province, every prefecture and county. A system of responsibility should be set-up not only for large infrastructure investment but also based on modern science and technology.
8. Based on self reliance, international cooperation should be actively carried out.

REFERENCES

Anonymous, (1990). To Carry Out "International Decade of Disaster Mitigation" - A Report to the Secretary-General of the UN. Beijing, Earthquake Publishing House, pp34-65.

Anonymous, (1992a). The Economic Development of Shandong Province and the Probability the Economic Exchange and Development between Shandong and Japan. The Dept. of Economics, Yamaguchi University. May, 1992, P. 1-26, P. 60-65

Anonymous, (1992b). The Feasibility Report on the Economic Cooperation between Shandong Province and Western Japan. The Task-Tackling Group of the Economy College of Shandong University (by Chinese Partner), Dec. 1992, P. 1-10

Anonymous, (1994). Agenda for China in the 21st Century. Beijing Chinese Environmental Science Publishing House.

Gao Q., (1991). On the Systematic Management of Disasters on the Coast and Disaster Mitigation and Development. Earthquake Publishing House, PP 178-187.

Gao Q, and Ma Z., (1990). Forecast and Countermeasures of Natural Disasters in China in the Coming 10 years Study on Disaster Mitigation in China. Chinese Science & technology Publishing House, pp6-10.

Ji M., (1992). Regional environment and geohazards in the Shandong Coastal Zone. Science Press, Beijing, pp216-246.

Ji M., (1993). Disaster About Storm Surge in Shandong Province Area and the Corresponding Strategy. *Disaster Reduction in China*, 3: 40-42.

Ji M., (1995). Coordination of Exploitation and Ecological Culture Establishment: Choice of the Strategies for Continuous Economic Development in the Delta of the Yellow River. Paper presented at the International Conference on Resource, Environment and Economic Growth, Wuhou, China.

Ji M. and Geng X., (1991). Some Ecological changes Caused by Seawater Intrusion Along the Coast of Shandong. *Coastal Engineering*, 11: 61-67.

Li D., (1992). To Speed up the Overall Development of Yellow River Delta on the Basis of Practice and Scientific Policy-making. *Shandong Administration*, 2: 23-26

Luo Z, (1993). Impact of the Ascending of Sea Level on Economic Development of Chinese Coastal Areas. *The Guide of Science & Technology*, 6: 61-62.

Wang J., (1991). Strategy for Development and Natural Disaster Reduction in the Coastal Areas in Shandong Province. China Population, Resources and Environment, 1: 35-40.

Wang J., (1992). The Calamity of Seawater Intrusion in Laizhou Bay Region and the Administering Measures. Paper presented at the International Symposium on Underground Water and the Environment, Beijing, P. R. China.

MARINE DISASTERS IN CHINA AND THE STRATEGY FOR THEIR MITIGATION

LE KENTANG
Institute of Oceanology, Chinese Academy of Sciences, 7 Nan-Hai road, Qingdao 266071, P.R.China

Abstract: China is one of those countries in the world which are frequently hit by marine natural disasters, which cause the economic losses of about 1 billion US dollars a year on an average during 1949-1990. The direct economic losses caused by marine disasters have had a rapidly increasing tendency year after year since the 1980's. This paper reviews the situation of marine disasters such as storm surge, sea wave, red tide, sea water encroachment and sea ice, and the successes and failures of their reduction. To mitigate losses caused by marine disasters, the strategy for their mitigation, which includes countermeasure, information, scientific and technological research, organization and training is also suggested.

1. INTRODUCTION

Marine disaster is one of the serious disasters in China, which includes storm surge, sea wave, red tide, sea ice, sea level rising, coastal erosion and silting, Tsunami, etc., and some man-made disasters such as sea water pollution, sea oil spill and sea water encroachment (Zhao et al., 1992).

China is one of those countries in the world, which are frequently hit by natural disasters every year. Also, China is a developing country in which people's living standard is lower. Thus, disaster reduction is an important strategic measure for developing Chinese economy (Murakami, 1976).

This paper includes three sections. Section 2 reviews the situation of major marine disasters in China, which frequently hit China and cause heavy losses to the life and property of the people since 1949. Section 3 analyzes examples of the success and failure of marine disaster reductions in China. And section 4 reviews the national system for marine disaster mitigation and suggests the strategy for it.

M. I. El-Sabh et al. (eds.), Land-based and Marine Hazards, 265–277.

2. NATURAL MARINE DISASTERS IN CHINA

China is frequently hit by natural disasters, which cause heavy losses to the life and property of the people. The direct economic losses in 1989, 1990 and 1991 were 52.5, 61.6 and more than 100.0 billion yuan (RMB, the same below and the rate of exchange between the RMB yuan and the U.S. dollar is about 5.5:1 in 1993), respectively (Murakami, 1976). The statistics (Sun, 1991) show that the economic losses caused by marine disasters is 5-6 billion yuan a year on an average during 1949-1990. It is worth to pay attention that the direct economic losses caused by marine disasters have had a rapidly increasing tendency year after year since the 1980's. The coastal area in China is an intensive economic development zones where building density caused by the growing concentration of people and economic assets in urban areas and rural and township enterprises have increased quickly since then. However, since a great many buildings are more highly exposed to natural marine disasters, this is one of the main reasons for the increase of economic loss caused by natural marine disasters. The other reason is that marine economy also has been developing quickly without appropriate preventive measures against disasters induced by storm surges and sea waves since the 1980's.

2.1 Storm Surge

Chinese coastal areas frequently suffer from the storm surges caused by tropical cyclones, extratropical cyclones and cold fronts every year. The heavy losses due to storm surge are mainly caused by typhoon and tropical storm. Table 1 lists the statistical results of typhoon and tropical storm landing in China during 1949-1990 (Wang and Xia, 1993). We define the storm surge with the maximal increment, $h \geq 1.0$m of water level caused by storm as a serious storm surge. Table 2 shows the statistics of severe storm surges caused by typhoon, which hit Chinese coast during 1949-1992. Fig. 1 shows a map of the coastal stations mentioned in Table 2 and five typical tracks of tropical storm. The statistics of the serious storm surges caused by extratropical cyclones are given in Table 3.

An incomplete statistics shows that there were 95 instances of storm surges caused by typhoon and tropical storm, among which the most severe 40 storm surges occurred in the past 40 years (1949-1990). Occurrence of the disaster generated by severe storm surge not only cause heavy losses to the life and property of the people, but also restrict the economic development of coastal areas. Although a storm surge incident causing ten thousands of mankind casualties has not occurred since 1949, the 1956 storm surge caused by Typhoon Wanda made more than 4000 citizens drowned (Bao, 1991; Wang et al., 1992).

From August 28th until September 1st, 1992, an extraordinarily serious storm surge generated by the combined effect of Typhoon Polly and astronomical spring tide hit the eastern coast of China, which caused the most severe storm surge disaster since 1949. The region affected by the disaster included provinces Fujian, Zhejiang, Jiangsu, Shandong,

Table 1.(a) Statistics of typhoon and tropical storm landing in China (1949-1990)

Year	Tropical depression	Tropical storm	Typhoon	Total
1949	2	4	3	9
1950	5	3	2	10
1951	1	2	4	7
1952	3	4	8	15
1953	2	2	7	11
1954		3	4	7
1955	4		5	9
1956	5	2	5	12
1957	1	2	4	7
1958	1	6	3	10
1959	3	2	4	9
1960	2	1	7	10
1961	3	3	9	15
1962	1		7	8
1963	2	1	6	9
1964		1	7	8
1965	1	4	5	10
1966	1	2	8	11
1967	1	3	9	13
1968	1		6	7
1969	1	2	3	6
1970	3	4	2	9
1971			12	12
1972		2	5	7
1973		3	8	11
1974		6	6	12
1975	2	1	6	9
1976	1	1	4	6
1977		4	4	8

Hebei and Liaoning, and Shanghai, which is the most widely-spread area suffered by a storm surge disaster since 1949. Governments at all levels in the disaster-affected region had taken anti-storm surge measures to prevent and to struggle against the disaster before and during its occurrence and did their best in disaster relief, which had greatly mitigated losses to the life and property of the people. Nevertheless, the economical losses and people casualties due to this disaster are still very heavy. For example, the disaster-affected population was more than 20 million, in which 193 persons died, about 1170 km long seawall was destroyed, the disaster-affected farmland reached up to 1.933 million ha, in which 333 thousand ha suffered from the storm surge. The direct economic loss due to the disaster reached up to 9.3 billion yuan.

Table 1.(b) Statistics of typhoon and tropical storm landing in China (1949-1990)

Year	Tropical depression	Tropical storm	Typhoon	Total
1970	3	4	2	9
1971			12	12
1972		2	5	7
1973		3	8	11
1974		6	6	12
1975	2	1	6	9
1976	1	1	4	6
1977		4	4	8
1978	1	3	4	8
1979		3	3	6
1980	1	6	4	11
1981	2	6	4	12
1982		1	3	4
1983		3	3	6
1984	2	5	3	10
1985	3	3	7	13
1986	1	3	5	9
1987			6	6
1988	2	2	6	10
1989	1	5	7	13
1990	1	2	8	11
Total	60	110	226	396

2.2 Sea Wave

The very rough sea with wave height over 4m can cause serious marine accidents. If wave height (h) exceeds more than 6 m, the high sea will destroy bank protector, offshore engineering and ships. Table 4 shows number of occurrences of high seas with wave height $h \geq 6m$ in the China seas from 1966 to 1990 (Bao, 1991) based on data taken at coastal stations and buoys. Maximum wave height in each year is also shown. It can be seen from Table 4 that the annual mean frequency of high sea with wave height over 6 m is 47.8. The annual mean of marine accidents is about 50. Bao (1991) shows that the perils of the ocean by ocean waves (from rough sea to high sea) in the China seas are 70 times on the annual average, which cause economic losses of 0.1 billion yuan or more and human casualties of more than 500 every year. In some years the perils of ocean waves generated by typhoon are startling. For instance, in 1989, several strong typhoons such as Brenda, Dot, Gordon, Sarah, Brian and Elsie struck the China coast one after the other. The extreme waves associated with these typhoons damaged and sunk 3408 ships, damaged seawalls at 10854 places, ruined 900 km of seawalls, which caused direct economic losses of 0.5 billion yuan and more than 200 people died and about thousand people wounded due to the high sea disasters.

Table 2: Statistics of the extraordinarily serious storm surges in China (1949-1992)

Date	Station	Maximal water level by storm surge(m)	Highest tidal levl(m) (above MSL)	disaster-affected area	Name of typhoon
Aug. 01,1956	Ganpu	5.02	4.37	Zhejiang P. and Shanghai	WANDA
July 15,1965	Nandu	2.87	3.33	Leizhou Peninsula and Guangxi P.	FRED
July 28,1969	Shantou	3.02	3.28	Shantou C.	VIOLA
Sep. 27,1969	Baiyantan	2.38	3.62	Mouth of Minjiang R. and Taiwan P.	ELSIE
Aug. 20,1974	Ganpu	2.56	5.83	Zhejiang P. and Shanghai	MARY
July 22,1980	Nandu	5.94	5.92	Leizhou Penin and Hainan I.	JOE
Sep. 01,1981	Lusi	2.03	4.29	Zhejiang P. and Shanghai	AGNES
July 21,1986	Shitouhu	1.17	3.96	Gulf of Beibuwan	8609
Sep. 05,1986	Nandu	3.52	3.37	Leizhou Penin and Hainan P	WAYNE
July 18,1989	Beijin	2.04	2.26	Mouth of Pearl R.	GORDON
Sep. 15,1989	Haimen	1.46	4.67	Taizhou C.	ANGELA
Sep. 08,1990	Wenzhou	2.41	3.87	Mouth of Minjian R.	DOT
Aug. 16,1991	Nandu	3.84	2.72	Leizhou Penin and Hainan P.	FRED
Aug. 30,1992	Aojiang	4.48		Zhejiang, Fujian, Shandong etc.	POLLY (9216)

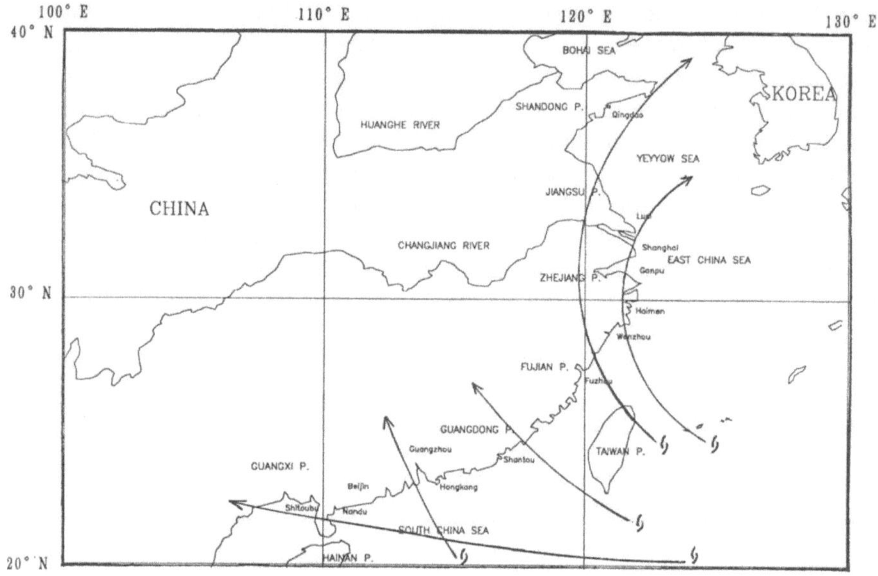

Figure 1: Five typical tracks of typhoon landings in China.

Table 3: Statistics of the remarkable storm surges caused by extratropical cyclone in China (1949-1990)

Date	Station	Maximal Water by storm surge(m)	Height tidal level (m)	Disaster- affected area
Apr. 5-6,1964	Tanggu	2.24		Bohaiwan Bay
	Yangjiaogou	3.21	3.16	Laizhouwan Bay
Nov. 7,1965	Tanggu	1.94	3.19	Bohaiwan Bay
Apr. 23,1969	Yangjiaogou	3.55	3.66	Laizhouwan Bay
Apr. 5,1980	Yangjiaogou	3.18	2.93	Laizhouwan Bay
Nov. 26,1987	Xiaying	2.46	2.90	Laizhouwan Bay

2.3 Red Tide

Red tide is a kind of phenomenon of ecological abnormality, which is characterized by making sea water coloured. It is caused by phytoplanktons, bacteria and/or protozoon, which suddenly become rich under some environmental conditions. So far the causes of the red tide formation is not so clear besides marine pollution (Murakami, 1976).

Table 4: Number of occurrences of high seas with wave height h>6m in the China Sea (1966-1990)

	Bohai Sea		**Yellow Sea**		**East China Sea**		**Taiwan Strait**		**South China Sea**		**Area east of Taiwan Island and Bashi Channel**	
Year	Maximal wave height (m)	N	Maximal wave height (m)	N	Maximal wave height (m)	N	Maximal wave height (m)	N	Maxima l wave height (m)	N	Maximal wave height (m)	N
1966	7.0	1	9.6	6	11.0	9	11.0	6	10.0	9	10.0	13
1967	9.0	3	9.8	10	9.0	8	9.5	7	9.5	14	9.5	14
1968	7.8	3	8.5	11	8.5	10	8.0	4	8.0	8	8.5	18
1969	7.3	1	7.5	4	15.0	10	10.0	6	15.0	14	15.0	14
1970	7.3	1			11.0	7	12	8	9.0	13	12.0	8
1971	7.3	1	11.0	8	10.0	13	10.0	8	11.0	27	10.0	6
1972	9.0	1	9.0	18	15.0	13	8.5	3	8.0	7	15.0	6
1973	7.3	1	9.3	6	10.0	4	10.0	1	11.0	15	11.0	5
1974			9.0	3	9.0	6	12.0	8	12.0	14	12.0	13
1975			9.8	6	15.0	8	9.0	3	15.0	6	9.0	9
1976			9.8	3	10.0	6	10.0	3	10.0	12	10.0	9
1977			10.0	5	10.0	7	10.0	6	10.0	12	10.0	7
1978			11.0	3	11.0	12	8.0	7	11.0	14	7.0	6
1979	6.9	1	7.5	4	9.6	10	8.0	3	8.0	11	9.0	9
1980	9.0	4	9.0	3	13.0	17	9.0	7	9.0	19	13.0	8
1981			10.0	12	10.0	11	10.0	4	10.0	16	10.0	10
1982			10.0	4	14.0	10	14.0	4	14.0	12	14.0	14
1983			12.0	5	12.0	7	10.0	6	10.0	16	12.0	9
1984			8.0	1	10.0	14	8.0	9	9.0	19	10.0	13
1985	7.5	2	11.0	7	12.0	16	12.0	11	12.0	11	10.0	8
1986			12.0	4	12.0	10	10.0	6	12.0	19	12.0	21
1987	8.5	1	12.0	4	12.0	11	12.0	7	12.0	9	12.0	14
1988			6.8	1	8.0	4	10.0	4	10.0	19	12.0	15
1989	6.5	1	7.0	3	9.0	7	9.0	7	10.0	20	10.0	10
1990	7.5	2	7.5	7	12.0	14	10.0	15	10.0	17	10.0	16
Total		23		148		244		153		153		275
Average		0.92		5.92		9.76		6.12		14.12		11.0

The disasters of red tide frequently occur in China's coastal areas in recent years because of the increase of pollutant substance emptying into the sea. 250 events of red tide disasters were reported in China's seas during 1970-1990, of which 178 occurred in the area around Hongkong and 72 in the coastal area off the China's mainland. It is worth to note that among 72 disasters mentioned above, there were 9 events in the 1970's, 17 events from 1980-1988, 12 events in 1989 and 34 events in 1990.

In August, 1977, a red tide event of Prorocentrum minimum which spread to 560 sq. km. and lasted 20 days, appeared off the Dagukou area near Tianjin and resulted in many dead fish floating on the sea surface.

In December, 1986, 136 persons were poisoned by eating Ruditapes philippinarum in the Cigu Village of County Dongshan, Province Fujian. It was concluded from the investigation and analysis that this was caused by Paralytic shellfish poison generated by a kind of red tide organisms ---- *Gymnodinium sp.*

In April, 1989, a red tide event occurred in the offshore area off County Qing, Province Fujian, causing severe damage to the 1300 mu cultivated Sinonovacula constricta and had over a hundred million post larvae of Penaeus penicillatus Alocou died, resulting in economic losses of 31 million yuan.

In August-September 1989, a widespread red tide occurred in the coastal waters of Province Hebei, city Tianjin and Province Shangdong. It affected about 1.5x104 hectares of shrimp pounds, resulting in a reduction of approximately 1x104 tons of shrimp yields, an economic loss of 2x108 yuan. Up to now, this was the largest, longest and most harmful red tide in the history of North China.

2.4 Sea Water Encroachment

Sea water encroachment is a kind of natural disaster caused by human activities in coastal area. This type of disaster is characterized by intrusion of sea water into underground aquifers through different ways (Lee and Cheng, 1974). In China it is mainly caused by over-extraction of underground water. Since the 1970's, this disaster has appeared in the coastal economic area of Provinces Shandong, Hebei and Liaoning around the Bohai Sea, particularly, around the Bohaiwan Bay and Laizhouwan Bay.

In the coastal economic area around the Laizhouwan Bay, the rate of sea water intrusion into coastal aquifers increased rapidly since the mid 70's. as shown in Table 5. (Wang and Xia, 1993). Table 6 shows the disaster statistics caused by sea water encroachment in the coastal area around the Laizhouwan Bay.

2.5 Sea Ice

In winter some parts of the Bohai Sea and northern Huanghai (Yellow) Sea are covered by sea ice every year (from mid-November to late February). Fig. 2 shows the typical sea-ice distributions in the normal and heavy ice year Historical data show that in the past 60 years sea ice disasters occurred 9 times, of which severe ice covers appeared 5 times (i.e. in 1936, 1947, 1957, 1969 and 1977) and the other 4 times (in 1966,1968, 1980 and 1986) occurred belong to less severe ice cover. In February-March, 1969, the most severe ice cover occurred in the Bohai Sea, when the whole Bohai Sea was ice-bounded, which

Table 5: Rate of sea water intrusion in City Leizhou

Year	Encroached Area (km^2)			Rate (m/a)
	Increment	Total	Annual Average	
1976 - 1979	15.8	15.8	4	46
1980 - 1982	23.4	39.2	7.8	92
1983 - 1984	31.9	71.1	16	177
1984 - 1990	167.1	283.2	31.3	348

Table 6: Statistics of sea water encroachment disasters in the Shandong Province (by the end of 1990)

Classification	County	Affected Area (km^2)	Affected Farmland (10^3mu)	Affected Villages	Population lacking fresh water (10^3 person)
Sea water Encroachment	Laizhou	238.2	271	132	107
	Luongkou	88.7	100	35	30
	Zhaoyuan	11.3			
	Penglai	5.4	5	18	36
	Changdao	5.3	6	13	5
	Fushan	40.0	40	20	20
	Muping	2.0	5	2	1
	Haiyang	2.0	5	2	1
	Jimuo	1.8	2	7	10
	Laoshan	15.0	18	20	60
	Jiaozhou	15.0	18	20	60
	Kianonan	4.5	5	10	4
	Huangdao	2.0	2	2	1
	Sum	431.2	447	281	335
Brackish water Encroachment	Pingdu	80.0	6	6	15
	Changyi	90.0	76	40	35
	Hanting	61.5	33	16	15
	Shouguang	54.0	70	51	40
	Guangrao	14.5	13	11	5
	Sum	299.5	198	123	110
Total		730.7	675	404	445

made more than 50 vessels surrounded and damaged the oil drilling platform "Sea Well No.2", causing economic loss of several billion yuan.

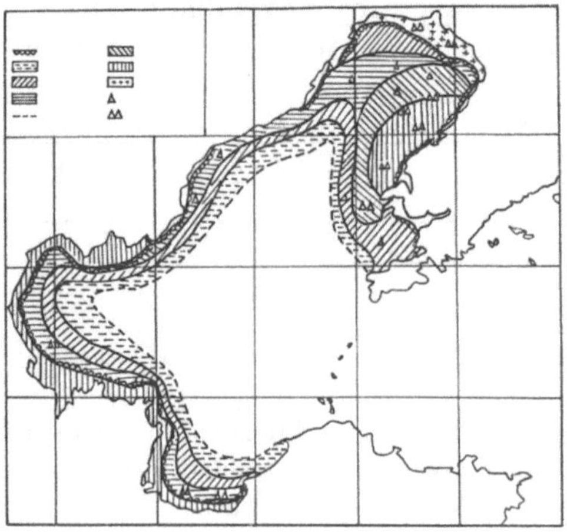

Figure 2a: Typical sea-ice distribution in the normal ice year.

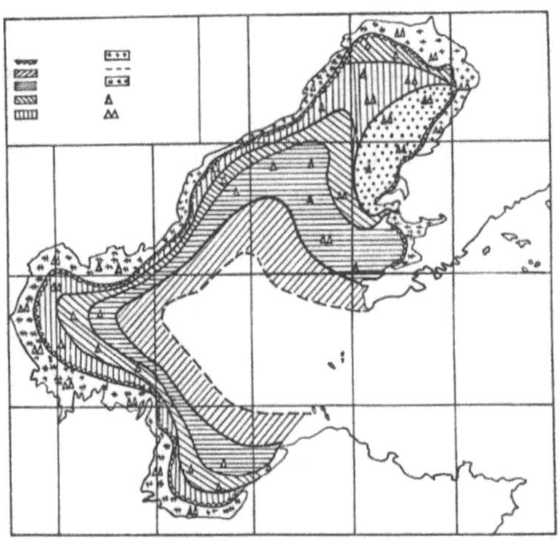

Figure 2b: Typical sea-ice distribution in the heavy ice year.

3. SUCCESSES AND FAILURES OF MARINE DISASTER REDUCTION IN CHINA

Since the founding of new China in 1949, Chinese Government has always paid great importance to the education against natural disasters, formulated a series of strategies and policies that conform to the realities in China. In the early 1960's, China started to establish a network for monitoring storm surges. Since then, the network which includes 60 tidal observatories for monitoring sea level variations has been set up. Up to now, a primary system for watching, monitoring, predicting and warning marine disasters has been established. The system together with the State Meteorological System has played a important role in marine disaster reduction. As a result many successful examples in mitigating marine disasters in China appeared.

The following 2 events of storm surge disasters caused by typhoon form a sharp contrast. On August 3rd, 1922, a severe typhoon hit the City Shantoe of Province Guangdong without any forecasting and warning in advance and the subsequent storm surge resulted in the drowning of nearly 70,000 people. After 47 years, Typhoon Viola (6903) struck the City Shantou on July 28th, 1969, but due to forecasting and early warning of the typhoon and storm surge, the governments and departments at different levels had time to take anti-natural disaster measures. Although the maximum level of the storm surge and the inundated area are almost the same as that of 1922, the people killed were reduced to one thousand.

In 1986, storm surge disasters occurred 4 times, causing the death of 76 people and 946 people wounded, in addition to economic losses of 0.96 billion yuan. Although the concerned departments had forecasted and warned the 4 storm surges in advance, the storm surges still caused heavy losses to the life and property of the people. This indicates that there are several shortcomings in current anti-natural marine disaster system, which force us to examine closely the strategy of the marine disaster reduction.

4. OUTLINES OF STRATEGY FOR MARINE DISASTER REDUCTION

To minimize the effects of marine hazards and reduce their drastic increase of losses to the life and property with the development of coastal and marine economy, it is necessary to establish a strategic plan for marine disaster mitigation in China. Such a plan should at least include the following aspects.

4.1 Countermeasure

The countermeasure against marine disasters should focus our attention on mitigating the above-mentioned disasters.

(1) Better design of engineering project against storm-surge in the vulnerable areas affected by storm-surge and sea wave.

(2) Control pollutants emptying into the Chinese rivers and sea.

(3) Restrict pumping of ground water to avoid their over-extraction.

4.2 Information

The key point of the information strategy for natural disaster reduction is the collection of disaster signals as early as possible, to transmit them to decision-making authorities and to rapidly provide feedback. Up to now, there still is a gap in this regard in China as compared with the other countries.

To close the gap as soon as possible, it is necessary to carry out the followings including improvement of existing disaster networks, establishment of multichannel, all-weather and comprehensive information system, and renewal of equipments to rapidly transmit the disaster information.

4.3 Modern Scientific and Technological Research

The research works for marine disasters in China have been carried out since the 1970's. A professional system for predicting marine disasters has been established. This includes the improvement of prediction methods in order to meet the needs of marine disaster reduction. So far the numerical models for predicting typhoon-induced storm surges and ocean waves should be improved not only due to the shortcomings of ocean models but also due to numerical simulation model of typhoon available.

One of the key projects for marine disaster reduction, carried out from 1991 to 1995, is the study on the pre-warning system against marine disasters, which is conducted by many scientists of the State Oceanic Administration (SOA), the Chinese Academy of Sciences (CAS) and the National Educational Committee (NEC).

Strategically, the research for marine disaster reduction should include not only the study on improving marine disaster forecasting and warning but also the study on how to assess marine disaster.

4.4 Organization

As mentioned above, CAS, SOA and NEC conduct marine disaster research project. SOA is a professional organization for preparing marine forecast and disaster warning and the governments and departments at different levels decide how to take precautionary measures and how to carry out disaster relief before, during and after occurrence of marine disasters. Although there have been many successful examples for marine disaster reduction in China since the 1960's, there are still some failures as mentioned earlier. From practical point of view, there should be a special organization system to improve the link between the above two systems, as suggested by Wang et al. (1992). Following Wang's suggestion, such an organization could be named as the "State Centre of Disaster

Reduction" (SCDR). SCRD should have a department to deal with marine disasters and have some branches at different levels.

4.5 Training

The goal of the training strategy is to raise the awareness of natural disaster reduction nationwide. To do this we suggest the following actions:

(1) Bring mass media into full play.

(2) Strengthen the basic skill training on natural disaster reduction.

(3) Add the contents of disaster reduction to the text books of primary and high schools.

(4) Run the special class on the disaster reduction management to train the leaders to raise their capabilities, decision-making and conduct.

REFERENCES

Bao C. (editor), 1991. Marine Disaster and its forecast. China Ocean Press, 205pp. (in Chinese).

Lee C. H. and Cheng R. T., 1974. On Seawater Encroachment in Coastal Aquifers, *Water Resour. Res.,* 10(5): 1039-1043.

Murakami, A., 1976. Red tide and eutrophication. Envir. Hazard Control Tech. Ass. Publ., 207pp.

Sun G., 1991. Estimate of the losses caused by natural disaster in China and countermeasure to reduce them. *Disaster Reduction in China* (trial issue) 1(1):37-40. (in Chinese).

Wang A. and Xia D., 1993. Tropical Cyclone Disasters in China and the Strategy for Disaster Reduction, *Natural Disaster Reduction in China*, 2(1):8-12.

Wang X., Ye L. and Bao C., 1992. Disaster of Typhoon-induced Storm Surge and its forecast in China, Study on Natural Disaster Reduction in China, China Meteorology Press, 300-306. (in Chinese).

Zhao D. et al., 1992. Environment and Disaster in the Coastal areas of the Shandong Province. China Science Press. 373 pp. (in Chinese).

STORM SURGE DISASTERS IN BEIBU BAY (CHINA) AND COUNTERMEASURES FOR THEIR REDUCTION

YU RIXIN
Guangxi Bureau of Hydrology and water Resources , 12 Yuan Hu Road, Nanning 530023, P.R.China

Abstract: Storm surge disasters in the Beibu Bay China are very serious. The historical documents and in-site investigations show that the years of 1501, 1906, 1934, 1965 and 1986 are those with quite major disaster due to storm surges. The conditions of the historical disasters remind us that the consciousness of disaster prevention and reduction of the whole nation must be strengthened. The technical measures for performing the study on the countermeasure to disaster prevention and reduction due to storm surges in the Beibu Bay can be regrouped into three aspects. The first one is the overall plan to accelerate the establishment of "four lines of defence". The second is to reinforce the monitoring measures and set up the warning and defence system. The third is to stagger the occurrence time of flood peaks and reduce the peak discharges by using scientific and engineering techniques, the disaster prevention by high buildings and the insurance of tide prevention.

1. INTRODUCTION

Beibu Bay (Figure 1) is a semi-enclosed shallow sea located in the Chinese subtropical zone. It is surrounded by Guangxi in the north, Vietnam in the west, the Leizhou Peninsula in the east, and the Hainan Island in the northeast..

It is one of the regions where the tropical cyclone occurs frequently in the shallow sea region of China. For example, 153 tropical cyclones affected the coast of Guangxi between 1949 and 1987. That is an average of 4 events per year. The occurrence time is mostly from June to October, with the main peak representing 28.8 % of the total occurs in August. Those occurred in July and September represent 20.3 %. The tropical cyclone and associated storm surges cause huge losses of lives and property of the people living in the coastal regions. It becomes a serious limitation to the economic development and external opening of the coastal regions. Therefore, the study on the countermeasures of disaster reduction is a strategic deployment concerning the development and external opening for the coastal regions.

2. HISTORICAL STORM SURGE DISASTER IN THE BEIBU BAY

In accordance with the historical literature and incomplete statistics, 49 serious disasters caused by windstorm occurred between the years from 983 to 1949, and 34 events from

M. I. El-Sabh et al. (eds.), Land-based and Marine Hazards, 279–285.

1950 to 1990. Most of the windstorm events were accompanied by natural disasters such as floods, water logging and storm surges.

In July 1501, Lianzhou and Linshan were hit by serious storm, where houses and trees were fully destroyed, and more than 150 persons were drowned. On August 3 (the lunar calendar) 1906, Hepu was hit by hurricane, houses were collapsed, trees were uprooted, boats were overturned, persons and animals were blown about by the wind, numerous

Figure 1: Location of Beibu bay, China

boats were sunk in the sea and inland rivers, nearly 1,000 persons were drowned. In Junanbu, 277 persons were drowned, objects were blown long distances, large ships were grounded. It was really an unusual disaster. In 1968, the Typhoon No. 11 hit the Beibu Bay, where 50 ships were sunk and 70 persons died. The disasters caused by surges are also quite serious. According to the historical document, the years of 1501, 1906, 1934, 1965, 1986 were those with serious disaster due to storm surges. For example, the Beibu

Bay was hit by typhoon on September 20,1906 and the water elevation in the coastal area was raised 2.5m and a relatively large storm surge occurred, only in Hepu and Beihai, more than 1,000 persons died, boats were sunk, houses were collapsed, seawall was broken and a lot of farmland was flooded. On July 29, 1934, the coastal area of the Beibu Bay was hit by the storm surges caused by tropical cyclone, the Tuanhe embankment in Kangxilin, Qingzhou was fully flooded. The tropical cyclone No. 6508 in July 1965 hit the Leizhou Peninsula, with maximum wind speed exceeding 110 km/hr.. The storm surge disaster was very serious. 733 ha of farmland were disastrous and 1925 tons of paddy were lost, 49 ships were sunk and 98 were destroyed, 250 km of sea wall were damaged, 112 villages were flooded, 1039 houses were collapsed, 10 persons died and 131 were wounded, the loss was very heavy. In July 1986, the tropical storm No. 8609 landed at Nankang, Hepu. It was just the period of spring tides. The sea level elevation observed at the coastal stations in the coast of the Beibu Bay was 0.26 to 0.65 m higher than the maximum values measured earlier. The flood peaks caused by rainfall and surges encountered, the high sea level superposed the flood caused by storm, so the disaster of historical significance occurred. most of agricultural embankment in the coastal region were overflowed by tidal water, 4,735 places of bank were broken, with the total length being 364 km, nearly half of the total length of sea wall. The Jianxin Embankment in Qinzhou City was constructed before 1949. It was named "Jianxin Embankment" because it was solid. But it was also destroyed this time, more than 53333 ha of farmland were flooded and the reduction of grains was more than 100,000 tons, 55693 houses were destroyed, 37 persons died and 2,060,000 persons were homeless, 273 ha seabeach breeding in the Fangchen Ge Autonomous county were fully damaged, all fishes and shrimps were flushed away from their breeding ponds. It was estimated, that the direct economic loss reached more than 0.14 billion dollars U.S., including more than 0.0375 billion dollars for the loss of embankment.

3. COUNTERMEASURES FOR DISASTER PREVENTION AND REDUCTION

At present, mankind hasn't the ability and the means to completely control and eliminate the disaster resulting from storm surges. It is possible, however, to mitigate the effect of these disasters and reduce the loss by adopting corresponding measures according to their frequencies and characteristics. At present, firstly it is necessary to strengthen the consciousness of disaster prevention and reduction of the whole nation. For decision-making leaders, it is necessary to overcome the short-term behaviour and the idea of not taken decision in the hope that the higher authorities would allocate funds one day. At the planning and management level, it is necessary to correct the tendency of eliminating the projects that deal with protections of flood, wind and surges in the overall planning for town and country, eliminate the phenomena of disputing over trifles, even shirking responsibility and scrambling profit among various departments. It is also necessary, for enterprises and inhabitant, to strengthen the consciousness of disaster prevention and reduction. Secondarily, the government enterprises and local inhabitants should increase their investment in disaster prevention measures. Only in this way, it is possible to realize the countermeasures for disaster reduction.

3.1 Performing overall planning and accelerating the establishment of "four lines of defence"

In order to reduce the negative impacts of flooding due to storm surges, it is necessary to establish a series of countermeasures including both engineering and biological measures, such as mangrove swamps, sea wall and enbankment, shelterbelt and shelter network of farmland. Such measures will contribute to the peoples safety and protect the shoreline fish breeding ponds and agricultural lands.

3.1.1 Performing well the planning for and strengthening the construction of sea wall

The total length of agriculture sea wall and embankment along the coast of Guangxi is about 730 km, most of which are earth bank. The actual measures for bank prevention are relatively bad, and hence their defence ability against floods and surges is low. In the past, the overall planning of seawall and embankment was not carried out and the investment wasn't enough. After the occurrence of the tropical storm No. 8609, the Guangxi government allocated 0.625 million U.S. dollars every year to block up the gap and maintain the embankment. This is considered far from the minimum requirement for consolidating, renovating the seawall and eliminating dangerous situations. Some of the seawalls are aging, and no repairs were carried out in a long time. This is a serious hidden danger. At present, the Department of Water Resources and Power of Guangxi is performing the planning of "Consolidation and Renovation of Seawall and Development of Beach in the Coastal Region of Guangxi". Such plan should be examined, approved and performed as soon as possible in order to ensure the safety of seawall and fully develop and utilize the resources of sea beach.

3.1.2 Construction of forest belt of mangrove and prevention of sea wall

The seawall, as a permanent hard option engineering measure, must be closely combined in concert with the biological measures such as mangrove swamps to absorb shock. Otherwise, the well constructed seawall cannot maintain for a long time. The disaster caused by the tropical storm occurred on 9 July, 1986 shows that the sea wall with prevention of mangrove on the outer slope was generally not destroyed by wave and tidal bore. This fact indicates that the luxuriant branches and leaves of mangrove forest can reduce the force and power of wave and can prevent well the seawall. The mangrove forest in Yinglu Bay of Hepu County is one of those growing well in the whole country, and has been classified as the national protection region of marine nature. The mangrove forests should be really protected. It is proposed to recover and construct mangrove forest in all segments of embankment or beaches wherever it is convenient to plant them. For the enclose surges land for cultivation, where it is not convenient for rice growing and marine breeding, the embankment should be demolished, and mangrove forest should be constructed there in order to recover the ecological equilibrium.. The relevant departments of forestry, water conservancy, marine and aquatic production should cooperate to actively help the different administrative units and inhabitants in coastal regions to better construct these mangrove forests.

3.1.3 Construction of sand-fixation forest belt and shelter network

Shelter belt can play a major role as windbreak and sand fixation. For example, since the 1950's, the peoples of Jing nationality who lived in three islands in the Ge Autonomous County of the Fang City have constructed a coastal shelter belt 19.8 km in length and 500 meters wide. It basically prevents the possible harm of strong wind and drift of sand to farmland, and grain yield keeps in high level steadily. Since the founding of new China, great achievements have been made in constructing such coastal shelter belts in Guangxi, 1, 667 ha of wind-proofing and sand fixation shelter belt mainly composed of beechwood in the coastal waste beach. The length of forested coastline reaches 554 km. But at present, the area of shelter-belt is still less and their construction should be strengthened. Eucalyptus and Taiwan jequirity all can be accepted as the main species of trees for shelter forest. The GuangXi Institute of Forestry Survey and Design prepared a plan which should be approved and performed in the near future..

In coastal region, frequent strong winds often cause the lodging of crops, flying of dust, and the sandization of soil. According to the initial statistics, the sandized area reaches 4,800 ha. Therefore, shelter-network of farmland in coastal region should be constructed to reduce wind speed and prevent the production of farmland. It is favorable to form the "microclimate" of agricultural production.

3.2 Enhancement of monitoring and establishment of warning and defence system

The guiding principle of disaster reduction should be "putting prevention first, and combining it with disaster mitigation". . The prerequisite of prevention is to enhance the ability of forecasting and precaution before the occurrence of disaster. Therefore, it is necessary to depend upon modern science and technology and to set up reliable warning system and facilities for disaster mitigation and protection. The department in charge of disaster response has the necessity to set up the information network system for monitoring information, forecasting, warning and defence related to storm surge disasters. This will heighten the ability of rapid response and scientific decision making for storm surge disaster and will certainly raise the effectiveness of disaster reduction.

Marine, hydrological and meteorological observations are essential basis for the construction and management of seawall, and hence the prevention of storm surges disaster. At present, such stations in coastal region of Guangxi are scarce. Therefore, it is urgent to increase wind-wave and tidal level observational stations in harbours and bays of coastal regions along the Beibu Bay. Furthermore the information, warning and forecasting of storm surges is an important non-structural measure to prevent disasters from severe wind and surges. It is also a measure with less cost with significant and rapid effectiveness. The departments of meteorology, hydrology and seas and oceans should strengthen the interdepartmental cooperation, closely concert and coordinate their efforts. Storm surge occurs rapidly and hence, the effectiveness in communication and action is very important. The information should be ensured to be received in time and communication should be improved continuously. Forecasting methods should also be improved.

Suggestion is also made to include meteorological aspects under the responsibility of the headquarters of flood control and drought mitigation in Beihai city. This will also include research institutions of seas and oceans, hydrology, meteorology, water conservation, in charge of the preparation and management of the warning and civil defence system. This system is a synthetic engineering, which is composed of automatic measurement and transmission, forecasting, information, warning, regulation and command subsystems. Various subsystems could incorporate either the local network or combined together to form the network. This system is a systematic engineering concerning many departments. Unified leadership and command and close cooperation are necessary. It is expected that this integrated system will play great role to disaster prevention and reduction in Beibu Bay.

3.3 Scientific regulation

(1) There are 125 large and small rivers flowing directly into the sea in the coastal regions of Guangxi. These include among others, the Nanliu River, the Qingjiang River, the Maoling River, the Fangchen River and the Beilun River. They have resulted in many serious disasters due to their floods. In many cases floods from rivers and storm surges occur at the same time as water logging. The flood-surges disaster that occurred in July 1986 is an example. To give full play of large and medium scale reservoirs, such as the Xiaojiang, Wanshenjiang and Hongchaojiang, we propose to stagger the occurrence time of flood peaks and reduce peak discharges, and also to consider the factor of backwater and optimally regulate water volume..

(2) The rise of high buildings in cities provides favorable condition for transfer and evacuation during flood, and greatly reduce casualties. For example, in 1985, a two-storied building of a primary school in the Hepu County saved more than 200 persons. Storied buildings are rare in seaside villages. During floods, these villages become a boundless ocean, so it is required to construct the central schools, government offices,

hospitals, special observational stations and bus stations on the places with higher elevation in designing village plan. It is convenient to construct three-storied buildings and radial roads among villages, which can be used as the points of evacuation during disasters of flood and surges.

(3) Insurance itself can neither prevent nor reduce the loss caused by storm surge disasters. But it however changes the burden for disaster loss. In this way, the loss is not borne one but shared by numerous persons who participate in the insurance during a relatively long period. It greatly improves the psychological pressure and pessimism of affected families and victims, and helps in recovering production and rebuilding of property following the disaster.

ACKNOWLEDGEMENT

I am very grateful to the following persons for their valuable advice and help: Prof. Zhang Hailun from the Nanjing Institute of Hydrology and water Resources, Senior Engineers Su Zhu and Liao Shijie from the Department of Water Resources and Power of Guandxi Autonomous Region and Engineer Su Zhukuan from the Bureau of Water Management. Thanks to Mr. Li Jiren of Hohai University for translating the manuscript to the English Language.

MANAGING FOREST FIRES : AN AUTOMATIC FIRE WEATHER STATION NETWORK IN CHINA

CHENG BANGYU AND JIN XIAOZHONG
Research Institute of Forest Protection, 75 Hapin Road, Harbin 150040, China

Abstract: Following huge forest fire of May 1987, the Chinese government has paid more attention to combat forest fires using modern science and technology measures. The Ministry of Forestry, using a World Bank loan, constructed several fire management and engineering projects. One of these projects is an Automatic Fire Weather Station Network system which has been designed for Daxinganling forest area in the northwestern corner of northeast China. It include 46 automatic stations, 20 store-and-forward Ultrahigh Frequency repeaters and 30 base stations linked to microwave main line by modems. The data can be collected and used for decision making on firefighting and prevention.

1. INTRODUCTION

During period 1950-1990 there were about 640,000 wildfires in China resulting in the burning of an area of 2,400,000 ha. and in the loss of life of more than 4000 persons (Bi 1993). The Daxinganling (DXAL) area is one of the most exposed regions to fire with damage from forest fires amounting to almost one-third that for the whole country.

Forest fire is influenced by environmental conditions and human activities. Like other natural disasters , it has random and accidental characters. In order to reduce forest fire hazards, we must not only strengthen the fire prevention programs, the fire response management and the establishment of firefighting teams, but also rely on modern scientific and technological measures.

Following the huge fire in 1987, the Chinese government paid more attention to the rehabilitation and protection of Daxinganling forest area. A world bank loan was used to fund several fire management engineering projects (Fuchs, 1988). One of these is the Automatic Fire Weather Station Network (AFWSN) system. It has been designed for hourly monitoring and collecting of weather data and proved to be an important tool for fire management decision making. This paper describes the AFWSN system as an example of countermeasures against fire hazards.

M. I. El-Sabh et al. (eds.), Land-based and Marine Hazards, 287–293.

2. OUTLINE OF STATION NETWORK

The DXAL forest area is located in extreme northeastern part of China between 46°57'N and 53°33'N, 119°01' E and 127°12'E, with an area 191100 km². It stretches across Heilongjiang and Inner Mongolian provinces. Its management is under the responsibility of the Ministry of Forestry DXAL (east), Bureau of Forestry Management (BFM) and Inner Mongolian DXAL (west) BFM.

The vegetation belongs to the cool climatic deciduous and coniferous forests. In the south there are many over-cutting and weeds area. Strong winds and drought during spring and fall seasons make for a serious fire weather condition.

Originally weather stations in DXAL were very rare and situated all along the railway. It was far from the needs of fire management. To overcome such a weakness in fire management, a World Bank loan was used to establish a microwave communication system to be fully utilized in DXAL's forest (Jin and Cheng, 1990). This system consists of 46 automatic weather stations and 2 Forest Fire Forecasting Offices (FFFO). It uses two kinds of communication - microwave and UHF link. It must set up 20 store-and-forward UHF repeaters and 30 base stations. Figure 1 shows the whole station network system.

3. STRUCTURE OF THE NETWORK

The data transmission system for fire weather station network is built according to DXAL's administration region. So, the Information Processing Centre (IPC) is located at East (Jagdaqi) BFM and West (Yakeshi) BFM respectively. The IPC has a Vax 4000/200 mini computer system with long-range communication computer terminals (COMPAQ 486 personal computer). They will collect and use the data for decision making related to firefighting and prevention.

The base station includes a long-range terminal COMPAQ 486 and a set of Data receive/transmit and control. It is set up in the Forest Fire Management Office of forestry bureau, which is considered as a primary user of data. Every base station can control 1-3 weather stations.

These weather stations can be divided into two classes and three kinds as well. Figure 1 shows its spatial distribution. In establishing this network, the following criteria were considered: (a) ensure a reasonable distance between the stations, (b) some sensors, such as the fuel moisture, are expensive, it was the first time to establish such a network in such a large forest region in China and so one has to consider our limited experience in maintenance and the fact that the sensor is still in the development stage and (c) the

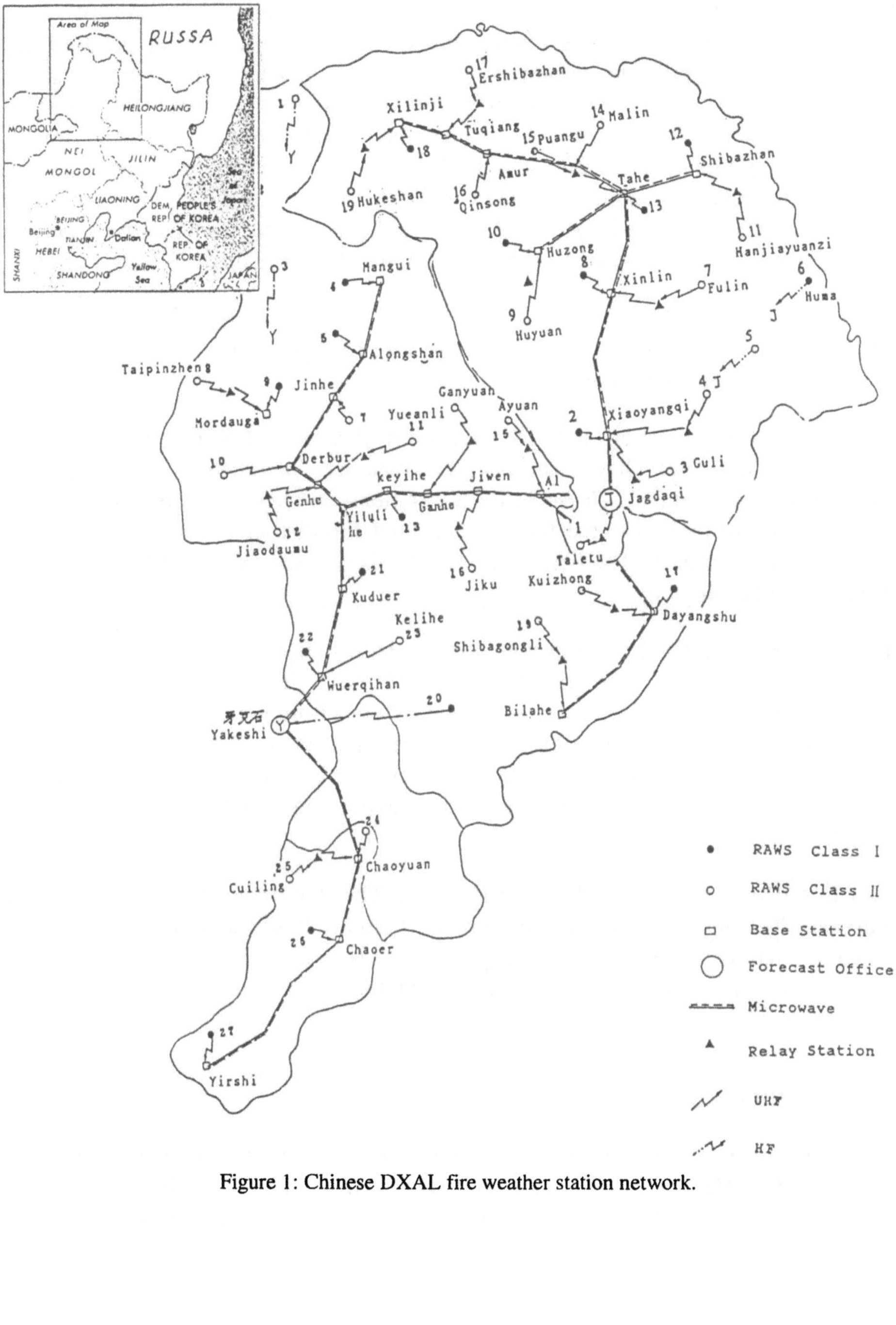

Figure 1: Chinese DXAL fire weather station network.

necessity for the workers at the base and FWS I stations to work together and monitor these sensors.

(1) There are 16 FWS I stations in DXAL. The station is Remote Automatic Weather Station (RAWS) with nine sensors - wind direction, wind speed, air temperature, air relative humidity, precipitation, barometric pressure, soil moisture/temperature, battery voltage etc. A few stations have sensors for measurement of the fuel moisture /temperature. The forest fuel moisture sensor consists of a standard pine fuel stick (dowel) machined to house a relative humidity sensor. The DCP calculates the corresponding fuel moisture from the relative humidity measured within the stick. This measuring system was first experimented in the United States of America. The measuring stations are located no more than 10 km from base station, and no additional data transmission relay station is needed.

FWS 1 is set up in the field where it is close to the forestry bureau. The working condition of this station is convenient for a technician.

(2). There are 30 FWS 2 stations in DXAL. Each has six sensors. Compared with FWS 1, the FWS 2 does not have sensors to measure barometric pressure, soil moisture/temperature and fuel. These 30 FWS 2 stations are well distributed in the whole DXAL area, and are located in the forestry farm and fire district station in order to be convenient for maintenance. The FWS 2 has two kinds of instruments: RAWS and AWS (Automatic Weather Station).

The RAWS's data of FWS 2 are transmitted to the base station by a relay station. There are 26 stations of this type. The other type is TRANSMIT. It is difficult to transmit the data to base station with just one repeater. So the data is auto-recorded and is transmitted manually using HF (High Frequency) transmission. The data are sent to long-range terminal at base station, then transferred to IPC. There are 4 stations of this kind. Most of the 20 relay stations are located on fire lookout towers and some are on microwave towers or buildings.

The structure of the network is shown in Figure 2. The figure shows the route of data transmission. The data transmission from weather station to base station is done by UHF, then into the mini computer of IPC through a microwave link. Data can be exchanged between the East and the West via the IPC mini computer.

4. SYSTEM PROPERTIES

The main equipment of this system was purchased by international competitive bidding and is provided by Handar CO. It has a large user in USA. The Boise Interagency Fire Centre, Bureau of Land Management and Forest Service operate more than 350 stations

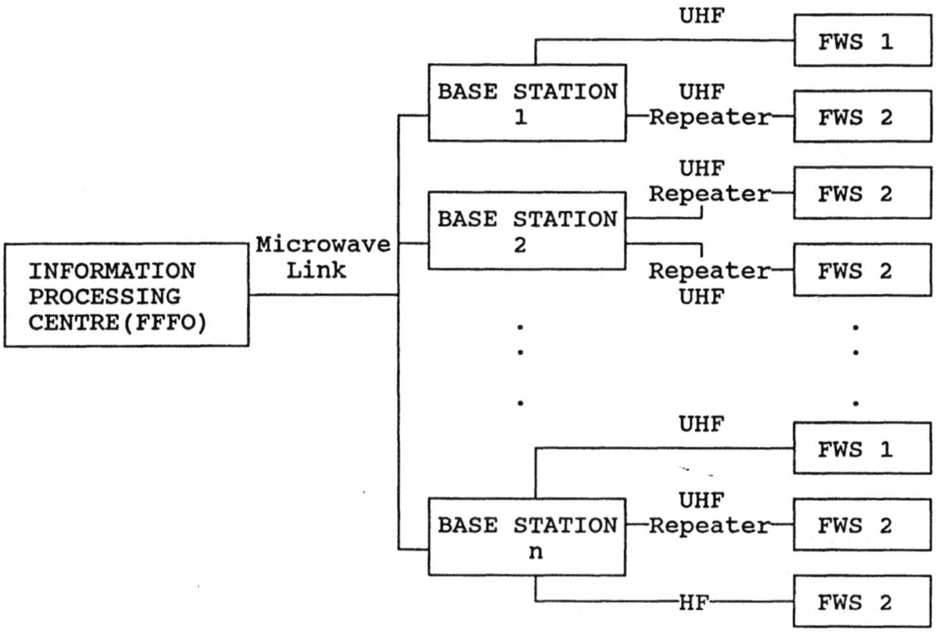

Figure 2: Configuration of station network.

as part of the system which has been in operation sine 1982 . The important difference between BIFC and DXAL system is the data transmission mode (Warren et al 1981). The former is by satellite (GOES) and the latter uses the ground repeater. We must consider the signal path loss by near ground atmosphere attenuation for RF signal. So the high quality communication devices are demanded in some field stations and the HF should be used.

The system has three working modes: monitoring, polling and dialogue. Its function is very flexible. It can provide the required data for fire management and can also provide the extreme values on local fire weather factors.

The DCP has 128 K RAM expansion memory and can be used to save data in case something goes wrong with the system. It is protected by stainless steel cover and the cable of sensors is protected by metal cover. They have the ability to resist severe environmental conditions.

The repeater and the RAWS use 18v/1.2A solar battery supplemented by a 26 Ampere Hour maintenance free battery to provide energy. The technical specification of the system is determined by two factors. One is to meet the WMO's standard for RAWS, and the other is to satisfy the practical needs of fire management. The equipment specifications are given in Appendix 1.

5. CONCLUSION

The AFWSN in DAXL forest area is an important engineering measure for fire management. It is still in the developmental stage. There is still much to be done to maximize its use for forest fire management. This system can also be used for flood and forest management.

REFERENCES

Bi, Z., (1993). How to protect the forest in China. *Fire Safety Science*, 2: 1-5.
Fuchs, F.F., (1988). Fire Protection Project In China. *Fire Management Notes*, 49: 3-7.
Jin, X. and CHENG B., (1990). The Theory of Fire Weather Station Network. *Forest Fire Prevention*, 3: 15-16.
Warren, J.R and Vance, D.L., (1981). Remote Automatic Weather Station For Resource and Fire Management Agencies, USDA. FS, BIFC Boise, Idaho GTR INT-116 11.

APPENDIX 1

System Specifications

Operating Ambient Temperature	-40°C-60°C (field equipment)
Storage Temperature	-55°C-70°C
Relative Humidity	0-100% (filed equipment)
Timing Accuracy	±15 min/month
DCP 's Quiescent Current	< 35 mA
Solar Panel Power Supply	20 AH/12V
Battery Voltage output	< 15 V
DCP's Standard RAM	16 K byte
DCP's RAM Expansion	128K byte
Data Measurement Interval	1 hr
Programming Input	1 Connector
MTBF	>7000 hrs

Sensor Specifications

Air Temperature	-40°C-50°C ± 0.5°C
Air Relative Humidity	0-80 % ± 2 %
	> 80 % ± 5 %
Wind Direction	0-360° ± 5°
Wind Speed	0-60 m/s ± 5 %RMSE
	(threshold 0.76 m/s)
Precipitation	< 20 mm < ±0.5mm
	> 20 mm < ±3% of FS
Barometer Pressure	800-1100mb ± 0.7 mb
Soil Temperature(10 cm)	-40°C-50°C ± 0.5 °C
Soil moisture	0-2000mb
Fuel Temperature	0-60°C
Fuel Moisture	0-25 % ±10 %
Response Time:	10hrs/63% final value

Data transmit Specifications

UHF Frequency	423.525 MHz
Rate of Data Transmission	1200 baud
Modem	FSK
BER(Microwave)	10_ (n=-6)
RF Power Output	1.6-25 W
Sensitivity of Receiver	-110dBm (BER=10_ n=-5)

HAZARDS-93 : CONCLUDING PLENARY SESSION AND RECOMMENDATIONS

FRANK THOMAS
Federal Emergency Managment Agency / MIT, 500 C Street, SW, Washington, D.C 20472, U.S.A.
HELÈNE DENIS
Dept. Industrial Eng., Ecole Polytechnique, C.P. 6079, Succ. A, Montreal H3C 3A7, Canada
ALAN EASTON
Dept. Mathematics, Swinburne University of Tech., P.O. Box 218, Hawthorn 3122, Australia
M. SEMIH YÜCEMEN
Dept. Civil Eng., Middle East Technical University, Ankara 06531, Turkey
and
WANG ANGSHENG
Inst. Atmospheric Phys., Chinese Acad. Sci., Qijiahuozi Deshengmenwai, Beijing 100029, P.R. China

HAZARDS-93 brought together scientists and government officials from 20 countries for the purpose of engaging in a free exchange of knowledge, experience and ideas regarding the scientific and socio-economic aspects of mitigating losses from natural and man-made disasters. Over 175 papers presented a broad range of scientific disciplinary and disaster management based issues.

The objective of the Concluding Plenary session was to capture the main conference themes and engage the conferees in discussion to identify action recommendations for the International Decade for Natural Disaster Reduction and hazard management community. The Plenary Committee Members presented synoptic discussions of themes drawn from the conference. Discussions were focussed on the need to unify government programs, improve education and communication, and exchange information and data. Specific action recommendations were addressed to the International Decade for Natural Disaster Reduction, International Society for the Prevention and Mitigation of Natural and Man-Made Hazards, national governments, and scientific disciplines.

Organizational partitioning of authority for disaster management functions is a widespread source of problems affecting national, provincial and local level governments. This fragmentation of authority requires increased commitment of resources to assure communication, coordination and cooperation among government managers, but also between scientific disciplines and the natural and man-made hazard management programs. Rarely does fragmentation align resources with the disaster management cycle of preparedness, response, recovery, and mitigation. Hence the challenge: How can/should authority for disaster management be aligned with the anatomy of disasters?

M. I. El-Sabh et al. (eds.), Land-based and Marine Hazards, 295–297.

What actions should be recommended to increase the effectiveness and efficiency of disaster management?

The following recommendations are intended to build upon existing organizations, programs and knowledge whenever possible.

ACTION RECOMMENDATIONS: SECRETARIAT FOR THE INTERNATIONAL DECADE FOR NATURAL DISASTER REDUCTION

1. Prepare guidance documents to assist individual nations address natural and man-made hazards in developing strategic plans for:

    (a) Systemic scaling and evaluation of disaster impacts of global, regional and national events,
    (b) Managing internal and regional hazards,
    (c) Achieving executive level hazard management awareness
    (d) Achieving effective communication between the scientific community and the media;

2. Facilitate development of a process and standards for:

    (a) A post-disaster survey system
    (b) Free exchange of scientific data among nations;

3. Establish a $1 Billion Fund for Loss Reduction analogous to the International Oil Spill Fund.

ACTION RECOMMENDATIONS: INTERNATIONAL SOCIETY FOR PREVENTION AND MITIGATION OF NATURAL AND MAN-MADE HAZARDS

1. Act as a catalyst by using work groups from Society membership to enhance cooperation and coordination for :

    (a) Exchange of data within the scientific communities and between scientists and government disaster management officials,
    (b) Integration of agency studies of hazard phenomena and the disaster management process,
    (c) Formation of interdisciplinary teams (including ecologists) to evaluate hazard studies and establish guidance for effectively addressing cross-cutting mitigation issues;

2. Plan the 1996 international symposium to:

(a) Utilize a disaster scenario to create a realistic focus on the interdependence of scientific and disaster manager roles in disaster mitigation,
(b) Assure high visibility for policy level issues and encourage attendance by policy makers.

ACTION RECOMMENDATIONS: NATIONAL GOVERNMENTS

1. Develop a unified national program for disaster management including:

(a) A multi-hazard management system,
(b) Strategic plans for data management, hazard identification, assessment of disaster impacts, and mitigation of risks.

2. Develop a management plan and agreement for scientific data and information exchange with neighboring nations.

3. Initiate development of a Southeast Asia cooperative organization for multi-hazard management based on experience with the Southeast Asia Tropical Cyclone Monitoring Organization.

ACTION RECOMMENDATIONS: SCIENTIFIC DISCIPLINES

1. Establish networks:

(a) Linking ecological sciences with disaster management community,
(b) Linking interdisciplinary teams to facilitate understanding, assessment, modeling and mitigation of risk.

2. Establish coordination mechanisms to:

(a) Foster development of inter-scientific hazard management programs,
(b) Focus research on commonalities of major catastrophic disaster events and build upon commonalities,
(c) Integrate ecological frameworks into mitigation studies.

3. Develop scientific research from the ground up through the operational-technical-process levels to the strategy level of senior managers.

LIST OF CONTRIBUTORS

AKKAYA, AY_EN
Department of Statistics, Middle East Technical University, 06531 Ankara, Turkey

BEGUM, SELINA
Water Resources Engineering Department,Bangladesh University of Engineering and Techonology, Dhaka-1000, Bangladesh.

BENOUAR, DJILLALI
USTHB, Civil Engineering Institute, BP 32 El-Alia, Bab Ezzouar, Alger, ALGERIA

CAMPESAN, G.
Institute of Marine Biology, CNR, Riva 7 Martiri, 1364/A, 30122 Venice, Italy

CANDIA-GALLEGOS, MARIO A.
Vector Engineering, Inc., 12438 Loma Rica Dr., Suite C, Grass Valley, CA 95945, USA.;

CHENG, BANGYU
Research Institute of Forest Protection, 75 Hapin Road, Harbin 150040, China

CHRISTIE, JENNY
School of Architecture, Victoria University of Wellington, P.O. Box 600, Wellington, New Zealand

DALLA VENEZIA, L.
Institute of Marine Biology, CNR, Riva 7 Martiri, 1364/A, 30122 Venice, Italy

DENIS, H.
Department of Industrial Engineering, Ecole Polytechnique, University of Montreal, Montreal, Quebec H3C 3A7, CANADA

DUAN, YIHONG
Shanghai Typhoon Institute,166 Puxi Road, Shanghai 200030, P.R. China

EASTON, ALAN
Dept. Mathematics, Swinburne University of Tech., P.O. Box 218, Hawthorn 3122, Australia

EL-SABH, MOHAMMED I.
Dept. D'Oceanographie, Universite du Quebec a Rimouski, 300 Allee des Ursulines, Rimouski, Quebec G5L 3A1, Canada

JI, MINGCHUAN
Senior Environmentalist, Station of Agriculture Environment Protection and Monitoring of Shandong Province, 17 Minziqian Road, Jinan city 250100, P. R. China

JIN, XIAOZHONG

Research Institute of Forest Protection, 75 Hapin Road, Harbin 150040, China

KAMEL, AHMED FAROUK
Nuclear Materials Authority, Maadi, P. O. Box. 530, Cairo, Egypt

KHANDEKAR, MADHAV L.
Atmospheric Environment Service, Downsview, Ontario, CANADA, M3H 5T4

LE, KENTANG
Institute of Oceanology, Chinese Academy of Sciences, 7 Nan-hai Road, Qingdao,P.R.China

LIU, XINGQUAN
Institute of Oceanology, Chinese Academy of Sciences, 7 Nan-hai Road, Qingdao,P.R.China

LTDA, INGECO, S.R. .
Ingenieros Consultores, P.O. Box 865, Cusco, Perú.

MENETTO, A.
Institute of Marine Biology, CNR, Riva 7 Martiri, 1364/A, 30122 Venice, Italy

MURTY, T.S.
National Tidal Facility, Flinders University of South Australia, GPO Box 2100, Adelaide SA 5001, Australia

NERALLA, V.R.
Atmospheric Environment Service, 4905 Dufferin Street, Downsview, Ontario M3H 5T4, Canada

PEREZ, J. CARLOS
GEOMEC Consultores S.R.L., Cusco, Perú.

QIN, ZENGHAO
Shanghai Typhoon Institute,166 Puxi Road, Shanghai 200030, P.R. China

SHI, JIUXIN
Institute of Oceanology, Chinese Academy of Sciences, 7 Nan-hai Road, Qingdao,P.R.China

SPRENKE, KENNETH F.
Department of Geological Engineering, University of Idaho, Moscow, ID 84843, USA.

STRADELLA, S.
Institute of Marine Biology, CNR, Riva 7 Martiri, 1364/A, 30122 Venice, Italy

THOMAS, FRANK
Federal Emergency Management Agency/MIT, 500 C Street, SW, Washington, D.C. 20472, U.S.A.

VENKATESH, S.
Environment Canada, 4905 Dufferin Street, Downsview, Ontario M3H 5T4, Canada

WANG, ANGSHENG
Inst. Atmospheric Phys., Chinese Acad. Sci., Qijiahuozi Deshengmenwai, Beijing 100029, P.R. China

WANG, JIANGONG
Vice Governor of shandong Province, Senior Environmentalist, 250011 Jinan city, P. R. China

YANG, GUISHAN
Nanjing Institute of Geography and Limnology, the Chinese Academy of Sciences. Nanjing 210008, People's Republic of China

YANG, HUATING
China Institute for Marine Development Strategy, State Oceanic Administration, No 8, Dahuisi Str., Haidian District, 100081 Beijing, China

YU, RIXIN
Guangxi Bureau of Hydrology and water Resources , 12 Yuan Hu Road, Nanning 530023, P.R.China

YÜCEMEN, M. SEM_H
Department of Civil Engineering and Earthquake Engineering Research Center, Middle East Technical University, 06531 Ankara, Turkey

ZHAO, KUIYI
Changchun Institute of Geography, Chinese Academy of Sciences, 10 Gongnong Road, Changchun 130021, P.R. China

Advances in Natural and Technological Hazards Research

Series Editor: Prof. Dr. Mohammed I. El-Sabh, *Département d'Océanographie, Université du Québec à Rimouski, 310 Allée des Ursulines, Rimouski, Québec, Canada G5L 3A1*

Publications

1. S. Tinti (ed.): *Tsunamis in the World.* Fifteenth International Tsunami Symposium (1991). 1993 ISBN 0-7923-2316-5
2. J. Nemec, J.M. Nigg and F. Siccardi (eds.): *Prediction and Perception of Natural Hazards.* Symposium Perugia, Italy (1990). 1993 ISBN 0-7923-2355-6
3. M.I. El-Sabh, T.S. Murty, S. Venkatesh, F. Siccardi and K. Andah (eds.): *Recent Studies in Geophysical Hazards.* 1994 ISBN 0-7923-2972-4
4. Y. Tsuchiya and N. Shuto (eds.): *Tsunami: Progress in Prediction, Disaster Prevention and Warning.* 1995 ISBN 0-7923-3483-3
5. A. Carrara and F. Guzzetti (eds.): *Geographical Information Systems in Assessing Natural Hazards.* 1995 ISBN 0-7923-3502-3
6. V. Schenk (ed.): *Earthquake Hazard and Risk.* 1996 ISBN 0-7923-4008-6
7. M.I. El-Sabh, S. Venkatesh, H. Denis and T.S. Murty (eds.): *Land-based and Marine Hazards.* Scientific and Management Issues. 1996 ISBN 0-7923-4064-7
8. J.M. Gutteling and O. Wiegman: *Exploring Risk Communication.* 1996 ISBN 0-7923-4065-5

Kluwer Academic Publishers – Dordrecht / Boston / London

MIX
Papier aus verantwortungsvollen Quellen
Paper from responsible sources
FSC® C105338
FSC
www.fsc.org

If you have any concerns about our products,
you can contact us on
ProductSafety@springernature.com

In case Publisher is established outside the EU,
the EU authorized representative is:
Springer Nature Customer Service Center GmbH
Europaplatz 3, 69115 Heidelberg, Germany

Printed by Libri Plureos GmbH
in Hamburg, Germany